T0340889

Introduction to Numerical Analysis and Scientific Computing

Introduction to
Numerical Analysis
and
Scientific Computing

Nabil Nassif
Dolly Khuwayri Fayyad

CRC Press
Taylor & Francis Group
Boca Raton   London   New York

CRC Press is an imprint of the
Taylor & Francis Group, an **informa** business

A CHAPMAN & HALL BOOK

CRC Press
Taylor & Francis Group
6000 Broken Sound Parkway NW, Suite 300
Boca Raton, FL 33487-2742

© 2014 by Taylor & Francis Group, LLC
CRC Press is an imprint of Taylor & Francis Group, an Informa business

No claim to original U.S. Government works

Printed on acid-free paper
Version Date: 20130618

International Standard Book Number-13: 978-1-4665-8948-3 (Hardback)

Library of Congress Cataloging-in-Publication Data

Nassif, Nabil, author.
   Introduction to numerical analysis and scientific computing / Nabil Nassif, Dolly Khuwayri Fayyad.
      pages cm
   Includes bibliographical references and index.
   ISBN 978-1-4665-8948-3 (hardback)
   1. Numerical analysis--Textbooks. 2. Computer science--Mathematics--Textbooks. I. Fayyad, Dolly Khuwayri, author. II. Title.

QA297.N37 2013
518--dc23
                                                                        2013012520

Visit the Taylor & Francis Web site at
http://www.taylorandfrancis.com

and the CRC Press Web site at
http://www.crcpress.com

Dedication

To the dear and supporting members of our respective families:

Norma, Nabil-John and Nadim Nassif

Georges, Ghassan and Zeina Fayyad

Contents

Preface

This work is the result of several years of teaching a one semester course on numerical analysis and scientific computing, addressed primarily to students in mathematics, engineering, and the sciences. Our purpose is to provide those students with fundamental concepts of numerical mathematics and at the same time stir their interest in the art of implementing and programming numerical methods.

The *learning objectives* of this book are mainly to have the students:

1. Understand floating-point number representations, particularly those pertaining to IEEE simple and double precision standards as being used in the scientific computer environment such as MATLAB® version 7. Please note that:

   MATLAB ® is a registered trademark of The Math Works, Inc.

   For product information, please contact:

   The Math Works Inc.
   3 Apple Hill Drive
   Natick, MA 01 760-20098 USA
   Tel: 508 647 7000
   Fax: 508 647 7001
   E-mail: info@mathworks.com
   Web: www.mathworks.com

2. Understand computer arithmetic as a source for generating round-off errors and be able to avoid the use of algebraic expressions that may lead to the loss of significant figures.

3. Acquire concepts on iterative methods for obtaining accurate approximations to roots of nonlinear equations. In particular, students should be able to distinguish between globally convergent and locally convergent methods as well as the order of convergence of a method.

4. Understand basic concepts of numerical linear algebra, such as: Gauss elimination, with or without partial pivoting used to solve systems of linear equations, and obtain the LU decomposition of a matrix and consequently compute its determinant value and inverse matrix.

5. Learn the basic Lagrange interpolation theorem and acquire the ability to use local polynomial interpolation through spline functions.

6. Learn the basic formulae of numerical differentiation and integration with the ability to obtain error estimates for each of the formulae.

7. Understand the concept of the order of a numerical method to solve an ordinary differential equation and acquire basic knowledge in using one step Runge-Kutta methods.

These objectives can be easily achieved in one semester by covering the core material of this book: the first five chapters, in addition to sections 7.1, 7.3 and 7.4 of Chapter 7.

Additional Topics

In addition to the core material, Chapter 6 provides additional information on numerical integration, specifically:
- One-dimensional adaptive numerical integration using Simpson's rule.
- Two-dimensional numerical integration on rectangles and polygons.
- Monte Carlo methods in 1 and 2 dimensions.
Also, in Chapter 7 on ordinary differential equations, specific sections discuss:
- Existence of the solutions that features Picard's iteration.
- Adaptive numerical integration based either on one Runge-Kutta method or on a pair of embedded Runge-Kutta methods.
- Multi-step methods of Adams types and backward difference methods.

Algorithms and MATLAB Programs

Special attention is given to algorithms' implementation through the use of MATLAB's syntax. As a matter of fact, each of the numerical methods explained in any of the seven chapters is directly expressed either using a pseudo-code or a detailed MATLAB program.

Exercises and Computer Projects

Each chapter ends with a large number of exercises. Answers to those with odd numbers are provided at the end of the book.
Throughout the seven chapters, several computer projects are proposed. These aim to test the students' understanding of both the mathematics of numerical methods and the art of computer programming.

Recommended sections for teaching a one semester course from the book: 1.1 to 1.5; 2.1 to 2.5; 3.1 to 3.5; 4.1 to 4.6; 5.1 to 5.7; 7.1, 7.3 and 7.4.

Nabil Nassif and Dolly Fayyad

About the Authors

Nabil Nassif received a Diplôme-Ingénieur from the Ecole Centrale de Paris and earned a master's degree in applied mathematics from Harvard University, followed by a PhD under the supervision of Professor Garrett Birkhoff. Since his graduation, Dr. Nassif has been affiliated with the Mathematics Department at the American University of Beirut, where he teaches and conducts research in the areas of mathematical modeling, numerical analysis and scientific computing. Professor Nassif has authored or co-authored about 50 publications in refereed journals and directed 12 PhD theses with an equal number of master's theses. During his career, Professor Nassif has also held several regular and visiting teaching positions in France, Switzerland, U.S.A. and Sweden.

Dolly Khoueiri Fayyad received her BSc and master's degrees from the American University of Beirut and her PhD degree from the University of Reims in France under the supervision of Professor Nabil Nassif. After earning her doctorate degree and before becoming a faculty member in the Mathematics Department of the American University of Beirut, she taught at the University of Louvain-la-Neuve in Belgium and then in the Sciences Faculty of Lebanon National University. Simultaneously, Dr. Fayyad has conducted research on the numerical solution of time-dependent partial differential equations and more particularly on semi-linear parabolic equations. She has also supervised several master's theses in her research areas.

List of Figures

List of Tables

Chapter 1

Computer Number Systems and Floating Point Arithmetic

1.1 Introduction

The main objective of this chapter is to introduce the students to modes of storage of **users' numbers** in a computer memory and as well providing the readers with basic concepts of **computer arithmetic**, referred to also as **Floating Point Arithmetic**. Although the principles covered are general and can apply to any finite precision arithmetic system, we apply those principles only to **Single** and **Double Precision IEEE** *(Institute of Electrical and Electronics Engineers)* systems. For additional detailed references, we refer to [8], [14], [19] and [23].

In this view, we start by describing computer number representation in the binary system that uses 2 as the base. Since the usual decimal system uses base 10, we discuss therefore methods of conversion from one base to another.

The octal and hexadecimal systems (respectively, base 8 and base 16 systems) are also introduced as they are often needed as intermediate stages between the binary and decimal systems. Furthermore, the subsequent hexadecimal notation is used to represent internal contents of stored numbers.

Since all machines have limited resources, not all real numbers can be represented in the computer memory; only a finite subset $\mathbb{F}$ of $\mathbb{R}$ is effectively dealt with. More precisely, $\mathbb{F}$ is a proper subset of the rationals, with $\mathbb{F} \subset Q \subset \mathbb{R}$. We shall therefore define first in general, normalized floating point systems $\mathbb{F}$ representing numbers in base $\beta \in \mathbb{N}$, $\beta \geq 2$ with a fixed precision p, and analyze particularly the standard IEEE single precision $\mathbb{F}_s$ and double precision $\mathbb{F}_d$ binary systems.

Moreover, the arithmetic performed in a computer is not exact; $\mathbb{F}$ is characterized by properties that are different from those in $\mathbb{R}$. We present therefore floating point arithmetic operations in the last sections of this chapter.

Note that IEEE stands for "Institute for Electrical and Electronics Engineers." The IEEE standard for floating point arithmetic (IEEE 754) is the most widely used standard for floating point operation and is followed by many hardware and software implementations; most computer languages allow or require that some or all arithmetic be carried out using IEEE formats and operations.

For any base $\beta \in \mathbb{N}$, $\beta \geq 2$, we associate the set of symbols S_β, which consists of β distinct symbols. To illustrate, we have the following examples:

$$S_{10} = \{0, 1, .., 9\},$$

$$S_2 = \{0, 1\},$$

$$S_{16} = \{0, 1, .., 9, A, B, C, D, E, F\}.$$

The general representation of $x \in \mathbb{R}$ in base β is given by:

$$x = \pm(a_N\beta^N + ... + a_1\beta + a_0 + a'_1\beta^{-1} + ... + a'_p\beta^{-p}) = \pm(a_N a_{N-1}...a_1 a_0 \cdot a'_1...a'_p)_\beta \tag{1.1}$$

where $0 \leq N < \infty$, $1 \leq p \leq \infty$ and a_i, $a'_i \in S_\beta$, with $a_N \neq 0$ being the most significant digit in this number representation.

The number x is thus characterized by its sign $\pm$, its integral part $E(x) = \sum_{i=0}^{N} a_i\beta^i$ and its fractional part $F(x) = \sum_{i=1}^{p} a'_i\beta^{-i}$, leading to the following general expression of x:

$$x = \pm(E(x) + F(x))$$

or also equivalently: $x = \pm(E(x).F(x))$

Note that in case $p = \infty$, the fractional part of x is said to be infinite.

Example 1.1 *The octal representation of* 0.36207 *is:*

$$(0.36207)_8 = 3 \times 8^{-1} + 6 \times 8^{-2} + 2 \times 8^{-3} + 7 \times 8^{-5}$$

The decimal representation of 57.33333... *is :*

$$(57.33333...)_{10} = (57.\overline{3})_{10} = 5 \times 10 + 7 + 3 \times 10^{-1} + 3 \times 10^{-2} + ...$$

The hexadecimal representation of $4.A02C$ *is :*

$$(4.A02C)_{16} = 4 + A \times 16^{-1} + 2 \times 16^{-2} + C \times 16^{-3}$$

1.2   Conversion from Base 10 to Base 2

Assume that a number $x \in \mathbb{R}$ is given in base 10, whereby:

$$x = \pm(d_N 10^N + ... + d_1 10 + d_0 + d'_1 10^{-1} + .. + d'_p 10^{-p}) = \pm(d_N d_{N-1}...d_1 d_0.d'_1...d'_p)_{10},$$

where $d_i, d'_i \in S_{10}\ \forall i$, $d_N \neq 0$, and $p \leq \infty$. We seek its conversion to base 2, in a way that:

$$x = \pm(b_M 2^M + ... + b_1 2 + b_0 + b'_1 2^{-1} + ... + b'_l 2^{-l}) = \pm(b_M b_{M-1}...b_1 b_0.b'_1...b'_l)_2,$$

where $b_i, b'_i \in S_2\ \forall i$, $b_M \neq 0$, $l \leq \infty$.
We convert successively the integral and fractional parts of x.

1.2.1   Conversion of the Integral Part

Starting with the integral part of x, $E(x)$ and writing:

$$E(x) = d_N 10^N + ... + d_1 10 + d_0 = b_M 2^M + ... + b_1 2 + b_0, \qquad (1.2)$$

one has to find the sequence $\{b_i | i = 0, ..., M\}$ in S_2, given the sequence $\{d_i | i = 0, ..., N\}$ in S_{10}. Both sequences are obviously finite. The conversion is done using the successive division algorithm of positive integers based on the Euclidean division theorem stated as follows:

Theorem 1.1 *Let D and d be two positive integers. There exist 2 non-negative integers q (the quotient) and r (the remainder), such that $r \in \{0, 1, 2, ..., d-1\}$, verifying:*

$$D = d \times q + r.$$

For notation purpose, we write $q = D \operatorname{div} d$ and $r = D \bmod d$.

Remark 1.1 *When $D < 0$ and $d > 0$, one has:*

$$D = q \times d + r, \quad with\ q = \lfloor \frac{D}{d} \rfloor < 0.$$

where $\lfloor r \rfloor : \mathbb{R} \to \mathbb{Z}$ designates the "floor function" of the real number r.

On the base of (1.2), if $E(x) = D$, then one seeks:

$$D = E(x) = (b_M 2^{M-1} + ... + b_1) \times 2 + b_0$$

where

$$(b_M 2^{M-1} + ... + b_1) = D \, div \, 2 \quad and \quad b_0 = D \, mod \, 2.$$

Thus if D is divided once by 2, the remainder in this division is b_0. We can repeat this argument taking then $D = b_M 2^{M-1} + ... + b_1$ to find b_1, then following a similar pattern, compute successively all remainders $b_2, ..., b_M$. The process is stopped as soon as the quotient of the division is identical to zero.

The corresponding MATLAB function can then be easily implemented as follows:

Algorithm 1.1 Integer Conversion from Base 10 to 2

```
% Input: D an integer in decimal representation
% Output: string  s of  binary symbols (0's and 1's)
%                  representing D in base 2
% All arithmetic is based on rules of the decimal system
function s = ConvertInt10to2(D)
s=[ ];
while D>0
%Divide D by 2, calculate the quotient q and the remainder r,
% then add r in s from right to left
        q=fix(D/2);
        r= D - 2*q ;
        s=[r s];
        D=q;
end
```

As an application, consider the following example.

Example 1.2 *Convert the decimal integer $D = 78$ to base 2.*

Using the above algorithm, we have successively:

$78 = 39 \times 2 + 0$
$39 = 19 \times 2 + 1$
$19 = 9 \times 2 + 1$
$9 = 4 \times 2 + 1$
$4 = 2 \times 2 + 0$
$2 = 1 \times 2 + 0$
$1 = 0 \times 2 + 1.$

Hence, one concludes that $(78)_{10} = (1001110)_2$.   ∎

We can now introduce base 8 in order to shorten this procedure of conversion. The octal system is particularly useful when converting from the decimal system to the binary system, and vice versa. Indeed, if

$$E(x) = b_M 2^M + ... + b_3 2^3 + b_2 2^2 + b_1 2 + b_0, \text{ with } b_i \in \{0, 1\},$$

Octal symbol	Group of 3 bits $o_i = b_{3i+2}b_{3i+1}b_{3i}$
0	0 0 0
1	0 0 1
2	0 1 0
3	0 1 1
4	1 0 0
5	1 0 1
6	1 1 0
7	1 1 1

TABLE 1.1: Table of conversion of octal symbols into base 2

we can group the bits 3 by 3 from right to left (supplying additional zeros if necessary), then factorize successively the positive powers of 8, i.e., $8^0, 8^1, 8^2, ...$ to have:

$$E(x) = ... + (b_5 2^5 + b_4 2^4 + b_3 2^3) + (b_2 2^2 + b_1 2 + b_0)$$

then equivalently:

$$E(x) = ... + (b_8 2^2 + b_7 2 + b_6)8^2 + (b_5 2^2 + b_4 2 + b_3)8^1 + (b_2 2^2 + b_1 2 + b_0)8^0$$

$$= \sum_{i=0}^{l} (b_{3i+2} 2^2 + b_{3i+1} 2 + b_{3i})8^i$$

Letting $o_i = b_{3i+2} 2^2 + b_{3i+1} 2 + b_{3i}$, one writes then the integral part as follows:

$$E(x) = \sum_{i=0}^{l} o_i 8^i$$

Note that for all values of i, $0 \le o_i \le 7$, implying that o_i is an octal symbol. The conversion is set up according to Table 1.1. Thus, to convert from base 2 to base 8, groups of 3 binary digits can be translated directly to octal symbols according to the above table. Conversion of an octal number to binary can be done in a similar way but in reverse order; i.e., just replace each octal digit with the corresponding 3 binary digits.

Hence, to convert an integer from base 10 to base 2, we can therefore start by converting it to base 8:

$$(E(x))_{10} \rightarrow (E(x))_8 \rightarrow (E(x))_2$$

We can then implement an algorithm that simulates this conversion process. Similarly to the previous one that converts integers from base 10 to base 2, this algorithm uses successive applications of the Euclidean division by 8.

Algorithm 1.2 Integer Conversion from Base 10 to Base 8

```
% Input: D an integer in decimal representation
% Output: string  s of  octal symbols (0 1 ... 7)
%                  representing D in base 8
% All arithmetic is based on rules of the decimal system
function s = ConvertInt10to8(D)
s=[ ];
while D>0
%Divide D by 8, calculate the quotient q and the remainder r,
% then add r in s from right to left
        q=fix(D/8);
        r= D - 8*q ;
        % or r=rem(D, 8);
        s=[r s];
        D=q;
end
```

In the preceding example, using this algorithm we have successively:
$78 = 9 \times 8 + 6$
$9 = 1 \times 8 + 1$
$1 = 0 \times 8 + 1.$
Hence, $(78)_{10} = (116)_8$ through 3 successive divisions by 8.
Referring to Table 1.1 that converts octal symbols to binary, we obviously deduce that:

$$(78)_{10} = (116)_8 = (001\,001\,110)_2 = (1001110)_2.$$

1.2.2   Conversion of the Fractional Part

To convert the fractional part $F(x)$ of the decimal x, we introduce the **successive multiplication algorithm**. Its principle runs as follows: given the sequence $\{d_i'\} \in S_{10}$, we seek the sequence $\{b_i'\} \in S_2$ with:

$$F(x) = d_1' 10^{-1} + .. + d_p' 10^{-p} = b_1' 2^{-1} + ... + b_l' 2^{-l} \tag{1.3}$$

Let $f = F(x)$. Note then the following identity:

$$2f = b_1' + b_2' 2^{-1} ... + b_l' 2^{1-l} = b_1' \cdot b_2' \cdot ... \cdot b_{l-1}'.$$

Obviously through one multiplication of f by 2, the integral and fractional parts of $2f$ are respectively:

$$E(2f) = b_1' \text{ and } F(2f) = b_2' 2^{-1} ... + b_l' 2^{1-l}$$

We can therefore repeat the same procedure, of multiplication by 2, to find successively b_2', then b_3', ... ,b_l'. The corresponding algorithm is the following:

Algorithm 1.3 Fraction Conversion from Base 10 to Base 2

```
% Input: F, fractional part of a decimal number 0<F<1
%              k, required maximum number of binary bits
% Output:  string s  (up to k bits) representing F in base 2
function s=ConvertFrac10to2(F,k)
s=[ ]  ;
i=1;
while F>0 & i<=k
     G=2*F;
     b=fix(G);
     F=G-b;
     s = [ s b ] ;
     i=i+1;
end
```

Note that if f has an infinite representation in base 10, its representation in base 2 will also be infinite. However, we could have situations where f is finitely represented in base 10 and infinitely represented in base 2. To illustrate, consider the following examples.

Example 1.3 *Convert* $(0.25)_{10}$ *to base 2.*

We apply the above algorithm to get successively:
$2 \times 0.25 = 0 + 0.5$
$2 \times 0.5 = 1 + 0.0$
Thus $(0.25)_{10} = (0.01)_2$. ∎

Example 1.4 *Convert* $(0.1)_{10}$ *to base 2.*

Applying the same non-terminating procedure, we have:

$2 \times 0.1 = 0 + 0.2$
$2 \times 0.2 = 0 + 0.4$
$2 \times 0.4 = 0 + 0.8$
$2 \times 0.8 = 1 + 0.6$
$2 \times 0.6 = 1 + 0.2$
$2 \times 0.2 = 0 + 0.4$
$2 \times 0.4 = 0 + 0.8$
$2 \times 0.8 = 1 + 0.6$
$2 \times 0.6 = 1 + 0.2$
...

Thus $(0.1)_{10} = (0.0001100110011...)_2 = (0.0\overline{0011})_2$.   ∎

We end up with an example where both representations are infinite.

Example 1.5 *Convert* $\frac{1}{3}$ *to base 2.*

Let us apply the successive multiplication algorithm to this fraction:
$2 \times \frac{1}{3} = 0 + \frac{2}{3}$

$2 \times \frac{2}{3} = 1 + \frac{1}{3}$
....   ∎

Hence: $\frac{1}{3} = (0.\overline{3})_{10} = (0.0101...)_2 = (0.\overline{01})_2$

Of course, base 8 can also be used as an intermediate stage:

$$(F(x))_{10} \rightarrow (F(x))_8 \rightarrow (F(x))_2$$

By grouping the bits 3 by 3 from left to right, supplying additional zeros if necessary, then factorizing successively negative powers of 8: 8^{-1}, 8^{-2}, ... one establishes through these steps the following identities:

$$F(x) = (b_1 2^{-1} + b_2 2^{-2} + b_3 2^{-3}) + (b_4 2^{-4} + b_5 2^{-5} + b_6 2^{-6}) + ...$$

$$= (b_1 4 + b_2 2 + b_3)8^{-1} + (b_4 4 + b_5 2 + b_6)8^{-2} + ... = o_1 8^{-1} + o_2 8^{-2} + ...$$

We can then have a new version of the successive multiplication by 8 algorithm converting a fractional decimal to octal, followed by a final conversion to a binary fractional using the table of conversion.
To illustrate, consider the following examples.

Example 1.6 *Convert* $(0.75)_{10}$ *to base 2, using base 8 as intermediate.*

A straightforward application of the procedure above yields: $8 \times 0.75 = 6 + 0.00$. Hence:

$$(0.75)_{10} = (0.6)_8 = (0.110)_2 = (0.11)_2$$

Example 1.7 *Convert* $x = (0.12)_{10}$ *to base 2, using base 8 as intermediate. Do not exceed 21 bits for the representation of x in base 2.*

Getting 21 bits in base 2 means reaching 7 digits in base 8. Therefore one only needs to apply 7 successive multiplications by 8. This yields:

$8 \times 0.12 = 0 + 0.96$

$8 \times 0.96 = 7 + 0.68$

$8 \times 0.68 = 5 + 0.44$

$8 \times 0.44 = 3 + 0.52$

$8 \times 0.52 = 4 + 0.16$

$8 \times 0.16 = 1 + 0.28$

$8 \times 0.28 = 2 + 0.24$

...

Hence $(0.12)_{10} = (0.0753412...)_8 = (0.000\,111\,101\,011\,100\,001\,010\,...)_2$.   ∎

1.3   Conversion from Base 2 to Base 10

We consider in this section inverse procedures that convert numbers from base 2 (or 8) to base 10. For a real number x, this is performed as previously on the integral part $E(x)$ first, then on the fractional part $F(x)$. Of course, the successive division and multiplication algorithms can be applied. However, this would mean dividing or multiplying successively by 10 and performing the arithmetic operations in base 2 (or 8). Instead, we follow up a straightforward **polynomial evaluation** process, with the arithmetic being performed in base 10. We start by discussing this last issue.

1.3.1   Polynomial Evaluation

Consider the polynomial $p_n(y)$ of degree n, with real coefficients $\{a_i | i = 0, 1..., n\}$ and $a_n \neq 0$:

$$p_n(y) = a_0 + a_1 y + ... + a_{n-1}y^{n-1} + a_n y^n \quad ; \quad y \in \mathbb{R}$$

A first way to evaluate $p_n(y)$ is by using a straightforward sum of products, as indicated in the following algorithm:

Algorithm 1.4 Direct Polynomial Evaluation

```
function p=EvaluatePolyStraight(a,y)
% Input a=[a(1),...,a(n+1)] and y
% Output Value of p(y)=a(n+1)*y^n+a(n)*y^{n-1}+...+a(2)*y+a(1)$
n=length(a)-1;
t=y;p=a(1);
for i=2:n+1
   p=p+a(i)*t;
   t=t*y;
end
```

This algorithm requires n additions and $2n$ multiplications.

A more efficient algorithm, called **Horner's algorithm**, uses **nested evaluation**. One starts by writing the given polynomial in nested form as shown below:

$$p_n(y) = a_n y^n + a_{n-1}y^{n-1} + ... + a_1 y + a_0 = (a_n y + a_{n-1})y^{n-1} + ... + a_1 y + a_0$$

$$= ((a_n y + a_{n-1})y + a_{n-2})y^{n-2} + ... + a_1 y + a_0$$

$$= (((a_n y + a_{n-1})y + a_{n-2})y + a_{n-3})y^{n-3}... + a_1 y + a_0$$

$$= (...(((a_n y + a_{n-1})y + a_{n-2})y + a_{n-3})y + ... + a_1)y + a_0$$

This method can be implemented as follows:

Algorithm 1.5 Nested Polynomial Evaluation

```
% Input a=[a(1),...,a(n+1)] and y
% Output Value of p(y)=a(n+1)*y^n+a(n)*y^{n-1}+...+a(2)*y+a(1)$
function p=EvaluatePolyNested(a,y)
n=length(a)-1;
p=a(n+1);
for i=n:-1:1
   p=p*y+a(i);
end
```

Such procedure requires n multiplications and n additions, i.e., a total of $2n$ operations, that is $2/3$ of the number of arithmetic operations in the previous algorithm. Thus, to minimize the number of arithmetic calculations, polynomials should always be expressed in nested form before performing an evaluation.

Example 1.8 *Write $f(x) = 5x^3 - 6x^2 + 3x + 1$ in nested form.*

$$f(x) = 5x^3 - 6x^2 + 3x + 1 = ((5x - 6)x + 3)x + 1$$

1.3.2 Conversion of the Integral Part

Rewriting identity (1.2) as:

$$E(x) = b_M 2^M + \ldots + b_1 2 + b_0 = d_N 10^N + \ldots + d_1 10 + d_0,$$

one seeks now to find the sequence $\{d_i\}$ in S_{10} given the sequence $\{b_i\}$ in S_2. Indeed, note that $E(x) = p_M(2)$, where p_M is the polynomial of degree M given by:

$$p_M(y) = b_M y^M + \ldots + b_1 y + b_0.$$

Hence finding $E(x)$ in base 10 reduces to the evaluation, using decimal arithmetic of the polynomial $p_M(y)$, for $y = 2$. In case one wants to use the octals as intermediates, the bits are first grouped 3 by 3 to write $E(x)$ as a polynomial in powers of 8, based on the table of conversion. That is:

$$E(x) = o_L 8^L + \ldots + o_1 8 + o_0 = q_L(8),$$

where q_L is a polynomial of degree L given by $q_L(y) = o_L y^L + \ldots + o_1 y + o_0$. Using decimal arithmetic, one computes then $q_L(y)$ for $y = 8$.

Example 1.9 *Convert the binary integer $D = (01110101110011)_2$ to base 10, using base 8 as intermediate.*

We first convert D to base 8 using the table of conversion:

$$D = (01110101110011)_2 = (001\,110\,101\,110\,011)_2 = (16563)_8$$

$$= 1 \times 8^4 + 6 \times 8^3 + 5 \times 8^2 + 6 \times 8 + 3.$$

Thus, using nested polynomial evaluation, one gets:

$$D = (((8+6)8+5)8+6)8+3 = (7539)_{10}.$$

1.3.3 Conversion of the Fractional Part

Given the sequence $\{b'_i\} \in S_2$, we seek now the sequence $\{d'_i\} \in S_{10}$, such that:

$$F(x) = f = b'_1 2^{-1} + \ldots + b'_l 2^{-l} = d'_1 10^{-1} + \ldots + d'_p 10^{-p}$$

Using decimal arithmetic, the evaluation of f is based on the following steps:

$$f = b'_1 2^{-1} + \ldots + b'_l 2^{-l} = 2^{-l}(b'_1 2^{l-1} + \ldots + b'_l)$$

that is, using nested polynomial evaluation:

$$f = 2^{-l} p_{l-1}(2),$$

where obviously:
$$p_{l-1}(y) = b_1' y^{l-1} + b_2' y^{l-2} ... + b_l'.$$

Clearly then, to use base 8 as an intermediate, through grouping the bits 3 by 3, then referring to the table of conversion, one gets a polynomial expression in negative powers of 8, specifically:

$$f = o_1' 8^{-1} + ... + o_{k-1}' 8^{-k+1} + o_k' 8^{-k}$$

Equivalently,

$$f = 8^{-k}(o_1' 8^{k-1} + ... + o_{k-1}' 8 + o_k') = 8^{-k} q_{k-1}(8),$$

with $q_{k-1}(y) = o_1' y^{k-1} + ... + o_{k-1}' y + o_k'$.
To illustrate consider the following example.

Example 1.10 *Convert the fractional octal $f = (0.00111000111)_2$ to base 10. Use base 8 as intermediate.*

We start by converting f to base 8, yielding:

$$f = (0.1616)_8 = 1 \times 8^{-1} + 6 \times 8^{-2} + 1 \times 8^{-3} + 6 \times 8^{-4} = 8^{-4}(1 \times 8^3 + 6 \times 8^2 + 1 \times 8 + 6)$$

Through nested evaluation,

$$8^3 + 6 \times 8^2 + 8 + 6 = ((8+6)8 + 1)8 + 6 = 910.$$

Thus:

$$f = 8^{-4} \times 910 = \frac{910}{4096} = 0.2221679$$

■

1.4   Normalized Floating Point Systems

1.4.1   Introductory Concepts

Recall that a standard way to represent a real number in decimal form is with a sign ($+$ or $-$), an integral part, a fractional part and a decimal point in between, for example: $+32.875$ or -0.0082.

Another standard computer notation called the **normalized floating point representation**, is obtained by shifting the decimal point and supplying appropriate powers of 10. Thus the preceding numbers have an alternate representation respectively as $+3.2875 \times 10^1$, or -8.2×10^{-3}.

In general, a non-zero real number x in the base β is written in the standard normalized floating point form:

$$x = \pm\, m\, \times \beta^e$$

where m is called the **mantissa**, with $1 \leq m < \beta$ and e the **exponent**, being a positive or negative integer. These parameters are obtained from (1.1) by writing:

$$x = \pm(a_N \beta^N + a_{N-1}\beta^{N-1} + ... + a'_p\beta^{-p}) = \pm(a_N + a_{N-1}\beta^{-1} + ... + a'_p\beta^{-(p+N)}) \times \beta^N$$

where $a_N \neq 0$, thus leading to

$$m = a_N + a_{N-1}\beta^{-1} + a_{N-2}\beta^{-2} + ... + a'_p\beta^{-(p+N)}, \text{ and } e = N$$

Remark 1.2 *If the number x has a non-terminating fractional part, in some cases the mantissa m can reach the value β.*

For example, consider the following decimal number x:

$$x = 0.9999999... = 9 \times 10^{-1} + 9 \times 10^{-2} + ...$$

The normalized floating point representation of x is:

$$x = (9 + 9 \times 10^{-1} + 9 \times 10^{-2} + ...) \times 10^{-1} = 9.99999999... \times 10^{-1}$$

Thus, the mantissa is infinite with

$$m = 9.\bar{9} = 9(1 + \tfrac{1}{10} + \tfrac{1}{10^2} + \tfrac{1}{10^3} + ...) = 9\tfrac{1}{1 - 1/10} = 10 = \beta$$

Example 1.11 *Base 10, 2 and 8 representations of $\frac{1}{3}$ in normalized floating point notations.*

1. In the normalized floating point notation, $\frac{1}{3}$ in base 10 is expressed as follows:
$$\frac{1}{3} = (0.\bar{3})_{10} = 3.\bar{3} \times 10^{-1}.$$

   Thus, in such system, the mantissa $m = 3.\bar{3}$ and the exponent $e = -1$.

2. However in base 2 (Example 1.5), it becomes:

$$\frac{1}{3} = (0.\overline{01})_2 = (0.0101010101...)_2 = 1.01010101... \times 2^{-2} = 1.\overline{01} \times 2^{-2},$$

   i.e., the mantissa is $m = 1.\overline{01}$ and the exponent $e = -2$.

3. Finally, to convert $\frac{1}{3}$ to base 8:

$$\frac{1}{3} = (0.010101010101...)_2 = (0.2525...)_8 = 2.\overline{52} \times 8^{-1}.$$

   where $m = 2.\overline{52}$ and $e = -1$.

■

Example 1.12 *Write the binary number* $x = (11001.0111)_2$ *in the normalized floating point notation.*

$$x = (11001.0111)_2 = 1.10010111 \times 2^4$$

■

Note that every computer system has a finite total capacity and a finite word length. Numbers used in calculations within a computer system must conform to the format imposed in that system; only real numbers with a finite number of digits can be represented, leading then to a strictly limited degree of precision. Real numbers representable in a computer are called **machine numbers**, and are written in a standard format.

A **floating point system** $\mathbb{F}$ consists of machine numbers and is defined as follows:

Definition 1.1 *A **normalized floating point system** $\mathbb{F} = \mathbb{F}(\beta, p, e_{\min}, e_{\max})$ is the set of all real numbers written in normalized floating point form $x = \pm\, m \times \beta^e$ where m is the mantissa of x and e, the exponent, such that:*

1. *If $x \neq 0$, then $m = m_0 + m_1\beta^{-1} + \ldots + m_{p-1}\beta^{-(p-1)}$; with $m_i \in S_\beta$, $m_0 \neq 0$, and $e_{\min} \leq e \leq e_{\max}$*

2. *If $x = 0$, then $m = 0$, while e could take any value or be selected according to other criteria.*

The main parameters of a floating point system $\mathbb{F} = \mathbb{F}(\beta, p, e_{\min}, e_{\max})$ are:

1. The **base** β

2. The **number of significant digits** p, called the **precision** of the system which is a finite positive integer that could be given a specific value (IEEE systems) or be defined by the user (MATHEMATICA or MAPLE)

3. The **range of the exponent** $[e_{\min}, e_{\max}]$, with $e_{\min} < 0$ and $e_{\max} = |e_{\min}| + 1$

4. A **convention for representing zero**

Note that since there is a complete symmetry with respect to zero, between the positive and negative elements of $\mathbb{F}$, we will analyze and prove in what follows properties of the positive elements only.

Theorem 1.2 *Let $x \in \mathbb{F} = \mathbb{F}(\beta, p, e_{\min}, e_{\max})$, with $x = +\, m \times \beta^e$ and $x \neq 0$.*

1. $1 \leq m < \beta$,

2. $x_{\min} \leq x \leq x_{\max}$, *where*

$$x_{\min} = \beta^{e_{\min}}$$

and

$$x_{\max} = (\beta - 1)(1 + \beta^{-1} + \dots + \beta^{-p+1})\beta^{e_{\max}} < \beta \times \beta^{e_{\max}}.$$

3. *If* $x = +m \times \beta^e \in \mathbb{F}$ *with* $x_{\min} \leq x < x_{\max}$, *then the successor of* x *is given by*

$$succ(x) = x + \beta^{1-p}\beta^e$$

leading to:

$$\frac{succ(x) - x}{x} \leq \beta^{-p+1}.$$

Proof.

1. The first part of the theorem is obtained straightforwardly from the definition.

2. It is enough to note that the minimum value of m is reached when $a_0 = 1$ and $a_i = 0$, for $1 \leq i \leq p - 1$, i.e., $m = 1$, while the maximum is obtained when $a_i = \beta - 1$ for all $0 \leq i \leq p - 1$. In this case $m = (\beta - 1)(1 + 1/\beta + \dots + (1/\beta)^{p-1}) = \beta(1 - (1/\beta)^p) < \beta$.

3. As for the third part, if $x = (m_0 + m_1\beta^{-1} + \dots + m_{p-1}\beta^{-(p-1)})\beta^e$, then the successor of x is obtained by adding 1 unit to the least significant digit of its mantissa, leading to the following identity:

$$succ(x) = x + \beta^{-p+1}\beta^e = (m + \beta^{-p+1})\beta^e \qquad (1.4)$$

Thus $succ(x) - x = \beta^{-p+1}\beta^e$ and

$$\frac{succ(x) - x}{x} = \frac{\beta^{-p+1}\beta^e}{m \times \beta^e} = \frac{\beta^{-p+1}}{m} \leq \beta^{-p+1} \qquad (1.5)$$

since $m \geq 1$. ∎

Definition 1.2 *In a floating point system* $\mathbb{F}(\beta, p, e_{\min}, e_{\max})$, *the* **system epsilon** *or* **epsilon machine** *is defined by the parameter* ϵ_M:

$$\epsilon_M = \beta^{-p+1}.$$

Clearly ϵ_M is a measure of the precision of the system, since according to (1.5) it is a maximum bound on the relative distance between two consecutive numbers in $\mathbb{F}(\beta, p, e_{\min}, e_{\max})$. Furthermore, note that equation (1.4) can be written as:

$$succ(x) = (m + \beta^{-p+1})\beta^e$$

from which one concludes that ϵ_M also represents the difference between the mantissas of two successive positive numbers in F.

As a direct application, we consider the following example:

Example 1.13 *Display the elements of the floating point system* $\mathbb{F} = \mathbb{F}(10, 3, -2, +3)$.

For non-zero numbers, we shall display only the positive elements; the negative ones being deduced by symmetry. This is done in Table 1.2. In this decimal floating point system, the following parameters in $\mathbb{F}$ are easily computed:

- $x_{\min} = 1.00 \times 10^{-2}$

- $x_{max} = 9.99 \times 10^3$

- $\epsilon_M = 10^{-2} = 0.01$.

- To represent zero, one might consider ± 0. For that purpose, we adopt a convention whereby ± 0 is represented by a 0 mantissa, regardless of the exponent. Therefore zero $\in \mathbb{F}(10, 3, -1, 2)$, and it is represented by $\pm\ 0.00 \times 10^e$ for any value of e.

- The total number of elements in $\mathbb{F}$ is

$$card(\mathbb{F}) = 2 \times [(9 \times 10^2) \times 6] + 2 = 10802$$

Moreover, the absolute distances between 2 successive or neighboring floating point numbers in $\mathbb{F}$, increase and are computed as in Table 1.3. ■

These results can be generalized and extended to any floating point system $\mathbb{F} = \mathbb{F}(\beta, p, e_{min}, e_{max})$. Absolute distances decrease towards zero, on intervals that are subset of $(0, \beta)$ and in contrast these distances increase on intervals in $[\beta, x_{max}]$ towards x_{max}, with

$$\max_{x \in (-\beta, +\beta) \cap \mathbb{F}} |x - \text{succ}(x)| \le \epsilon_M,$$

We note also that the ϵ-machine $\epsilon_M = \beta^{1-p}$ being the smallest upper bound of **relative distances** in $\mathbb{F}$ coincides with the smallest absolute distance between successive points **only** on the interval $[1, \beta)$. The following table summarizes such fact. Thus, when computing in $\mathbb{F}$, criteria for "numerical convergence" should be preferably established in terms of relative errors and not absolute ones.

Positive numbers in $\mathbb{F}(10, 3, -2, 3)$
1.00×10^{-2}
1.01×10^{-2}
....
9.98×10^{-2}
9.99×10^{-2}
1.00×10^{-1}
1.01×10^{-1}
....
9.98×10^{1}
9.99×10^{-1}
1.00×10^{0}
1.01×10^{0}
....
9.98×10^{0}
9.99×10^{0}
1.00×10^{1}
1.01×10^{1}
....
9.98×10^{1}
9.99×10^{1}
1.00×10^{2}
1.01×10^{2}
....
9.98×10^{2}
9.99×10^{2}
1.00×10^{3}
1.01×10^{3}
....
9.98×10^{3}
9.99×10^{3}

TABLE 1.2: Display of the elements in $\mathbb{F}(10, 3, -2, 3)$

Interval	Neighboring numbers distance
$[10^{-2}, 10^{-1})$	$\epsilon_M \times 10^{-2} = 10^{-4}$
$[10^{-1}, 1)$	$\epsilon_M \times 10^{-1} = 10^{-3}$
$[1, 10^{1})$	$\epsilon_M \times 10^{0} = 10^{-2} = \epsilon_M$
$[10^{1}, 10^{2})$	$\epsilon_M \times 10^{1} = 10^{-1}$
$[10^{2}, 10^{3})$	$\epsilon_M \times 10^{2} = 1$
$[10^{3}, 10^{4})$	$\epsilon_M \times 10^{3} = 10$

TABLE 1.3: Absolute distances between successive numbers in the floating point system $\mathbb{F}(10, 3, -2, 3)$

Interval	Neighboring numbers distance
....	....
$[1/\beta^3, 1/\beta^2)$	β^{-p-2}
$[1/\beta^2, 1/\beta)$	β^{-p-1}
$[1/\beta, \beta)$	β^{-p}
$[1, \beta)$	$\beta^{-p+1} = \epsilon_M$
$[\beta, \beta^2)$	β^{-p+2}
$[\beta^2, \beta^3)$	β^{-p+3}
....	....

TABLE 1.4: Absolute distances between successive numbers in a general floating point system $\mathbb{F}(\beta, p, e_{\min}, e_{\max})$

4 bytes, a total of 32 bits		
t sign	biased exponent c	f part of mantissa m
1 bit	8 bits	23 bits

FIGURE 1.1: A word of 4 bytes in IEEE single precision

1.4.2 IEEE Floating Point Systems

A computer operating in binary normalized floating point mode represents numbers as described earlier except for the limitation imposed by the finite word length. In this section, we shall describe the **internal representation and storage** of numbers for IEEE floating point systems. Addressable words of 4 bytes (32 bits or digits) and 8 bytes (64 bits) are used respectively in single and double precision floating point systems referred to as $\mathbb{F}_s$ and $\mathbb{F}_d$. In what follows, we analyze some properties of these systems successively.

1. **IEEE single precision floating point system**

   By single-precision IEEE floating point numbers, we mean all acceptable numbers belonging to the normalized floating point system $\mathbb{F}_s = \mathbb{F}(2, 24, -126, +127)$, where a non-zero number x stored in a word of 4 bytes is organized as follows:

   $$x = \pm(1.f)_2 \times 2^e = (-1)^t (1.f)_2 \times 2^{c-127}$$

   according to Figure 1.1. Note the following:
   (i) In $\mathbb{F}_s$, if $x \neq 0$, the first bit in the mantissa is always 1, so that this bit does not have to be stored. The stored mantissa consists of the rightmost 23 bits and contains the fractional part f with an understood binary point. So the mantissa actually corresponds to 24 binary digits since there is a **hidden bit**. Moreover, the mantissa of each non-zero positive number is restricted by the mantissas of x_{min} and x_{max}, satisfying the

following inequality:

$$1.000...000 \leq (1.f)_2 \leq 1.111....11$$

(ii) In order to store positive numbers only, the **biased exponent** c is introduced, with $e = c - 127$. The values of c in $\mathbb{F}_s$ are bounded as follows:

$$(0)_{10} = (00\ 000\ 000)_2 < c < (11\ 111\ 111)_2 = (255)_{10}$$

The values $c = 0$ and $c = 255$ are reserved for special machine numbers obtained in calculations, that are not elements of $\mathbb{F}_S$.

Thus, the value $c = 0$ is reserved for ± 0 and the **subnormal or denormalized numbers** (in case of underflow in the computations), while the value $c = 255$ includes $\pm \infty$ (in case of overflow in the computations) and "undefined" **NaN** numbers as for example: $0/0, \infty/\infty, x_d/x_d, \infty - \infty, ...$ The sign of NaN has no meaning, but it may be predictable in some circumstances; most applications (as MATLAB for example) ignore its sign , and place such elements by "sort functions" at the high end of positive numbers. Note also that once generated, a NaN propagates through all subsequent computations.

The value of the biased exponent c in $\mathbb{F}_s$, $\forall x \neq 0$, is thus strictly restricted by the inequality:

$$(1)_{10} = (00\ 000\ 001)_2 \leq c \leq (11\ 111\ 110)_2 = (254)_{10}$$

or equivalently

$$-126 \leq e \leq 127.$$

We may then extend Definition 1.1 as follows to the IEEE single precision system.

Definition 1.3 *Let x be a machine number in $\mathbb{F}_s(2, 24, -126, +127)$, where the biased exponent $c = e + 127$, then:*

a- If $1 \leq c \leq 254$, i.e., $-126 \leq e \leq 127$: $x = (-1)^t (1.f) \times 2^{c-127}$. Moreover, if $t = 1$ then $x < 0$ and if $t = 0$ then $x > 0$.

b- The case $c = 0$ is reserved for special number representations: 0 and denormalized numbers:

- The case $c = f = 0$ is reserved for the zeros, where $|x| = 0$. By convention we write $x = \pm 0$.

- The case $c = 0$, and $f \neq 0$, is used to fill the gap between 0 and x_{min} (or $-x_{min}$ and 0), with **denormalized numbers**. By convention, we write $x = x_d = \pm 0.f \times 2^{-126}$.

c- $c = 255$ is reserved for representations of $\pm \infty$ and NaN numbers defined as follows:

c	f	e = c − 127	m	Number being represented
0	0	Not Applicable	0.0	±0
0	≠ 0	Not Applicable	0.f	$(-1)^t(0.f)2^{-126}$
0 < c < 255	any	−127 < e < 128	1.f	$(-1)^t(1.f)2^{c-127}$
255	0	Not Applicable	1.0	±∞
255	≠ 0	Not Applicable	1.f	NaN (Not a Number)

TABLE 1.5: The IEEE single precision system

c	Number	Representation in $\mathbb{F}(2, 24, -126, 127)$
c=0	0	0.00...00
c=1	x_{min}	$1.00..00 \times 2^{-126}$
....	....	....
c=127	1	$1.00..00 \times 2^0$
...	....	....
c=254	x_{max}	$1.11...11 \times 2^{127}$

TABLE 1.6: IEEE single precision positive elements

- The case $c = 255$ and $f = 0$ represents $x = \pm\infty$.

- The case $c = 255$ and $f \neq 0$ represents "Not a Number" written as $x = \textbf{NaN}$.

Table 1.5 provides all the elements of $\mathbb{F}_s$ while Table 1.6 gives some of its non-negative elements.

Table 1.7 gives the basic parameters of $\mathbb{F}_s$. Note that the machine epsilon $\epsilon_M = (2^{-23})_2 = (2^{1-24})_2 < (2 \times 10^{-7})_{10} < (10^{1-7})_{10}$. This implies that in a simple computation in base 10, approximately 7 significant decimal digits of accuracy may be obtained in single precision.

When more precision is needed, then **IEEE double precision** can be used. In that case each double precision floating number is stored in 2 computer memory words (8 bytes ≡ 64 bits).

Parameter	Expression(base 2)	Decimal value
x_{min}	2^{-126}	1.175494×10^{-38}
x_{max}	$(1.1...1)_2 \times 2^{127} = 2^{128}(1 - 2^{-24})$	3.402824×10^{38}
ϵ_M	2^{-23}	1.192093×10^{-7}
p	24=23+implicit bit	≈ 7

TABLE 1.7: x_{min}, x_{max}, ϵ machine and p in IEEE single precision

8 bytes, a total of 64 bits		
t sign	biased exponent c	f part of mantissa m
1 bit	11 bits	52 bits

FIGURE 1.2: A word of 8 bytes in IEEE double precision

c	f	$e = c - 1023$	m	**Number being represented**
0	0	Not Applicable	0.0	± 0
0	$\neq 0$	Not Applicable	$0.f$	$(-1)^t (0.f) 2^{-1022}$
$0 < c < 2047$	any	$-1023 < e < 1024$	$1.f$	$(-1)^t (1.f) 2^{c-1023}$
2047	0	Not Applicable	1.0	$\pm\infty$
2047	$\neq 0$	Not Applicable	$1.f$	NaN (Not a Number)

TABLE 1.8: Values in IEEE - double precision system

2. **IEEE double precision floating point system**

Definition 1.1 is also used to define the IEEE double precision system $\mathbb{F}_d = \mathbb{F}(2, 53, -1022, 1023)$, where a non-zero number in standard floating point representation corresponds to:

$$x = \pm(1.f)_2 \times 2^e = (-1)^t (1.f)_2 \times 2^{c-1023}$$

with $e = c - 1023$, and the biased exponent c verifying: $1 \leq c \leq 2046$. The system $\mathbb{F}_d$ uses a word of 8 bytes organized as indicated in Figure 1.2. On the basis of those concepts explained for $\mathbb{F}_s$, the number system $\mathbb{F}_d$ is displayed in Table 1.8 and the basic parameters for $\mathbb{F}_d$ are displayed in Table 1.9. Note that the epsilon machine $\epsilon_M = 2^{-52} \approx 2.2 \times 10^{-16} < 10^{1-16}$. This implies that in a double precision computation corresponds to approximately 16 significant decimal digits. Note that in the process of representing machine numbers in $\mathbb{F}_s$ or $\mathbb{F}_d$, it is convenient to use the **hexadecimal symbols** (base 16) to get a "compact" representation of binary contents of a computer word, whether 4 or 8 bytes. Considering the symbols A, B, C, D, E, F as representing 10, 11, 12, 13, 14, and 15, Table 1.10 provides the hexadecimal symbols representations in base 2. Representing then machine binary numbers with hexadecimal symbols

Parameter	**Expression(base 2)**	**Decimal value**
$x_{\min}$	2^{-1022}	$2.2250738507201 \times 10^{-308}$
$x_{\max}$	$(1.1...1)_2 \times 2^{1023} = 2^{1024}(1 - 2^{-53})$	$1.79769313486231 \times 10^{308}$
ϵ_M	2^{-52}	$2.220446049250313 \times 10^{-16}$
p	53=52+implicit bit	≈ 16

TABLE 1.9: $x_{\min}$, $x_{\max}$, ϵ machine and p in IEEE double precision

Hexadecimal	Binary
0	0000
1	0001
2	0010
3	0011
4	0100
5	0101
6	0110
7	0111
8	1000
9	1001
A	1010
B	1011
C	1100
D	1101
E	1110
F	1111

TABLE 1.10: Binary representations of hexadecimal symbols

is particularly easy. We need only regroup the binary digits from groups of 3 (as required in the octal system), to groups of 4. Note that the reverse procedure can also be used.

Example 1.14 *Determine the hexadecimal representation of the decimal number $d = -52.234375$ in both single precision and double precision.*

We start by converting the given number to binary, then normalize it:

- $E(x) = (52)_{10} = (64)_8 = (110\ 100)_2$
- $F(x) = (0.234375)_{10} = (0.17)_8 = (0.001\ 111)_2$
- Therefore:
  $(52.234375)_{10} = (110\ 100.001\ 111)_2 = (1.101\ 000\ 011\ 110\ )_2 \times 2^5$

In $\mathbb{F}_s(2, 24, -126, +127)$:
- The normalized mantissa of d is $m = 1.101\ 000\ 011\ 110$
- The exponent of d is $e = (5)_{10} = c - 127$ implying that the biased exponent is $c = (132)_{10} = (204)_8 = (10\ 000\ 100)_2$
The single precision machine representation of d is then:

$$[1100\ 0010\ 0101\ 0000\ 1111\ 0000\ 0000\ 0000]_2 = [C250F000]_{16}$$

In $\mathbb{F}_d(2, 53, -1022, +1023)$:
- The normalized mantissa of d is $m = 1.101\ 000\ 011\ 110$

- The exponent of d is $e = (5)_{10} = c - 1023$, and the biased exponent is therefore $c = (1028)_{10} = (2004)_8 = (10\,000\,000\,100)_2$
The double precision machine representation of d is:

$$[1100\ 0000\ 0100\ 1010\ 0001\ \ 1110\ 0000\ ...\ 00\,00]_2 = [C04A1E0000000000]_{16}$$

Example 1.15 *Determine the binary number x in $\mathbb{F}_s$ that corresponds to $[45DE4000]_{16}$, then find its decimal representation.*

The 32 bits string representation (or machine number representation) of x is:
$$[01000101110111100100000000000000]_2$$

The biased exponent is $c = (10\,001\,011)_2 = (213)_8 = (139)_{10}$, so $e = 139 - 127 = 12$. Therefore:

$$(x)_2 = +(1.101\ 111\ 001)_2 \times 2^{12}$$

Example 1.16 *Determine the machine number representation of the binary number $b = 2^{-128}$ in IEEE single precision.*

$b = 2^{-128} < 2^{-126} = x_{min}$, meaning that b is a denormalized number in single precision. Moreover, as $b = 2^{-2} \times 2^{-126} = 0.01 \times 2^{-126}$, its corresponding machine number is:

$$[00000000001000000000000000000000]_2$$

1.4.3 Denormalized Numbers in MATLAB

The default format for numbers in MATLAB is IEEE double precision. One can easily check out the denormalized numbers in the system, as indicated through the following set of commands.

```
realmin %2^(-1022)
ans =
  2.2251e-308
>> 0.5*2^(-1022)
ans =
  1.1125e-308
>> 0.25*2^(-1022)
ans =
  5.5627e-309
>> 0.125*2^(-1022)
ans =
  2.7813e-309
```

1.4.4   Rounding Errors in Floating Point Representation

Consider a general floating point system $\mathbb{F} = F(\beta, p, e_{\min}, e_{\max})$, with $\beta \geq 2$. For all $x \in \mathbb{R}$ with $x_{min} < |x| < x_{max}$, and $x \notin \mathbb{F}$, we seek for a procedure leading to the representation of x in $\mathbb{F}$. For such x, there exist x_1 and $x_2 = succ(x_1)$, with $x_1, x_2 \in \mathbb{F}$, such that $x_1 < x < x_2$. The process of replacing x by its nearest representative element in $\mathbb{F}$ is called **correctly rounding**, and the error involved in this approximation is called **round-off error**. We want to estimate how large it can be.

Definition 1.4 *The floating point representation of x in $\mathbb{F}$ is an application* $fl \colon \mathbb{R} \to \mathbb{F}$, *such that* $fl(x) = x_1$ *or* $fl(x) = x_2$ *following one of the rounding procedures defined below.*

1. **Rounding by Chopping**:
   $fl_0(x) = x_1$, if $x > 0$, (and $fl_0(x) = x_2$, if $x < 0$)
   (i.e., $fl_0(x)$ is obtained by simply dropping the excess of digits in x)

2. **Rounding to the closest**:

   (a) $fl_p(x) = x_1$ if $|x - x_1| < |x - x_2|$
   (b) $fl_p(x) = x_2$ if $|x - x_2| \leq |x - x_1|$

Remark 1.3

Note that to round $x < 0$, we could apply the above procedures to $|x|$ first, then multiply the result obtained by -1.

Remark 1.4 *Let* $x = (1.b_1..b_{23}b_{24}b_{25}...)_2$. *Rounding x in $\mathbb{F}_s$ to the closest stands as follows:*

- *If $b_{24} = 0$, then $fl_p(x) = x_1$.*

- *If $b_{24} = 1$ then $fl_p(x) = x_2$.*

Proof. To obtain this result, based on the definition above, simply note that if

$$x_1 = (1 \cdot b_1 b_2 ... b_{23})2^e, \text{ and } x_2 = succ(x_1) = x_1 + (2^{-23})2^e$$

then the midpoint of the line segment $[x_1, x_2]$ is

$$x_M = \frac{x_1 + x_2}{2} = x_1 + (2^{-24})2^e = 1 \cdot b_1...b_{23}1\,; \; (x_M \notin \mathbb{F})$$

■

Consequently, since in the general case $x_M = (x + \frac{\beta^{-p+1}}{2} \times \beta^e)$ is the midpoint of the line segment $[x, succ(x)]$, one easily verifies the following result graphically:

Theorem 1.3 *Let $x \in \mathbb{R}$ and $x \notin \mathbb{F} = F(\beta, p, e_{min}, e_{max})$, with $x_{min} < |x| < x_{max}$. Then:*

$$fl_p(x) = fl_0\left(x + \frac{\beta^{-p+1}}{2} \times \beta^e\right)$$

Example 1.17 *Let $x = (13.14)_{10}$. Find the internal representation of x using IEEE single precision notation (rounding to the closest if needed). Find then the hexadecimal representation of x.*

As a first step we convert x to a binary number:

$$x = (1101.0010001111010111000010100011111...)_2$$

We next normalize the number obtained:

$$x = (1.1010010001111010111000010100011111...)_2 \times 2^3$$

Hence, the 2 successive numbers x_1 and x_2 of $\mathbb{F}_s$ are:

$$x_1 = (1.10100100011110101110000)_2 \times 2^3$$

$$x_2 = (1.10100100011110101110001)_2 \times 2^3$$

Obviously, rounding x to the closest gives $fl_p(x) = x_2$.
Note also that $e = 3$ and $c = (130)_{10} = (10000010)_2$
is as follows:

4 bytes = 32 bits		
t	c	f
0	10000010	10100100011110101110001

or also equivalently:

01000001010100100011110101110001

with hexadecimal representation:

$$[4\ 1\ 5\ 2\ 3\ D\ 7\ 1]_{16}$$

We turn now to the error that can occur when we attempt to represent a given real number x in $\mathbb{F}$. As for relative error estimates we have the following.

Proposition 1.1 *Let $x \in \mathbb{R}$ with $x \notin \mathbb{F} = \mathbb{F}(\beta, p, e_{min}, e_{max})$ and $x_{min} < |x| < x_{max}$. Then, the representations of x in $\mathbb{F}$ verify the following relative error estimates:*

1. $\frac{|x - fl_0(x)|}{|x|} < \epsilon_M,$

2. $\frac{|x - fl_p(x)|}{|x|} \leq \frac{1}{2}\epsilon_M,$

where $\epsilon_M = \beta^{-p+1}$ is the epsilon machine of the system.

Proof. Without loss of generality, we shall prove the above properties for positive numbers. Let x_1 and x_2 be in $\mathbb{F}(\beta, p, e_{\min}, e_{\max})$, such that

$$x_1 < x < x_2 = succ(x_1).$$

Then,

$$|x - fl_0(x)| < (x_2 - x_1) \text{ and } |x - fl_p(x)| \leq \frac{(x_2 - x_1)}{2}.$$

Furthermore, given that $x_1 < x$, the estimates of the proposition are obviously verified since in both cases $\frac{x_2 - x_1}{x_1} \leq \epsilon_M$. ∎

Remark 1.5 *Note that Proposition 1.1 can be summarized by the following estimate:*

$$\frac{|x - fl(x)|}{|x|} \leq u \text{ where } u = \begin{cases} \epsilon_M, & \text{if } fl = fl_0 \\ \epsilon_M/2, & \text{if } fl = fl_P \end{cases}$$

This inequality can also be expressed in the more useful form:

$$fl(x) = x(1 + \delta) \text{ where } |\delta| \leq u \tag{1.6}$$

To see that, simply let $\delta = \frac{fl(x) - x}{x}$. Obviously $|\delta| \leq u$, with $fl(x)$ yielding the required result.

∎

Remark 1.6 *When computing a mathematical entity $E \in \mathbb{R}$ (for example, $E = \pi$, $\sqrt{2}$, $\ln 2$,..) up to r decimal figures, one seeks an approximation $\hat{E}$ to E such that $\hat{E} \in \mathbb{F}(10, r, e_{\min}, e_{\max})$, a user floating point system with a base of 10 and r significant digits. A rounding procedure to the closest would yield $\hat{E}$ satisfying the following error estimate:*

$$\frac{|E - \hat{E}|}{|E|} \leq \frac{1}{2} 10^{1-r}.$$

To illustrate, we give some examples.

Example 1.18   *1. Consider $E = \pi = 3.14159265358979... \in \mathbb{R}$. In seeking for the representative $\hat{E}$ of $\pi \in \mathbb{F} = \mathbb{F}(10, 6, e_{\min}, e_{\max})$, we first look for 2 successive numbers x_1 and x_2 in $\mathbb{F}$ such that*

$$x_1 \leq E \leq x_2.$$

Obviously $x_1 = 3.14159$ and $x_2 = 3.14160$. Rounding to the closest would select $\hat{E} = 3.14159$, with

$$\frac{|E - \hat{E}|}{|E|} \leq \frac{1}{2} \frac{|x_2 - x_1|}{x_1} = 1.59155077526 \times 10^{-6} \leq \frac{1}{2} 10^{1-6} = 5 \times 10^{-6} = \frac{\epsilon_M}{2}$$

2. *Similarly,* $\hat{E} = 1.4142136$ *approximates* $E = \sqrt{2}$ *up to 8 significant figures. Since*

$$x_1 = 1.4142135 < \sqrt{2} = 1.414213562373095... < x_2 = 1.4142136$$

and

$$\frac{|x_2 - x_1|}{2x_1} = \frac{7.071067628}{2} \times 10^{-8} = 0.35 \times 10^{-7} < 0.5 \times 10^{1-8} = \frac{\epsilon_M}{2}.$$

1.5   Floating Point Operations

For a given arithmetic operation $\cdot = \{+, -, \times, \div\}$ in $\mathbb{R}$, we define respectively in $\mathbb{F}$ the **floating point operations**: $\odot = \{\oplus, \ominus, \otimes, \oslash\}$, i.e.,

$$\odot : \mathbb{F} \times \mathbb{F} \to \mathbb{F}$$

Each of these operations is called a **flop** and, according to IEEE standards, is designed as follows.

Definition 1.5 *In the standards of floating point operations in IEEE convention:*

$$\forall x \ and \ y \in \mathbb{F}, \ x \odot y = fl(x \cdot y).$$

This definition together with (1.6) leads to the following estimate:

$$x \odot y = (x \cdot y)(1 + \delta), \text{ with } |\delta| \leq \mathbf{u},$$

where $u = \epsilon_M$ or $u = \frac{\epsilon_M}{2}$, depending on the chosen rounding procedure. Practically, Definition 1.5 means that $x \odot y$ is computed according to the following steps:

- First: correctly in $\mathbb{R}$ as $x \cdot y$

- Second: normalizing in $\mathbb{F}$

- Third: rounding in $\mathbb{F}$

Under this procedure, the relative error will not exceed u.

Remark 1.7 *Let* $x, y \in \mathbb{F} = \mathbb{F}(\beta, p, e_{min}, e_{max})$.

$$x \oplus y = fl(x + y) = (x + y)(1 + \delta) = x(1 + \delta) + y(1 + \delta)$$

meaning that $x \oplus y$ *is not precisely* $(x + y)$, *but is the sum of* $x(1 + \delta)$ *and* $y(1 + \delta)$, *or also that it is the exact sum of a slightly perturbed* x *and a slightly perturbed* y.

Example 1.19 *If x, y, and z are numbers in $\mathbb{F}_s$, what upper bound can be given for the relative round-off error in computing $z \otimes (x \oplus y)$, with rounding to the closest $(fl = fl_p)$.*
In the computer, the innermost calculation of $(x + y)$ will be done first:

$$fl(x + y) = (x + y)(1 + \delta_1) \,, |\delta_1| \leq 2^{-24}$$

Therefore:

$$fl[z \, fl(x + y)] = z \, fl(x + y)(1 + \delta_2) \,, |\delta_2| \leq 2^{-24}$$

Putting both equations together, we have:

$$fl[z \, fl(x + y)] = z(x + y)(1 + \delta_1)(1 + \delta_2) = z(x + y)(1 + \delta_1 + \delta_2 + \delta_1\delta_2)$$

$$= z(x + y)(1 + \delta_1 + \delta_2) = z(x + y)(1 + \delta),$$

where $\delta = \delta_1 + \delta_2$.

In this calculation, we neglect $|\delta_1\delta_2| \leq 2^{-48}$. Moreover, $|\delta| = |\delta_1 + \delta_2| \leq |\delta_1| + |\delta_2| \leq 2^{-24} + 2^{-24} = 2^{-23}$ ∎

Although rounding errors are usually small, their accumulation in long and complex computations may give rise to unexpected wrong results, as shown in the following example:

Example 1.20 *Consider the following sequence of numbers:*

$$I_1 = 1, \; I_n = \frac{2}{n}[(\frac{1}{n})^3 + (\frac{2}{n})^3 + \dots + (\frac{n-1}{n})^3 + \frac{1}{2}], \; n = 2, 3, \dots \qquad (1.7)$$

It can be proved that $\lim_{n \to \infty} I_n = 0.5$.

However, when we compute I_n in single precision MATLAB, we obtain the results displayed in Table 1.11, which clearly shows that the relative errors for $n = 2^p$, $p = 7, 8, 9, 10, 11, 12, 13, 14$. One can check that such relative errors decrease for $p \leq 11$ and stop following a decreasing pattern for $p > 11$, vastly because of round-off errors propagation.

A similar case regarding (non-)convergence due to rounding errors can be also found in [26], p. 7. We look now for specific problems caused by rounding errors propagation.

1.5.1   Algebraic Properties in Floating Point Operations

Since $\mathbb{F}$ is a proper subset of $\mathbb{R}$, elementary algebraic operations on floating point numbers do not satisfy all the properties of analogous operations in $\mathbb{R}$. To illustrate, let x, y, $z \in \mathbb{F}$. The floating point arithmetic operations verify the following properties:

n	I_n	$\frac{\lvert I - I_n \rvert}{\lvert I \rvert}$
128	5.0003052×10^{-1}	6.1035156×10^{-5}
256	5.0000769×10^{-1}	1.5377998×10^{-5}
512	5.0000197×10^{-1}	3.9339066×10^{-5}
1024	5.0000048×10^{-1}	9.5367432×10^{-7}
2048	4.9999997×10^{-1}	5.9604645×10^{-8}
4096	5.0000036×10^{-1}	7.1525574×10^{-7}
8192	4.9999988×10^{-1}	2.3841858×10^{-7}
16384	5.0000036×10^{-1}	7.1525574×10^{-7}

TABLE 1.11: Effects of round-off error propagation on the convergence of the sequence I_n defined in (1.7)

1. Floating point addition is commutative in $\mathbb{F}$

$$x \oplus y = fl(x + y) = fl(y + x) = y \oplus x$$

2. Floating point multiplication is commutative in $\mathbb{F}$

$$x \otimes y = y \otimes x$$

3. Floating point addition is not associative in $\mathbb{F}$

$$(x \oplus y) \oplus z \not\equiv x \oplus (y \oplus z)$$

4. Floating point multiplication is not associative in $\mathbb{F}$

$$(x \otimes y) \otimes z \not\equiv x \otimes (y \otimes z)$$

5. Floating point multiplication is not distributive with respect to floating point addition in $\mathbb{F}$

$$x \otimes (y \oplus z) \neq (x \otimes y) \oplus (x \otimes z)$$

Example 1.21 *Let* $x = 3.417 \times 10^0$, $y = 8.513 \times 10^0$, $z = 4.181 \times 10^0 \in \mathbb{F}(10, 4, -2, 2)$. *Verify that addition is not associative in* $\mathbb{F}$.

$x \oplus y = 1.193 \times 10^1$ and $(x \oplus y) \oplus z = 1.611 \times 10^1$,
while: $y \oplus z = 1.269 \times 10^1$ and $x \oplus (y \oplus z) = 1.610 \times 10^1$. ∎

Particularly, associativity is violated whenever a situation of overflow occurs as in the following example.

Example 1.22 *Let* $a = 1 * 10^{308}$, $b = 1.01 * 10^{308}$ *and* $c = -1.001 * 10^{308}$ *be 3 floating point numbers in* F_D *expressed in their decimal form.*

$$a \oplus (b \oplus c) = 1 * 10^{308} \oplus 0.009 * 10^{308} = 1.009 * 10^{308}$$

while

$$(a \oplus b) \oplus c = \infty$$

since $(a \oplus b) = 2.01 * 10^{308} \equiv \infty > x_{max} \approx 1.798 * 10^{308}$ in F_D ∎

1.5.2   The Problem of Absorption

Let x, y be two non-zero numbers $\in \mathbb{F}_s$, with

$$x = m_x \times 2^{e_x}, \, y = m_y \times 2^{e_y}$$

Assume $y < x$, so that:

$$x + y = (m_x + m_y \times 2^{e_y - e_x}) \times 2^{e_x}.$$

Clearly, since $m_y < 2$, if also $e_y - e_x \leq -25$, then

$$x + y < (m_x + 2^{-24}) \times 2^{e_x} = (x + succ(x))/2.$$

Hence using $fl = fl_p$, one gets:

$$x \oplus y = fl_p(x + y) = x,$$

although $y \neq 0$. In such a situation, we say that y is **absorbed** by x.

Definition 1.6 *(Absorption Phenomena) Let x and y be 2 non-zero elements in $\mathbb{F}(\beta, p, e_{\min}, e_{\max})$. y is said to be absorbed by x, if $x \oplus y = x$.*

Example 1.23 *Consider the sum of n decreasing positive numbers $\{x_i | i = 1, ..n\}$, with $x_1 > x_2 > ... > x_i > x_{i+1} > ... > x_n$, and let $S_n = \sum_{i=1}^{n} x_i$. There are two obvious ways to program this finite series; by increasing or decreasing index. The corresponding algorithms are as follows:*

Algorithm 1.6 Harmonic Series Evaluation by Increasing Indices

```
% Input : x=[x(1),...,x(n)]
% Output : sum of all components of x by Increasing index
function S=sum1(x)
S=0 ;
n=length(x) ;
for i=1:n
 S=S+x(i)
end
```

which leads then for example for $n = 4$ to the floating point number

$$S_1 = (((x_1 \oplus x_2) \oplus x_3) \oplus x_4).$$

Algorithm 1.7 Harmonic Series Evaluation by Decreasing Indices

```
function S=sum2(x)
% Input x=[x(1),...,x(n)]
% Output : sum of all components of x by Decreasing index
S=0 ;
n=length(x) ;
for i=n:-1:1
 S=S+x(i)
end
```

which gives for $n = 4$, the floating point number

$$S_2 = (((x_4 \oplus x_3) \oplus x_2) \oplus x_1)$$

Obviously, $S_1 \neq S_2$ and S_2 is more accurate than S_1 that favors the absorption phenomena.

Example 1.24 *Consider the following sequence of numbers in* $\mathbb{F}(10, 4, -3, 3)$, $x_1 = 9.999 \times 10^0$, $x_2 = 9.999 \times 10^{-1}$, $x_3 = 9.999 \times 10^{-2}$ and $x_4 = 9.999 \times 10^{-3}$.

The exact value of $\sum_{i=1}^4 x_i$ is $11.108899 = 1.1108899 \times 10^1$. Using rounding by chopping, for example, the first algorithm would give 1.108×10^1 while the second provides 1.110×10^1! ∎

Example 1.25 *Consider Euler's number* $e = 2.718217.......$ *It is given by the Taylor's series expansion of* e^x *for* $x = 1$:

$$e = 1 + \frac{1}{1!} + \frac{1}{2!} + ... \frac{1}{n!} + ...$$

Computing e up to 8 significant figures with rounding to the closest and using 11 terms, one gets, summing up by increasing n:

$$1 + \frac{1}{1!} + \frac{1}{2!} + ... \frac{1}{10!} = 2.7182820,$$

while summing by decreasing n, one obtains:

$$\frac{1}{10!} + \frac{1}{9!} + ... + \frac{1}{1!} + 1 = 2.7182817.$$

1.5.3   The Problem of Cancellation or Loss of Precision

A loss of significance can occur when computing in normalised floating point systems. This problem of cancellation occurs when subtracting two positive floating point numbers of almost equal amplitude. The closer the numbers are, the more pronounced is the problem. To start, consider the following example.

Example 1.26 *Let x_1, $x_2 \in \mathbb{F}(10, 5, -3, 3)$. To subtract $x_2 = 8.5478 \times 10^3$ from $x_1 = 8.5489 \times 10^3$, the operation is done in two steps:*

x_1	8.5489×10^3
x_2	8.5478×10^3
$x = x_1 - x_2$	0.0011×10^3
Normalized result	1.1000×10^0

Hence the result appears to belong to a new floating point system $\mathbb{F}(10, 2, -3, 3)$ that is less precise ($p = 2$) than the original one ($p = 5$). Three zeros have been supplied in the last three least significant fractional places. We are experiencing the phenomenon of **Cancellation** that causes **loss of significant figures** in floating point computation. This can be summarized by the following proposition.

Proposition 1.2 *Let $x, y \in \mathbb{F} = \mathbb{F}(\beta, p, e_{min}, e_{max})$. Assume x and y are two numbers of the same sign and the same order, $(|x|, |y| = O(\beta^e))$. Then there exists $k > 0$, such that $x - y$ is represented in a less precise floating point system $\mathbb{F}(\beta, p - k, e_{min}, e_{max})$.*

Proof. Assume the two numbers x and y are expressed as follows.

$$x = (a_0 + a_1\beta^{-1} + ... + a_k\beta^{-k} + ... + a_{p-1}\beta^{-p+1}) \times \beta^e$$

and

$$y = (a'_0 + a'_1\beta^{-1} + ... + a'_k\beta^{-k} + ... + a'_{p-1}\beta^{-p+1}) \times \beta^e$$

with $a_i = a'_i$ for $i \leq k - 1 < p - 1$. It is obvious that:
$x - y = ((a_k - a'_k)\beta^{-k} + ... + (a_{p-1} - a'_{p-1})\beta^{-p+1}) \times \beta^e = (c_k\beta^{-k} + ... + c_{p-1}\beta^{-p+1}) \times \beta^e$
Hence: $x - y = (c_k + ... + c_{p-1}\beta^{-(p-k-1)}) \times \beta^{e-k}$, with $c_k \neq 0$
Consequently, $x - y$ is represented in a system in which precision is $p - k$. ∎

Example 1.27 *Let $x_1 = 1.00000000000000000000011 \times 2^{-126}$ and $x_2 = 1.00000000000000000000010 \times 2^{-126}$ be 2 numbers $\in \mathbb{F}_S(2, 24, -126, +127)$. To subtract x_2 from x_1:*

x_1	$1.00000000000000000000011 \times 2^{-126}$
x_2	$1.00000000000000000000010 \times 2^{-126}$
$x = x_1 - x_2$	$0.00000000000000000000001 \times 2^{-126}$
Normalized result	$2^{-149} < x_{min}$

In that extreme case, rounding the result to the closest gives $fl_p(x) = 0$, although $x_1 \neq x_2$!

Example 1.28 *Alternate series and the phenomenon of cancellation.*

Consider the example of computing $\exp(-a)$, $a > 0$. For that purpose, we choose one of the following alternatives:

1. A straightforward application of the Taylor's series representation of $\exp(x)$, giving for $x = -a$, an alternating series:

$$\exp(-a) = 1 - a + \frac{a^2}{2!} - \frac{a^3}{3!} + \frac{a^4}{4!} + ... + (-1)^n \frac{a^n}{n!} + ..., \qquad (1.8)$$

2. On the other hand, computing first $\exp(a)$ for $a > 0$, using the same series representation, which however has all its terms positive,

$$\exp(a) = 1 + a + \frac{a^2}{2!} + \frac{a^3}{3!} + \frac{a^4}{4!} + ... + \frac{a^n}{n!} + ..., \qquad (1.9)$$

followed up by an inverse operation:

$$\exp(-a) = 1/\exp(a). \qquad (1.10)$$

would yield more accurate results.

Computing with the first power series for large negative values of a, leads to drastic cancellation phenomena, while the second alternative provides accurate results as the following example indicates.

Example 1.29 *Consider the computation of* $\exp(-20)$, **which exact value is** $2.061153622438558 \times 10^{-9}$.

The implementation of the following 2 algorithms is done in MATLAB, which uses double precision IEEE formats:

Algorithm 1.8 Implementing e^x: **Alternative 1**

```
function y=myexp(x)
tol=0.5*10^(-16);
y=1;
k=1;
T=x;
while abs(T)/y>tol;
    y=y+T;k=k+1;T=T*x/k;
end
```

Algorithm 1.9 Implementing e^x: **Alternative 2**

```
function y=myexp(x)
tol=0.5*10^(-16);
y=1;
k=1;
v=abs(x);
```

```
T=v;
while abs(T)/y>tol;
    y=y+T;k=k+1;T=T*v/k;
end
if x<0
    y=1/y;
end
```

The results came as follows.

First alternative (1.8)	Value
	-19
Second alternative (1.10)	Value
	$2.061153622438558 \times 10^{-9}$

Another example deals with the computation of the roots of a quadratic equation.

Example 1.30 *Consider the computation of the roots of $x^2 + 2bx + c = 0$, where c is a positive number "much smaller" than b^2.*

There are 2 ways for handling the numerical computation of the solutions to this obvious problem.

1. A straightforward application of the well-known formulae:

$$x_1 = -b - \sqrt{b^2 - c} \approx -2b; \quad x_2 = -b + \sqrt{b^2 - c} \approx 0. \qquad (1.11)$$

   There is obviously, in this way, loss of significant figures when computing x_2

2. However, computing first x_1 then using

$$x_2 = \frac{c}{x_1} \qquad (1.12)$$

   would not result in loss of digits.

1.6   Computing in a Floating Point System

Clearly in normalized floating point systems $\mathbb{F} = \mathbb{F}(\beta, p, e_{min}, e_{max})$, no irrational nor rational numbers that do not fit the finite format imposed by the computer can be represented, neither too large nor too small real numbers are. Thus the effective number system for a computer is not a continuum, but rather a non-uniformly distributed finite subset of the rational numbers, i.e., a "strange" set of rational numbers with irregular gaps. The total number of elements in $\mathbb{F}$ is easily computed and is given by:

$$card(\mathbb{F}) = 2(\beta - 1)(\beta)^{p-1}(e_{max} - e_{min} + 1) + 2 \qquad (1.13)$$

Note that this count excludes the denormalized numbers, but includes ± 0.
In what follows, we analyze particularly some cardinality and distribution properties of floating point systems $\mathbb{F}$, where the exponents are such that $e_{max} = |e_{min}| + 1$, as for example the cases of the IEEE single and double precision systems F_s and F_d.

1.6.1 Cardinality and Distribution of Floating Point Systems

Let $\mathbb{F} = \mathbb{F}(2, p, E_{min}, E_{max})$, with $E_{max} = |E_{min}| + 1$, and $E_{min} < 0$. Note that

$$card(\mathbb{F}) = 2 * card(\mathbb{F}_+) + 2,$$

where $\mathbb{F}_+$ is the set of all non-zero positive elements of $\mathbb{F}$. Based on (1.13), it can be easily shown that:

$$card(\mathbb{F}_+) = 2^{p-1}(E_{max} + |E_{min}| + 1).$$

Hence:

$$N_F = card(\mathbb{F}) = 2^p(E_{max} + |E_{min}| + 1) + 2$$

Since also $E_{max} = |E_{min}| + 1$, then:

$$N_F = 2^p(2E_{max}) + 2 = 2^{p+1}(E_{max}) + 2.$$

On the other hand, if we consider now $\mathbb{F}_0$, the subset of non-zero elements of $\mathbb{F}$ defined as follows:

$$\mathbb{F}_0 = \{x \in \mathbb{F} | x = \pm 1.f \times 2^e, \ E_{min} \le e \le 0\}$$

one finds that:

$$N_{F_0} = card(\mathbb{F}_0) = 2^p(E_{max})$$

since in that case the number of different values taken by the exponent in $\mathbb{F}$ is

$$|E_{min}| + 1 = E_{max}$$

Note now that N_{F_0} represents half of the total of the non-zero elements of $\mathbb{F}$, since:

$$\frac{N_{F_0}}{N_F - 2} = \frac{2^p(E_{max})}{2^{p+1}(E_{max})} = \frac{1}{2}. \qquad (1.14)$$

This leads to the following proposition:

Proposition 1.3 *In a floating point system* $\mathbb{F}(2, p, E_{\min}, E_{\max})$, *with* $E_{\max} = |E_{\min}|+1$, *half of the non-zero floating point numbers are located in the interval* $(-2, 2)$ *with the other half located in* $[-x_{\max}, -2] \cup [2, x_{\max}]$.
Proof. This follows from formula (1.14).   ∎

It is also worth noting that all floating point numbers $\pm 1.f \times 2^e$ become integers for $e \geq p-1$. These facts are visualized in the simulation that follows in the next section.

1.6.2   A `MATLAB` Simulation of a Floating Point System

The following function generates the non-negative numbers of a floating point system $\mathbb{F}(b, p, emin, emax)$.

Algorithm 1.10 Simulation of a Floating Point System

```
function x=float_v(b,p,emin,emax)
x=[];
epsm=b^(-p+1);
M=1:epsm:b-epsm;
E=1;
for e=0:emax
x=[x M*E];
    E=b*E;
end
E=1/b;
for e=-1:-1:emin
x=[M*E x];
    E=E/b;
end
x=[0 x];
```

As a result, we plot respectively in Figures 1.3 and 1.4 the distribution of non-negative numbers of $\mathbb{F}(2, 4, -6, 7)$ and $\mathbb{F}(2, 3, -3, 4)$.

1.6.3   Tips for Floating Point Computation

To conclude, we may set an ensemble of rules that could avoid situations where accuracy can be jeopardized by the propagation of rounding errors through all type of floating point operations and more particularly through absorption and cancellation. When programming in finite precision arithmetic requests, some safeguarding habits are useful **whenever possible**. For example:

1. Seek always algorithms that would solve numerically a problem with the least number of flops.

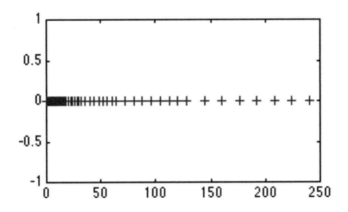

FIGURE 1.3: Distribution of numbers in $\mathbb{F}(2, 4, -6, 7)$

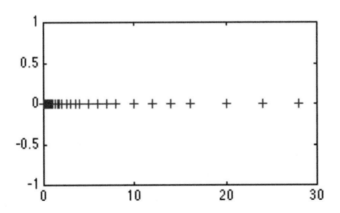

FIGURE 1.4: Distribution of numbers in $\mathbb{F}(2, 3, -3, 4)$

2. Use Taylor's series expansions to avoid loss of significant figures.

3. Avoid using alternating series in case the solution to the problem can be obtained using a series of positive (or negative) numbers.

4. Sum up positive elements of a series by adding from the smallest to the largest.

1.7 Exercises

1. Find the binary representation of the following decimal numbers.

    (a) $e \approx (2.718)_{10}$

    (b) $\frac{7}{8}$

    (c) $(792)_{10}$

2. Convert the following decimal numbers to octal numbers.

    (a) 37.1

    (b) 12.34

    (c) 3.14

    (d) 23.38

    (e) 75.231

    (f) 57.231

3. Convert the following binary numbers to octals and then to decimal numbers.

    (a) $(110\ 111\ 001.101\ 011\ 101)_2$

    (b) $(1\ 001\ 100\ 101.011\ 01)_2$

4. Convert the following numbers as required.

    (a) $(100\ 101\ 101)_2 = ($ $)_8 = ($ $)_{10}$

    (b) $(0.782)_{10} = ($ $)_8 = ($ $)_2$

    (c) $(47)_{10} = ($ $)_8 = ($ $)_2$

    (d) $(0.47)_{10} = ($ $)_8 = ($ $)_2$

    (e) $(51)_{10} = ($ $)_8 = ($ $)_2$

    (f) $(0.694)_{10} = ($ $)_8 = ($ $)_2$

    (g) $(110\ 011.111\ 010\ 110\ 110\ 1)_2 = ($ $)_8 = ($ $)_{10}$

    (h) $(351.4)_8 = ($ $)_2 = ($ $)_{10}$

    (i) $(45753.127664)_8 = ($ $)_2 = ($ $)_{10}$

5. Convert $x = (0.6)_{10}$ first to octal and then to binary. Check your result by converting directly to binary.

6. Prove that the decimal number $\frac{1}{7}$ does not have a finite expansion in the binary system.

7. Prove or disprove - by giving a counter example - the following statements:

   (a) Any real number that has a finite representation in the binary number system is of the form $\pm m/2^n$, where n and m are positive integers.

   (b) Any real number of the form $\pm m/2^n$ has a finite representation in the binary number system

   (c) Any number that has a finite representation in the binary system must have a finite representation in the decimal system.

   (d) Any number that has a finite representation in the decimal system must have a finite representation in the binary system.

   (e) A number has a finite representation in the octal system if and only if it has a finite representation in the binary system.

8. Display the positive elements of the floating point system $\mathbb{F} = \mathbb{F}(2, 3, -2, +3)$. Determine the cardinality of $|F$.

9. Determine the IEEE single precision representation of the decimal number 64.015625.

10. Determine the IEEE single and double precision representations of the following decimal numbers:

    (a) 0.5, −0.5

    (b) 0.125, −0.125

    (c) 0.03125, −0.03125

    (d) 1.0, −1.0

    (e) +0.0, −0.0

    (f) −987.0054321

    (g) 385.65

    (h) 10^{-2}

11. Identify the decimal floating point numbers corresponding to the following bit strings in the IEEE single precision system:

    (a) 0 00000000 00000000000000000000000

    (b) 1 00000000 00000000000000000000000

    (c) 0 11111111 11111111111111111111111

    (d) 1 11111111 11111111111111111111111

    (e) 0 00000001 00000000000000000000000

    (f) 0 10000001 11110000000000000000000

(g) 0 01111111 00000000000000000000000

(h) 0 01111011 11111001100110011001101

12. In the IEEE single precision system, what are the bit-string representation for the following sub-normal numbers?

  (a) $2^{-128} + 2^{-139}$

  (b) $2^{-132} + 2^{-145}$

  (c) $2^{-129} + 2^{-130}$

  (d) $\sum_{k=127}^{149} 2^{-k}$

13. Determine the decimal numbers that have the following IEEE single precision system representations:

  (a) $[3F27E520]_{16}$

  (b) $[CA3F2900]_{16}$

  (c) $[C705A700]_{16}$

  (d) $[494F96A0]_{16}$

  (e) $[4B187ABC]_{16}$

  (f) $[45223000]_{16}$

  (g) $[45607000]_{16}$

  (h) $[C553E100]_{16}$

  (i) $[437F0001]_{16}$

14. Convert the greatest positive element in single precision to an octal number "o" and write it in normalized floating point notation. Convert then the resulting "o" to a decimal number "d" and write it in normalized floating point notation.

15. (a) Identify the binary number x whose 32 bit-string representation in single precision is as follows:

  $$1 \quad 00000001 \quad 00000000000000000000000$$

  (b) Find the next largest and smallest machine numbers in single precision for the number x given above, then write their hexadecimal representation.

16. Consider the binary number $b = 1.01 \times 2^{+128}$.

  (a) Write the machine number representing b in IEEE double precision, then write its corresponding hexadecimal representation.

  (b) Write the machine number representing b in IEEE single precision, then write its corresponding hexadecimal representation.

(c) Let x_M be the midpoint of the interval $[0, b]$. Write the machine number representing x_M in IEEE single precision, then write its hexadecimal representation.

17. Consider the binary number $b = 2^{-127} + 2^{-130}$.

(a) Write the machine number representing b in IEEE single precision, then write its corresponding Hexadecimal representation.

(b) Find the successor of b ($y = succ(b)$) in IEEE single precision, then write its corresponding machine number and Hexadecimal representation.

(c) Write the machine number representing b in IEEE double precision, then write its corresponding hexadecimal representation.

(d) Find the predecessor of b ($x = pre(b)$) in IEEE double precision, then write its corresponding machine number and hexadecimal representation.

18. For some values of x, the following functions cannot be accurately computed by using the given formula. Explain and find a way around the difficulty.

(a) $f(x) = \sqrt{x^2 + 1} - x$

(b) $f(x) = \sqrt{x^4 + 4} - 2$

(c) $f(x) = \sqrt{x + 2} - \sqrt{x}$

(d) $f(x) = (\sqrt{x + 4})^{1/2} - (\sqrt{x})^{1/2}$

19. For some values of x, the following functions cannot be accurately computed by using the given formula. Explain and find a way around the difficulty.

(a) $f(x) = 1 - \sin x$

(b) $f(x) = 1 - \cos x$

(c) $f(x) = 2\cos^2 x - 1$

(d) $f(x) = (\cos x - e^{-x})/\sin x$

(e) $f(x) = e^x - \sin x - \cos x$

20. For some values of x, the following functions cannot be accurately computed by using the given formula. Explain and find a way around the difficulty.

(a) $f(x) = \tanh x = \frac{e^x - e^{-x}}{e^x + e^{-x}}$

(b) $f(x) = \frac{1}{x^3}(\sinh x - \tanh x)$

21. For some values of x, the following functions cannot be accurately computed by using the given formula. Explain and find a way around the difficulty.

   (a) $f(x) = \ln(x) - 1$

   (b) $f(x) = \ln x - \ln(1/x)$

   (c) $f(x) = x^{-2}(\sin x - e^x + 1)$

   (d) $f(x) = e^x - e$

22. Let $f(x) = \ln(x + \sqrt{x^2 + 1})$. Show how to avoid loss of significance in computing $f(x)$ when x is negative. Hint: Compute first $f(-x)$.

23. For some values of x, the function $f(x) = x + \sqrt{x^2 - 1}$ cannot be accurately computed by using the given formula.

   (a) What values of x are involved? What remedy do you propose?

   (b) Carry 3 decimal significant figures, for example in $\mathbb{F}(10, 3, -24, +25)$ with rounding to the closest, and compute $f(-10^2)$ directly, then using the suggested remedy. (The exact value of $f(-10^2)$ is -0.005000125006250).

24. Let $f(x) = \frac{(e^x - 1) - \sin x}{x^2}$

   (a) For some values of x the function $f(x)$ cannot be accurately computed by using the given formula. What are the non-negative values of x that cause the problem? What remedy do you propose?

   (b) Use the first 2 terms only of the Taylor's series derived in (a), to approximate $f(10^{-4})$ in $\mathbb{F}(10, 5, -15, +15)$, rounding to the closest.

   (c) Find the absolute relative error in this approximation if the exact value of $f(10^{-4})$ is $0.50003333807.....$

25. Let
$$f(x) = \frac{e^x - e^{-x}}{x}$$

   (a) For which value of x, the given function cannot be accurately computed. Explain and find a way around the difficulty.

   (b) Carry 3 significant digits with rounding to the closest to evaluate $f(0.1)$ directly.

   (c) Repeat part (b) using the suggested remedy.

   (d) The actual value is $f(0.1) = 2.003335000$. Find the relative error for the values obtained in parts (b) and (c).

26. Use the Taylor polynomial of degree 4 to find an approximation to e^{-3} by each of the following methods, carrying 3 significant digits with rounding to the closest:

(a) $e^{-3} = \sum_{i=0}^{5} \frac{(-1)^i 3^i}{i!}$

(b) $e^{-3} = \frac{1}{e^3} = \frac{1}{\sum_{i=0}^{5} \frac{3^i}{i!}}$

(c) An approximate value of e^{-3} is 6.74×10^{-3}. Compare this value with the results obtained in (a) and (b). Explain your answer.

1.8 Computer Projects

Exercise 1 : Conversion: Decimal - Binary

1. Write a MATLAB
   function [Ibase2, Fbase2, b] = Convert10to2(d, k)
   that takes as input a non-zero decimal number d and a positive integer
   k and converts d to a binary number b up to k fractional digits. Your
   function should output the 2 vectors Ibase2 and Fbase2 that represent
   respectively the integral and fractional parts of b, and the binary number
   b displayed with its sign and its integral and fractional parts.

2. Write a MATLAB
   function [Ibase10, Fbase10, d] = Convert2to10(Ibase2, Fbase2)
   that takes as input two vectors Ibase2 and Fbase2 that represent respectively the integral and fractional parts of a binary number, converts
   to base 10 and outputs the results as 2 numbers Ibase10 and Fbase10
   that are respectively the integral and fractional parts of the corresponding decimal number d and the decimal number d displayed with its sign
   and its integral and fractional parts.
   Hint: Use nested polynomial evaluation.

3. Write a MATLAB
   function [B, I] = ConvertFraction10to2Pattern(D,m) that takes
   as input a decimal integer D consisting of k digits where $m = 10^k$.
   This function converts the decimal fractional $f = \frac{D}{m}$ into a binary fractional number represented by the vector B, and identifies the repeating
   pattern in B (if there is any), starting at component I and ending at
   n=length(B). In case the converted fractional part is finite, then no repeating pattern occurs and the value of I should be zero.
   For example:

   (a) To convert $f = 0.1$: input $D = 1$ and $m = 10$. This function outputs
       $B = [00011]$ and $I = 2$, since $(0.1)_{10} = (0.0\ 0011\ 0011\ 0011\ ......)_2$

   (b) To convert $f = 0.25$: input $D = 25$ and $m = 100$. This function
       outputs $B = 01$ and $I = 0$, since $(0.25)_{10} = (0.01)_2$.

   Remark: To minimize rounding errors in case I is a "large" number, it
   is more efficient to express fractional numbers as a ratio of 2 integers
   (for example $f = $ D/m ...).

4. Test each one of the 3 functions above for 3 different cases and save the
   results in a Word document.

Exercise 2 : Conversion from Double to Single Precision

1. Write a MATLAB
   function [t e f] = GetVectorD(v)   which takes as input a binary
   vector v of 64 bits or components representing a machine number in
   IEEE double precision, and extracts the values of the sign (t), the
   exponent (e) and the fractional part of the mantissa (f).

2. Write a MATLAB
   function x = ConvertDoubletoSingle(v) which takes as input a binary vector v of 64 bits representing a machine number in the IEEE
   double precision system. Your function should convert v to a single precision <u>machine number</u> and should output the result as a vector x of
   32 bits, unless x represents a "denormalized number" or "Not a Number." In these 2 cases, your function should only display a message: ' x
   represents NaN ' or ' x represents a denormalized number ' .
   At the end, if x represents an element of $F_S(2, 24, -126, +127)$, your
   function should also display the corresponding number in normalized
   floating point form, i.e., $xs = \pm 1.f \times 2^e$ or $xs = \pm 0$. Note the following
   remarks:

   (a) Use rounding by chopping when needed: (fl_0).

   (b) The smallest single precision denormalized number is: 2^{-149}

   (c) For any exponent $e < -149$, the corresponding number in single
       precision is rounded to zero.

3. In Exercise 2, test function 1 for 3 different test cases, then function 2
   for 5 different test cases including: "NaN', denormalized numbers, ± 0
   and $\pm\infty$. Save the results in a Word document.

Call for previous functions when needed.

Exercise 3 : Conversion: Decimal - Octal - Binary

1. Write a MATLAB function [E8 , F8] = Convert2to8(E2, F2) that
   takes as input two binary vectors E2 and F2 that are respectively the
   integral and fractional parts of a positive binary number b, converts
   them to octals and outputs the results as 2 vectors E8 and F8 that are
   respectively the integral and fractional parts of a positive octal number
   o.

2. Write a MATLAB function [E10, F10, d] = Convert8to10(E8, F8)
   which takes as input two octal vectors E8 and F8 that represent respectively the integral and fractional parts of a positive octal number o,
   converts to base 10 and outputs the results as 2 decimal numbers, E10
   and F10 that represent respectively the integral and fractional parts of
   the positive decimal number d using Nested Polynomial Evaluation. As
   a last step, this function should also display d as a decimal number.

3. Test each one of the 2 functions above for 3 different test cases and save the results in a Word document.(Consider different lengths for all input vectors.)

Exercise 4 : Successors and Rounding Procedures
Let $x = +mx \times 10^{ex}$ be a positive decimal number in $\mathbb{F}(10, p, -20, +20)$, written in normalized floating point form, with $-20 \leq ex < +20$, and $p < 15$.

1. Write a MATLAB
   function [my, ey] = GetSuccessor(mx, ex, p)
   which takes as inputs:

   - mx: the mantissa of x in standard normalized floating point notation
   - ex: the exponent of x
   - p: the precision of the floating point system to which x belongs

   Let y be the successor of x in $\mathbb{F}(10, p, -20, +20)$. This function should output:

   - my: the mantissa of y displayed with a precision p (the non-significant digits of the fractional part need not be displayed).
     Hint: First compute my, then use num2str(my,p) for output of my in the required format.
   - ey: the exponent of y.

2. Let $m = +m_1.m_2m_3...m_p$ be a positive decimal number which integral part is m_1, and fractional part is $0.m_2m_3...m_p$.
   Write a MATLAB
   function [m] = ConvertVectortoDecimal(M)
   that takes as input a vector M of length p which i^{th} component is the decimal digit m_i, for $i = 1, ..., p$, and output is the decimal number m represented by M.
   Use format long g to display m in double precision, discarding the non-significant zeros of the fractional part.

3. Write a MATLAB function [mz, ez] = Round(Mx, ex, n, t) which takes as inputs:

   - Mx: a vector of length p which components represent the mantissa mx of the decimal number $x \in \mathbb{F}(10, p, -20, +20)$.
   - ex: the exponent of x.
   - n: a positive integer less than or equal to p ($n \leq p$), representing the precision to be reached.
   - t: a parameter taking the values 1 or 2.

   This function should compute z, the representative of x in $\mathbb{F}(10, n, -20, +20)$ by rounding x to the closest if $t = 1$ or by chopping if $t = 2$, and output:

- **mz**: the mantissa of z displayed with a precision n.
  Hint: First compute **mz**, then use **num2str(my,n)** to output **mz** in the required format (the non-significant zeros of the fractional part should be discarded).

- **ez**: the exponent of z

As a result, your function should also **display** z in normalized floating point representation in $\mathbb{F}(10, n, -20, +20)$.

4. Test each one of the 2 functions above for 3 different test cases and save the results in a Word document.

Remark: Call for previous functions when needed.

Chapter 2

Finding Roots of Real Single-Valued Functions

In this chapter we consider one of the most encountered problems in scientific computing, which is the problem of computing the **root or zero** of a real-valued function f of one variable. We focus on what we consider to be three basic methods: *the bisection, Newton's and the secant methods*. In short, any of these methods compute a solution of a nonlinear equation starting from one **initial data**, then adopting some **iterative method** that - under favorable conditions - will **converge** to a zero of the function f.

To study other methods, we refer to other textbooks such as [1], [4] [7], [9], [15], [21], [26] and [29].

2.1 Introduction

Let f be a real-valued function of a real variable admitting a specific regularity on its domain D, i.e., let f be k-times continuously differentiable, with $k \geq 1$ ($f \in C^k(D)$). We seek to find the roots of this function f, defined as follows:

Definition 2.1 *The set R of roots of the function $f(x)$ is defined as:*

$$R = \{r \in \mathbb{R} : f(r) = 0\}.$$

Given some computational tolerance $\epsilon_{tol} = \frac{1}{2}10^{1-m}$, $m = 1, 2, ...$, our objective is to compute one or more roots of f, within such ϵ_{tol}. Specifically, for any $r \in R$, we seek an approximation r_a to r, ($r_a \approx r$), such that:

$$\frac{|r - r_a|}{|r|} \leq \epsilon_{tol}. \tag{2.1}$$

(We say then, that r_a approximates r up to m decimal places)

The search for a specific root of a function requires two steps.

1. **Step 1: Locate the root,** i.e., seek an interval (a, b), with $O(|b - a|) = O(|r|)$, such that:

$$f(x) \in C([a, b]), \text{ (i.e., } f(x) \text{ is at least continuous)} \qquad (2.2)$$
$$r \in (a, b) \qquad (2.3)$$
$$f(a) \times f(b) < 0 \qquad (2.4)$$
$$\forall x \in (a, b), \ x \neq r \Rightarrow f(x) \neq 0 \qquad (2.5)$$

2. **Step 2:** Generate a sequential process leading to a **sequence** $\{r_n\}_{n \geq 0}$ the terms of which are in (a, b) for all values of n, and that converges to r, i.e., satisfying:

$$r_n \in (a, b) \, \forall n \text{ and } \lim_{n \to \infty} r_n = r. \qquad (2.6)$$

The generation of such a sequence is usually done through an **iterative** procedure (or method) where $r_n = g(r_{n-1}, ..., r_{n-k})$, $k \geq 1$.

We start by introducing some general properties verified by such methods.

Definition 2.2 *A numerical method is said to be a* **one-step** *method in case* $k = 1$, *the initial state of the sequence being determined by the only choice of* r_0; *otherwise, it is a* **multi-step** *method of order* k, *and its initial state is then determined by the choice of* $r_0, .., r_{k-1}$.

The **order of convergence** of a method measures the rate at which the sequence $\{r_n\}$ generated by the numerical process converges to the root r. It is defined as follows:

Definition 2.3 Order of Convergence of a Method
A method is of order $\alpha > 0$, *if there exists a sequence of positive numbers* $\{t_n\}_{n \geq 0}$, *such that* $\forall n \geq 1$:

$$|r - r_n| \leq t_n, \text{ with } t_n \leq C t_{n-1}^\alpha \qquad (2.7)$$

Equivalently, in the special case where $t_n = |r - r_n|$:

$$|r - r_n| \leq C|r - r_{n-1}|^\alpha \qquad (2.8)$$

The constants C and α are independent from n, with $C < 1$ for $n = 1$.

If $\alpha = 1$ the convergence is said to be **linear**, while if $\alpha > 1$ the convergence is **super-linear**. In particular, if $\alpha = 2$ the convergence of the method is **quadratic**. (Note also that the greater α is, the faster is the method.)

Definition 2.4 Global Convergence vs. Local Convergence
*A method is said to be **globally** convergent if the generated sequence $\{r_n\}_n$ converges to r for any choice of the initial state; otherwise it is **locally** convergent.*

When implemented, the process generating the elements of $\{r_n\}$ will be stopped as soon as the 1st computed element r_{n_0} satisfies some predefined "stopping criteria."

Definition 2.5 Stopping Criteria
Given some tolerance ϵ_{tol}, a standard stopping criterion is defined by the following relative estimates:

$$\frac{|r_{n_0} - r_{n_0-1}|}{|r_{n_0}|} \leq \epsilon_{tol} \text{ and } \frac{|r_n - r_{n-1}|}{|r_n|} > \epsilon_{tol} \text{ if } n < n_0. \tag{2.9}$$

The "remainder" $f(r_n)$ can also be used to set a stopping criterion since $f(r) = \lim_{n\to\infty} f(r_n) = 0$. Thus, one may use a relative evaluation of the remainder. Specifically, find the first element r_{n_0} of the sequence $\{r_n\}$ satisfying:

$$\frac{|f(r_{n_0})|}{|f(r_0)|} \leq \epsilon_{tol} \text{ and } \frac{|f(r_n)|}{|f(r_0)|} > \epsilon_{tol} \text{ if } n < n_0. \tag{2.10}$$

Note also that by using the Mean-Value Theorem one has:

$$0 = f(r) = f(r_n) + (r - r_n)f'(c_n), \text{ where } c_n = r + \theta(r_n - r), \theta \in (0,1).$$

Thus if f' is available (referring also to (2.9)), a more sophisticated stopping criterion would be:

$$\frac{|f(r_{n_0})|}{|r_{n_0}f'(r_{n_0})|} \leq \epsilon_{tol} \text{ and } \frac{|f(r_n)|}{|r_{n_0}f'(r_{n_0})|} > \epsilon_{tol} \text{ if } n < n_0. \tag{2.11}$$

In this chapter, we shall analyze successively three root finding iterative methods: the bisection method, Newton's method and the secant method

2.2 How to Locate the Roots of a Function

There are basically two approaches to **locate the roots** of a function f. The first one seeks to **analyze the behavior of** f analytically or through plotting its graph, while the second one transforms the problem of root finding into an equivalent **fixed point problem**. We illustrate this concept on some specific examples.

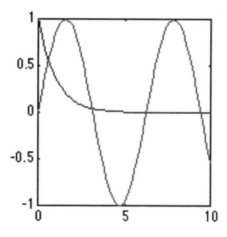

FIGURE 2.1: Roots of $e^{-x} - \sin(x)$, $x > 0$

Example 2.1 *Locate the roots of the function $f(x) = e^{-x} - \sin(x)$.*

Analyzing the behavior of the function
A first analysis for $x < 0$ indicates, since the exponential $e^{-x} > 1$ and $\sin(x) \leq 1$, one concludes that $f(x) > 0$ for $x < 0$. Furthermore as $f(0) = 1$, this implies that all the roots of the function lie in the interval $(0, \infty)$. For $x > 0$, we put the problem in a **fixed point problem**.
For that purpose, we let $g_1(x) = e^{-x}$ and $g_2(x) = \sin(x)$. Solving the problem $f(r) = 0$ can be made equivalent to solving the equation $g_1(r) = g_2(r)$, in which r becomes a "fixed-point" for g_1 and g_2. Hence plotting these 2 functions on the same graph, one concludes straightforwardly that g_1 and g_2 intersect at an infinite number of points with positive abscissa, that constitute the set of all roots of f. This is shown in Figure 2.1

Example 2.2 *Locate the roots of the quadratic polynomial $p(x) = x^4 - x^3 - x - 1$.*

To use the fixed point method, let $g_1(x) = x^4 - x^3$ and $g_2(x) = x + 1$. It is easy to verify in this case that these 2 functions intersect twice, implying consequently that f has 2 roots located respectively in the intervals $(-1, 0)$ and $(1, 2)$ as indicated in Figure 2.2.

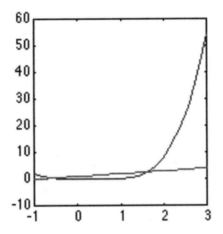

FIGURE 2.2: Roots of $p(x) = x^4 - x^3 - x - 1$

2.3 The Bisection Method

The **bisection method** is a procedure that repeatedly **"halves"** the interval in which a root r has been located. This "halving" process is reiterated until the desired accuracy is reached. Specifically, after locating the root in (a, b) we proceed as follows:

- Compute $r_1 = \frac{a+b}{2}$ the **midpoint** of (a, b) and $y = f(r_1)$. If it happens fortuitously that $f(r_1) = 0$ then the root has been found, i.e., $r = r_1$. Otherwise $y \neq 0$ and 2 cases may occur:

    – either $y \times f(a) < 0$, implying that $r \in (a, r_1)$
    – or $y \times f(a) > 0$, in which case $r \in (r_1, b)$.

    Let the initial interval $(a, b) = (a_0, b_0)$.
    Either way, and as a consequence of this first halving of (a_0, b_0), one obtains a new interval $(a_1, b_1) = (a_0, r_1)$ or $(a_1, b_1) = (r_1, b_0)$, such that one obviously has:

$$r \in (a_1, b_1), \text{ with } b_1 - a_1 = \frac{1}{2}(b_0 - a_0) \text{ and } |r - r_1| \leq (b_1 - a_1). \quad (2.12)$$

- Evidently this process can be repeated, generating a sequence of intervals $\{(a_n, b_n)| \, n \geq 1\}$ such that:

$$r \in (a_n, b_n) \text{ with } b_n - a_n = \frac{1}{2}(b_{n-1} - a_{n-1}) \quad (2.13)$$

and a sequence of iterates $\{r_n \,|\, n \geq 1\}$, with $r_n \in (a,b) \, \forall n$, and where

$$r_n = \frac{1}{2}(a_{n-1} + b_{n-1}) \text{ with } |r - r_n| \leq (b_n - a_n). \qquad (2.14)$$

- The process is achieved when the interval (a_n, b_n) is relatively small with respect to the initial interval, specifically when the least value of n is reached, for which:

$$\frac{b_n - a_n}{b_0 - a_0} \leq \epsilon_{tol} \qquad (2.15)$$

where ϵ_{tol} is a given computational tolerance.

At the end of this process, the best estimate of the root r would be the last computed value of r_n as in (2.14).

The bisection method is implemented through the following algorithm:

Algorithm 2.1 Bisection Method

```
function r=myBisection(f,a,b,tol,kmax)
% Inputs:  f, a, b, kmax, tol
% kmax: maximum acceptable  number of iterations; tol=0.5*10^(-p+1)
% S= [length last (a,b)] / [length initial (a, b)]
% Output: r, vector of converging midpoints.
fa=f(a);
% length of initial interval (a,b)
ab=abs(b-a);
% Initialize n and S
n=1;S=1;
while S>tol & n<kmax
  r(n)=(a+b)/2;y=f(r(n));
    if y*fa<0
        b=r(n);
    elseif  y*fa>0
        a=r(n);fa=y;
    elseif y*fa=0
        disp('r(n) is the root' )
        break
    end
S=(abs(b-a)/ab);
n=n+1;
end
%If n>=kmax, reconsider the values of a, b, S, kmax
if n>=kmax
    disp ('error  no convergence' );
else
    n=n-1;
end
```

The parameter $kmax$ is used as a programming safeguard. This eliminates the possibility of entering an infinite loop in case the sequence diverges, or also when the program is incorrectly coded. If k exceeds $kmax$ with $\frac{b_k - a_k}{b - a} > tol$, the written algorithm would then signal an error.

Thus, (2.12), (2.13) and (2.14) lead to the following result.

Theorem 2.1 *Under assumptions (2.2)-(2.5), the bisection algorithm generates 2 sequences $\{a_n\}_{n\geq 0}$ and $\{b_n\}_{n\geq 0}$ from which one "extracts" a sequence of iterates $\{r_n\}_{n\geq 1}$, with $r_n = a_n$ or $r_n = b_n$, such that:*

1. *$a_0 = a$, $b_0 = b$,*

2. *$r \in (a_n, b_n)$ with $a_n < r < b_n$, $\forall n \geq 0$,*

3. *The sequences $\{a_n\}$ and $\{b_n\}$ are respectively monotone increasing and decreasing,*

4. *$b_n - a_n = \frac{b_{n-1} - a_{n-1}}{2} = \frac{b-a}{2^n} \forall n \geq 1$, and $\lim_{n\to\infty} a_n = \lim_{n\to\infty} b_n = r$,*

5. *$|r - r_n| \leq b_n - a_n, \forall n \geq 1$.*

Proof. 1. and 2. are obtained by construction.
To prove 3., given (a_{n-1}, b_{n-1}) with $r \in (a_{n-1}, b_{n-1})$, then by definition of the method, $r_n = \frac{1}{2}(a_{n-1} + b_{n-1})$ will either be a_n or b_n. Therefore, in the case the process is reiterated, this implies that either $a_n = a_{n-1}$ and $b_n < b_{n-1}$ or $a_n > a_{n-1}$ and $b_n = b_{n-1}$ which proves the required result. (Note that neither of these sequences can "stagnate." For example, the existence of an n_0 such that $a_{n_0} = a_n$, $\forall n \geq n_0$, would imply that $r = a_{n_0}$, i.e., the process is finite and the root has been found after n_0 steps!)
4. follows from the "halving" procedure. It can be easily shown by induction, that $b_n - a_n = \frac{b-a}{2^n}$ and therefore $\lim_{n\to\infty} b_n - a_n = 0$, meaning that the sequences of lengths $\{(b_n - a_n)\}$ of the intervals $\{(a_n, b_n)\}$ converge to 0. Hence, the sequences $\{a_n\}$ and $\{b_n\}$ have the same limit point r.
Finally, to obtain 5., just note again that $r_n = a_n$ or b_n, with $r \in (a_n, b_n)$. ∎

A consequence of these properties is the linearity of the convergence and an estimate on the minimum number of iterations needed to achieve a given computational tolerance ϵ_{tol}. Specifically, we have the following result.

Corollary 2.1 *Under the assumptions of the previous theorem, one obtains the following properties:*

1. *The bisection method converges linearly, in the sense of definition (2.3), i.e.,*

$$|r - r_n| \leq t_n = b_n - a_n, \text{ with } t_n \leq \frac{1}{2}t_{n-1}$$

2. *The minimum number of iterations needed to reach a tolerance of $\epsilon_{tol} =$*
   $0.5 \times 10^{1-p}$ is given by

$$k = \left\lceil (p-1)\frac{\ln(10)}{\ln(2)} + 1 \right\rceil$$

Proof. The first part of the corollary is a direct result from the previous theorem. As for the second part, it is achieved by noting that the method reaches the desired accuracy, according to the selected stopping criteria, whenever n reaches the value k such that:

$$\frac{b_k - a_k}{b - a} \leq \epsilon_{tol} < \frac{b_{k-1} - a_{k-1}}{b - a} < \dots \frac{b_1 - a_1}{b - a} = \frac{1}{2} < \frac{b_0 - a_0}{b - a} = 1. \qquad (2.16)$$

From equation (2.16) and since $\frac{b_n - a_n}{b-a} = \frac{1}{2^n} \; \forall \; n \geq 0$, we can estimate the **least number of iterations** required (theoretically) to reach the relative precision $\epsilon_{tol} = \frac{1}{2}10^{1-p}$, p being the number of significant decimal figures fixed by the user. Such integer k satisfies then:

$$\frac{1}{2^k} \leq \frac{1}{2}10^{1-p} < \frac{1}{2^{k-1}}. \qquad (2.17)$$

Equivalently:

$$-k\ln(2) \leq (1-p)\ln(10) - \ln(2) < -(k-1)\ln(2),$$

from which one concludes that:

$$k\ln(2) \geq (p-1)\ln(10) + \ln(2) > (k-1)\ln(2),$$

leading to:

$$k \geq (p-1)\frac{\ln(10)}{\ln(2)} + 1 > k - 1, \qquad (2.18)$$

The integer k is computed then as:

$$k = \left\lceil (p-1)\frac{\ln(10)}{\ln(2)} + 1 \right\rceil.$$

∎

Note that such k is independent of a and b, since it estimates the ratio $\frac{b_k - a_k}{b-a}$, a measure of the <u>relative</u> reduction of the size of the interval (a_k, b_k) containing r. Table 2.1 provides the estimated number of iterations k with respect to a precision p required by the user. Obviously the method is slowly convergent! Nevertheless, since at each step the length of the interval is reduced by a factor of 2, it is advantageous to choose the initial interval as small as possible. In applying the bisection method algorithm for the above 2 examples, one gets the following results:

Precision p	Iterations k
3	8
5	15
7	21
10	31
15	48

TABLE 2.1: Estimated number of iterations with respect to a requested precision in the bisection method

Iteration	Iterate
1	$5.000000 \ 10^{-1}$
...	....
10	$5.888672 \ 10^{-1}$
11	$5.885009 \ 10^{-1}$
12	$5.886230 \ 10^{-1}$
13	$5.885010 \ 10^{-1}$
14	$5.885620 \ 10^{-1}$
15	$5.885315 \ 10^{-1}$
16	$5.885468 \ 10^{-1}$
17	$5.885391 \ 10^{-1}$
18	$5.885353 \ 10^{-1}$
19	$5.885334 \ 10^{-1}$
20	$5.885324 \ 10^{-1}$
21	$5.885329 \ 10^{-1}$

TABLE 2.2: Bisection iterates for the first root of $f(x) = e^{-x} - \sin(x)$

Iteration	Iterate
1	$1.500000 \ 10^0$
2	$2.250000 \ 10^0$
...	....
10	$1.620118 \ 10^0$
11	$1.618653 \ 10^0$
12	$1.617921 \ 10^0$
13	$1.618287 \ 10^0$
14	$1.618104 \ 10^0$
15	$1.618013 \ 10^0$
16	$1.618059 \ 10^0$

TABLE 2.3: Bisection iterates for one root of $f(x) = x^4 - x^3 - x - 1$

n	a_n	b_n	r_n	$f(a_n) \times f(b_n)$
0	0	1	0.5	+
1	0.5	1	0.75	+
2	0.75	1	0.875	-
3	0.75	0.875	0.813	-
4	0.75	0.813	0.782	-
5	0.75	0.782	0.766	-
6	0.75	0.766	0.758	+
7	0.758	0.766	0.762	+
8	0.762	0.766	0.764	-
9	0.762	0.764	0.763	+
10	0.763	0.764	0.763	

TABLE 2.4: Convergence of the intervals (a_n, b_n) to the positive root of $f(x) = \ln(1 + x) - \frac{1}{1+x}$

1. Let $f(x) = e^{-x} - \sin(x)$. Results of bisection iterates in finding the first root of f in the interval $[0, 1]$, with a tolerance $\epsilon = 0.5 \times 10^{-5}$ (6 significant figures rounded) are given in Table 2.2. The bisection method took 20 iterations to reach a precision of 6. The $21st$ was needed to meet the termination condition.

2. Let $f(x) = x^4 - x^3 - x - 1$. Search for the root of f in the interval $[0, 3]$ with $\epsilon = 0.5 \times 10^{-4}$ (5 significant figures rounded). The results of the bisection iterates are given in Table 2.3.

Table 2.4 illustrates the convergence of the sequence of intervals $\{(a_n, b_n) | n = 1, 2, ..., 10\}$, generated by the bisection method for the function $f(x) = \ln(1 + x) - \frac{1}{1+x}$, as proved in Theorem 2.1. Computations are carried out up to 3 significant figures. To conclude, the bisection is a **multi-step method** that, although conceptually clear and simple, has significant drawbacks since, as theory and practice indicate, it is a slowly convergent method. However it **globally converges** to the searched solution and can be used as a starter to more efficient **locally convergent** methods, notably both the Newton's and secant methods.

2.4 Newton's Method

Newton's (or **Newton-Raphson's**) method is one of the most powerful numerical methods for solving non-linear equations. It is also referred to as the **tangent method**, as it consists in constructing a sequence of numbers $\{r_n | r_n \in (a, b) \, \forall n \geq 1\}$, obtained by intersecting tangents to the curve $y = f(x)$ at the sequence of points $\{(r_{n-1}, f(r_{n-1})) | n \geq 1\}$ with the X-Axis. Constructing such tangents and such sequences requires additional assumptions to (2.2)-(2.5) as derived hereafter.

To start, let $r_0 \in (a, b)$ in which the root is located, and let $M_0 = (r_0, f(r_0))$ be the point on the curve

$$\{(\mathcal{C}) | y = f(x), \, a \leq x \leq b\}.$$

Let also $(\mathcal{T}_0)$ be the tangent to $(\mathcal{C})$ at M_0 with equation given by:

$$y = f'(r_0)(x - r_0) + f(r_0).$$

The intersection of $(\mathcal{T}_0)$ with the X-Axis is obtained for $y = 0$ and is given by:

$$r_1 = r_0 - \frac{f(r_0)}{f'(r_0)}. \tag{2.19}$$

To insure that $r_1 \in (a, b)$, r_0 should be chosen "close enough" to r. Specifically, since $f(r) = 0$, $\overline{(2.19)}$ is equivalent to:

$$r_1 - r = r_0 - r - \frac{f(r_0) - f(r)}{f'(r_0)} \tag{2.20}$$

Using Taylor's expansion of $f(r)$ about r_0 up to first order, one has:

$$f(r) = f(r_0) + f'(r_0)(r - r_0) + \frac{1}{2} f''(c_0)(r - r_0)^2, \; c_0 = r_0 + \theta_0(r - r_0), \; 0 < \theta_0 < 1,$$

thus leading to:

$$\frac{f(r) - f(r_0)}{f'(r_0)} = (r - r_0) + \frac{1}{2} \frac{f''(c_0)}{f'(r_0)} (r - r_0)^2, \text{ with } c_0 \in (a, b).$$

Hence, imposing on f and on the interval (a, b) the following additional assumptions:

$$f(x) \in C^2(a, b), \text{i.e., } f(x), f'(x), f''(x) \text{ are continuous on } (a, b) \tag{2.21}$$

$$f'(x) \neq 0 \; \forall x \in (a, b) \tag{2.22}$$

one concludes from (2.20):

$$|r_1 - r| = \frac{1}{2} \frac{|f''(c_0)|}{|f'(r_0)|}(r - r_0)^2 \tag{2.23}$$

Based on these additional assumptions, we define also the positive constant:

$$C = \frac{1}{2} \frac{\max_{x \in (a,b)} |f''(x)|}{\min_{x \in (a,b)} |f'(x)|}. \tag{2.24}$$

which will then lead to:

$$|r - r_1| \leq C|r - r_0|^2 \tag{2.25}$$

This gives a preliminary "closeness" result of r_1 with respect to the root r, in terms of the "closeness" of r_0, without however insuring yet the required location of r_1 in (a, b). In this view, letting now:

$$I_0 = \{x|\, |r - x| < \frac{1}{C}\} \cap (a, b) \tag{2.26}$$

and selecting initially r_0 in I_0, leads to the required result as shown hereafter.

Lemma 2.1 *If $r_0 \in I_0$ as defined in (2.26), then $r_1 \in I_0$ with*

$$|r - r_1| \leq |r - r_0| \tag{2.27}$$

Proof. Let $e_i = C|r - r_i|$, $i = 0, 1$, where C verifies (2.24). Multiplying (2.25) by C one obviously has:

$$e_1 \leq e_0^2$$

moreover, since $e_0 < 1$ and $C > 0$, the required result is reached.   ∎

Thus selecting $r_0 \in I_0$ and reaching r_1 satisfying (2.27), the process can be continued beyond that step. In fact one generates a sequence of **Newton's** iterates $\{r_n|\, n \geq 2\}$ with $r_n \in (a, b) \,\forall n$, given by a formula generalizing (2.19). Specifically, one has:

$$r_{n+1} = r_n - \frac{f(r_n)}{f'(r_n)}, \; n \geq 0. \tag{2.28}$$

with $(r_{n+1}, 0)$ being the intersection with the X-Axis of the tangent to the curve $(\mathcal{C})$ at the point $(r_n, f(r_n))$, as indicated in Figure 2.3. Clearly, Newton's method is a one-step iteration $r_{n+1} = g(r_n)$, with the **iteration** function $g(x)$ given by:

$$g(x) = x - \frac{f(x)}{f'(x)}. \tag{2.29}$$

We turn now to the analysis of the convergence of Newton's method, i.e., the convergence of Newton's iterates $\{r_n\}_{n \geq 0}$.

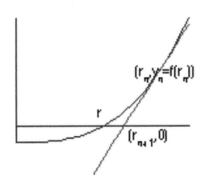

FIGURE 2.3: Intersection with the X-Axis of the tangent to $(\mathcal{C})$ at $(r_n, f(r_n))$

Theorem 2.2 *Let $f(x)$ satisfy assumptions (2.2)-(2.5), in addition to (2.21) and (2.22), then for $r_0 \in I_0$, with C as defined in (2.24), the sequence of Newton's iterates:*

$$r_{n+1} = r_n - \frac{f(r_n)}{f'(r_n)}, \quad n \geq 0,$$

is such that:

1. $r_n \in I_0, \ \forall n \geq 0$

2. $\lim_{n \to \infty} r_n = r$

3. $|r - r_{n+1}| \leq C|r - r_n|^2$, *meaning that Newton's method is quadratic with $\alpha = 2$.*

Proof. The proof of this theorem follows from arguments used to obtain Lemma 2.1. In fact, one derives as for (2.25) that:

$$e_{n+1} \leq e_n^2, \ \forall n \geq 0. \tag{2.30}$$

where $e_i = C|r - r_i|$, $i = n, n + 1$.
Moreover, it can be easily proved by induction on n, that (2.30) in turn implies that:

$$e_n \leq (e_0)^{2^n} \ \forall \ n \geq 1. \tag{2.31}$$

As $e_0 < 1$ then $r_n \in I_0$ with $\lim_{n \to \infty} e_n = 0$, proving parts 1 and 2 of the lemma. In addition to these results, and as derived in (2.23) and (2.25), one concludes that :

$$|r_{n+1} - r| = \frac{1}{2} \frac{|f''(c_n)|}{|f'(r_n)|}(r - r_n)^2 \leq C|r - r_n|^2, \tag{2.32}$$

with $c_n = r_n + \theta_n(r - r_n)$, $0 < \theta_n < 1$. Referring to (2.8) that result obviously implies that $\alpha = 2$. ∎

Note also that inequality (2.31) allows obtaining an estimate on the minimum number of iterations needed to reach a computational tolerance $\epsilon_{tol} = 0.5 \times 10^{1-p}$. Specifically, we prove now:

Corollary 2.2 *If $r_0 \in I_0$, the minimum number of iterations needed to reach $\epsilon_{tol} = 0.5 \times 10^{1-p}$ is given by:*

$$n_0 = \lceil \ln(1 + \frac{(p-1)\ln(10) + \ln(2)}{|\ln(e_0)|})/\ln(2)\rceil.$$

Proof. Note that ϵ_{tol} is reached whenever $n = n_0$ satisfies the following inequalities:

$$\frac{|r - r_{n_0}|}{|r - r_0|} \le 0.5 \times 10^{1-p} < \frac{|r - r_n|}{|r - r_0|}, \ \forall n < n_0.$$

Since also $\frac{|r-r_n|}{|r-r_0|} = \frac{e_n}{e_0}$, $\forall n \ge 1$, then from (2.31):

$$\frac{|r - r_{n_0}|}{|r - r_0|} \le (e_0)^{2^{n_0}-1}.$$

The sought for minimum number of iterations n_0 would thus verify:

$$(e_0)^{2^{n_0}-1} \le 0.5 \times 10^{1-p} < (e_0)^{2^n-1}, \ \forall n < n_0.$$

Since $e_0 < 1$, this is equivalent to:

$$2^{n_0} \ge 1 + \frac{(p-1)\ln(10) + \ln(2)}{|\ln(e_0)|} > 2^n, \ n < n_0.$$

This leads to n_0 satifying:

$$n_0 \ge \frac{\ln(1 + \frac{(p-1)\ln(10)+\ln(2)}{|\ln(e_0)|})}{\ln(2)} > n_0 - 1$$

and therefore:

$$n_0 = \lceil \ln(1 + \frac{(p-1)\ln(10) + \ln(2)}{|\ln(e_0)|})/\ln(2)\rceil,$$

which is the desired result. ∎

To illustrate, assume $e_0 = \frac{1}{2}$, then it results from this lemma that:

$$n_0 = \lceil \ln(2 + (p-1)\frac{\ln(10)}{\ln(2)})/\ln(2)\rceil.$$

Precision p	Iterations n_0
7	4
10	5
16	6

TABLE 2.5: Estimate of the number of iterations as a function of the precision in Newton's method

Table 2.5 provides values of n_0 relative to a precision p. Thus, one can assert that Newton's method is a **locally** and **quadratically convergent** method. When a root r of a function $f(x)$ is located in an interval (a, b), the first step is to insure finding a sub-interval $I_0 \subset (a, b)$ containing r, in which $|r - r_0| \le \frac{1}{C}$, with the constant C given in (2.24).

A rule of thumb would be to select r_0 after 1 or 2 applications of the bisection method. Such a step would make sure the initial condition r_0 is close "enough" to r.

In the following algorithm, the initial choice is being selected after one bisection iteration. Note that Newton's method requires the availability of the first derivative $f'(x)$ of $f(x)$. This is the "price" to pay in order to get a quadratic convergence.

Algorithm 2.2 Newton's Method

```
% Input f, df, a, b,Tol=0.5*10^(-p+1), kMAX
% Output: r is the sequence of approximations to the root up to Tol
% Find the first approximation by the Bisection rule
% The chosen stopping criteria is S=| r(n+1)-r(n) | / |r(n) |<= Tol
function r=myNewton(f,df,a,b,Tol,kmax)
r(1)=(a+b)/2;n=1; S =1;
while S >Tol & n< nMAX
      F=f(r(n)); DF=df(r(n));
      r(n+1)=r(n)-F/DF;
      S = abs[r(n+1)-r(n)]/ abs[r(n)];
      n=n+1;
end
if n>=kmax
     disp ('error  no convergence' );
else
     n=n-1;
end
```

The following example illustrates the general behavior of Newton's method.

Example 2.3 *Find the roots of $f(x) = \sin(x) - e^{-x}$ in the interval $(0, 2)$, using Newton's method.*

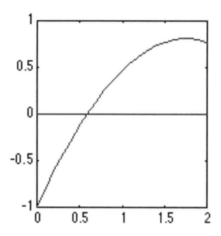

FIGURE 2.4: Finding a root of $f(x) = \sin(x) - e^{-x}$ using Newton's method

Iteration	Iterate
0	1.75
1	1.8291987×10^2
2	1.8206468×10^2
3	1.8221346×10^2
4	1.8221237×10^2
...	...

TABLE 2.6: A case of a diverging Newton's iteration

The graph of the function on that interval is shown in Figure 2.4.

Obviously, Newton's method is not applicable when the initial choice of the iteration r_0 is selected randomly in the interval $(0, 2)$. For example if r_0 is chosen in the interval $(1.5, 2)$, the generated sequence $\{r_n\}$ may not fall in the interval $(0, 2)$ and thus the method fails to converge, as is shown in Table 2.6 resulting from the application of Newton's algorithm with $r_0 = 1.75$. Obviously, the convergence is taking place to a root that **is not** in the interval $(0, 2)$. On the other hand, one application of the bisection method would start the iteration with $r_0 = 1$, leading to an efficiently convergent process as shown in Table 2.7. Obviously, about 4 iterations would provide 10 significant figures, a fifth one leading to 16 figures, i.e., a more than double precision answer.

However, there are cases, as in the first example below, where the convergence of the method is not affected by the choice of the initial condition, whereby Newton's method converges unconditionally.

Iteration	Iterate
0	1.0
1	$4.785277889803116 \times 10^{-1}$
2	$5.841570194114709 \times 10^{-1}$
3	$5.885251122073911 \times 10^{-1}$
4	$5.885327439585476 \times 10^{-1}$
5	$5.885327439818611 \times 10^{-1}$
6	$5.885327439818611 \times 10^{-1}$

TABLE 2.7: A case of a converging Newton's iteration

Example 2.4 The Square Root Function

Using Newton's method, we seek an approximation to $r = \sqrt{a}$, where $a > 0$. Clearly, such r is the unique positive root of $f(x) = x^2 - a$, with Newton's iterates satisfying the following identity:

$$r_{n+1} = r_n - \frac{f(r_n)}{f'(r_n)} \equiv \frac{1}{2}(r_n + \frac{a}{r_n}), \ \forall n \geq 0 \qquad (2.33)$$

(It is easy to check graphically that the sequence converges to $\sqrt{a}$ for any initial choice of $r_0 > 0$).
Based on the equation above:

$$r_{n+1} - r = \frac{1}{2}(r_n - 2r + \frac{a}{r_n})$$

Equivalently, since $a = r^2$:

$$r_{n+1} - r = \frac{(r_n - r)^2}{2r_n} \geq 0 \qquad (2.34)$$

The following results can therefore be deduced:

1. $r_n \geq r, \ \forall \ n \geq 1$

2. The generated iterative sequence $\{r_n\}$ is a decreasing sequence, since:

$$r_{n+1} - r_n = -\frac{f(r_n)}{f'(r_n)} = -\frac{(r_n^2 - r^2)}{2r_n} \leq 0$$

   based on the property 1 above.

3. The sequence $\{r_n\}$ converges to the root of f, i.e., $\lim_{n \to \infty} r_n = r$, since rewriting (2.34) as:

$$r_{n+1} - r = \frac{r_n - r}{2}(1 - \frac{r}{r_n})$$

in turn by induction leads to:

$$r_{n+1} - r < \frac{1}{2}(r_n - r) < \ldots < \frac{1}{2^{(n-1)}}(r_1 - r)$$

4. The convergence is notably quadratic, since from (2.34) and for all $n \geq 0$:

$$|r_{n+1} - r| = |\frac{(r_n - r)^2}{2r_n}| < C|r_n - r|^2 \text{ where } C = \frac{1}{2r}$$

As for IEEE standard notations, note that since

$$a = m \times 2^e \text{ with } 1 \leq m < 2$$

then the square root function is such that:

$$\sqrt{\ } : (m, e) \to (m', e') \text{ with } \sqrt{a} = m' \times 2^{e'}$$

The normalized mantissa and exponent of $\sqrt{a}$ are computed as follows:

1. If $e = 2k$, then $m' = \sqrt{m}$ with $1 \leq m' < \sqrt{2} < 2$, and $e' = k$, i.e.,

$$\sqrt{\ } : (m, e = 2k) \to (m' = \sqrt{m}, e' = k)$$

2. If $e = 2k+1$, then $a = 2m \times 2^{2k}$ and $m' = \sqrt{2m}$ with $1 < \sqrt{2} \leq m' < 2$, and $e' = k$, i.e.,

$$\sqrt{\ } : (m, e = 2k + 1) \to (m' = \sqrt{2m}, e' = k)$$

In either case, Newton's iteration in binary mode may start with $r_0 = 1$. ∎

The local character of convergence of Newton's method is well illustrated in the interesting case of the reciprocal function.

Example 2.5 The Reciprocal of a Positive Number

Assume $a > 0$. We seek an approximation to $r = \frac{1}{a}$, where r is the unique positive root of $f(x) = a - \frac{1}{x}$. Obviously, Newton's iterations satisfy the following identity:

$$r_{n+1} = r_n(2 - ar_n), \ \forall \, n \geq 0 \qquad (2.35)$$

Choosing restrictively the initial condition $r_0 \in (0, 2/a)$ leads to an iterative sequence $\{r_n\}$ where:

$$r_{n+1} > 0, \text{ whenever } r_n \in (0, 2/a) \qquad (2.36)$$

In such case, for all $n \geq 1$, it is left as an exercise to prove that:

1. $r_{n+1} - r = -\frac{(r_n - r)^2}{r}$

2. The generated sequence is an increasing sequence

3. The sequence $\{r_n\}$ converges to the root of f, i.e., $\lim_{n\to\infty} r_n = r$

4. Convergence of the sequence is quadratic.

Considering IEEE standard notations as for the square root function example, if

$$a = m \times 2^e, \text{ with } 1 < m < 2$$

then the inverse function is such that:

$$inv : (m, e) \to (m', e'), \text{ with } \frac{1}{a} = m' \times 2^{e'}$$

The normalized mantissa and exponent of $1/a$ are respectively:

$$m' = 2/m \text{ and } e' = -e - 1$$

since $\frac{1}{a} = \frac{1}{m} \times 2^{-e}$ or more adequately:

$$\frac{1}{a} = \frac{2}{m} \times 2^{-e-1}, \text{ with } 1 < \frac{2}{m} < 2$$

$\blacksquare$

2.5 The Secant Method

Recall that Newton's iteration satisfies formula (2.28):

$$r_{n+1} = r_n - \frac{f(r_n)}{f'(r_n)} \text{ where } f'(r_n) = \lim_{h\to 0} \frac{f(r_n + h) - f(r_n)}{h}$$

One drawback of Newton's method is the necessary availability of the derivative. In case such function is difficult to program, an alternative would be to avoid the calculation of $f'(r_n)$, and replace it by the backward divided difference approximation to the derivative:

$$f'(r_n) \approx [r_{n-1}, r_n] = \frac{f(r_n) - f(r_{n-1})}{r_n - r_{n-1}}$$

As indicated in Figure 2.5, obtaining r_{n+1} uses the secant to the curve $y = f(x)$ passing through the points $(r_{n-1}, f(r_{n-1}))$ and $(r_n, f(r_n))$, the equation of which is:

$$y = \frac{f(r_n) - f(r_{n-1})}{r_n - r_{n-1}}(x - r_n) + f(r_n)$$

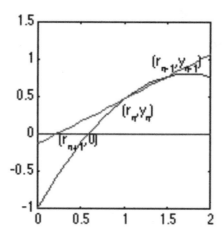

FIGURE 2.5: Intersection with the X-Axis of the secant passing by the points $(r_n, f(r_n))$ and $(r_{n-1}, f(r_{n-1}))$ on $(\mathcal{C})$

The intersection of this secant line with the X-Axis would provide the $(n+1)$-iterate **secant method** formula:

$$r_{n+1} = r_n - \frac{f(r_n)}{[r_n, r_{n-1}]} \equiv r_n - \frac{f(r_n)(r_n - r_{n-1})}{f(r_n) - f(r_{n-1})}, n \geq 2 \qquad (2.37)$$

The secant method is a two-step method of the form $r_{n+1} = g(r_n, r_{n-1})$, its processing requiring selection of r_0 and r_1. Of course, if the method is succeeding, the points r_n will be approaching a zero of f, so $f(r_n)$ will be converging to zero.

Practically, if a root r of the function f is located in the interval (a, b), one would suggest applying twice the bisection method in order to implement (2.37) as shown in the following algorithm.

Algorithm 2.3 Secant Method

```
% Input f, a, b,TOL, kMAX
% Find the first 2 approximations by the Bisection rule
function r=mySecant(f,a,b,TOL,kmax)
r(1)=(a+b)/2 ;
if f(a)*f(r(1)) < 0
    r(2)=(r(1)+a)/2 ;
else
    r(2)=(b+r(1))/2 ;
end
k=2; S = 1;
while S >TOL & k<=kMAX
```

```
        d=(f(r(k))-f(r(k-1))/r(k)-r(k-1));
        r(k+1)=r(k)-f(r(k))/d;
        S = abs (r(k+1)-r(k)]) abs (r(k)) ;
        k=k+1;
end
if n>=kmax
        disp ('error no convergence' );
else
        n=n-1;
end
```

The advantages of the secant method relative to the tangent method are that (after the first step) only one function evaluation is required per step (in contrast to Newton's iteration which requires two) and that it is almost as rapidly convergent. It can be shown that under the same assumptions as those of Theorem 2.2, the basic secant method is **superlinear** and has a **local character** of convergence.

Theorem 2.3 *Under the hypothesis of Theorem 2.2 and for r_0 and $r_1 \in I_0$ (defined in (2.26)), then one has:*

1. *$\lim_{n \to \infty} r_n = r$,*

2. *There exists a sequence $\{t_n|, \ n \geq 0\}$ such that:*

$$|r - r_n| \leq t_n, \ \text{with} \ t_n = O(t_{n-1})^\gamma \ and \ \gamma = \frac{1 + \sqrt{5}}{2} \qquad (2.38)$$

*i.e the **order of convergence** of the secant method is the **Golden Number** $\gamma \approx 1.618034$ in the sense of (2.7).*

Proof. Starting with the following identity (Theorem 4.5):

$$f(r) = f(r_n) + [r_{n-1}, r_n](r - r_n) + \frac{1}{2}(r - r_n)(r - r_{n-1})f''(c) \ ; \ c = r_n + \theta(r - r_n),$$

where $0 < \theta < 1$. As $f(r) = 0$, one deduces:

$$r_n - \frac{f(r_n)}{[r_{n-1}, r_n]} = r + \frac{1}{2}(r - r_n)(r - r_{n-1})\frac{f''(c)}{[r_{n-1}, r_n]}$$

Since $[r_{n-1}, r_n] = f'(c_1)$, then under the assumptions of Theorem 2.2, one concludes that:

$$|r - r_n| \leq C|r - r_{n-1}|.|r - r_{n-2}|, \ \forall n \geq 2 \qquad (2.39)$$

with C as defined in (2.24). Again, let $e_i = C|r - r_i|, \ i = n - 2, n - 1$, then (2.39) is equivalent to:

$$e_n \leq e_{n-1}.e_{n-2}, \ \forall n \geq 2. \qquad (2.40)$$

With the assumption that the initial conditions r_0, r_1 are selected so that:

$$\delta = \max(e_0, e_1) < 1 \qquad (2.41)$$

one obviously concludes that $e_2 \leq e_0 e_1 < \delta^2$ and that $e_3 \leq e_1 e_2 < \delta^3$. Let $\{f_n | n \geq 0\}$ be a Fibonacci sequence defined by:

$$f_0 = f_1 = 1, \; f_n = f_{n-1} + f_{n-2}, \; n \geq 2.$$

It is well known that the solution of this second order difference equation is given by:

$$f_n = \frac{1}{\sqrt{5}}((\frac{1+\sqrt{5}}{2})^{n+1} - (\frac{1-\sqrt{5}}{2})^{n+1}) = \frac{1}{\sqrt{5}}((\frac{1+\sqrt{5}}{2})^{n+1} + (-1)^{n+1}(\frac{\sqrt{5}-1}{2})^{n+1}).$$

Let $\gamma = \frac{1+\sqrt{5}}{2}$ be the Golden Number, then:

$$f_n = \frac{1}{\sqrt{5}}(\gamma^{n+1} + (-\frac{1}{\gamma})^{n+1})$$

As $n \to \infty$, the first term of f_n tends to $+\infty$ while the second tends to 0 so that $f_n = O(\gamma^{n+1})$.

Based on the choice of r_0 and r_1 in I_0, $e_0 < \delta^{f_0}$ and $e_1 < \delta^{f_1}$. By induction, assuming that $e_k < \delta^{f_k}$, $\forall k \leq n-1$, then using (2.40), one has:

$$e_n \leq e_{n-1} e_{n-2} < \delta^{f_{n-1}+f_{n-2}} = \delta^{f_n}, \; \forall n \geq 2. \qquad (2.42)$$

As $\delta < 1$, this last inequality proves the first part of the theorem, i.e., that

$$\lim_{n\to\infty} e_n = 0.$$

As for the second part of the theorem, given that:

$$|r - r_n| \leq \frac{1}{C} e_n < t_n = \frac{1}{C} \delta^{f_n},$$

then:

$$\frac{t_n}{t_{n-1}^\gamma} = C^{\gamma-1} \delta^{f_n - \gamma f_{n-1}}.$$

Note that

$$f_n - \gamma f_{n-1} = \frac{1}{\sqrt{5}}(\gamma^{n+1} + (-\frac{1}{\gamma})^{n+1} - \gamma^{n+1} - \gamma(-\frac{1}{\gamma})^n) = \frac{2}{\sqrt{5}}(-\frac{1}{\gamma})^{n+1}.$$

Hence $f_n - \gamma f_{n-1} \to 0$ and therefore there exists a constant K such that:

$$t_n \leq K(t_{n-1})^\gamma.$$

■

To illustrate the secant method, we consider the following example.

Iteration	Iterate
0	1.0
1	1.5000000000
2	0.21271008648
3	0.77325832517
4	0.61403684201
5	0.58643504642
6	0.58855440366
7	0.58853274398
8	0.58853274398

TABLE 2.8: Application of the secant method for the first root of $f(x) = \sin(x) - e^{-x}$

Example 2.6 *Approximate the root of $f(x) = \sin(x) - e^{-x}$ up to 10 decimal figures, in the interval (0, 2) using the secant method.*

Results of this process are given in Table 2.8. Besides computing the initial conditions, the secant method requires about 6 iterations to reach a precision $p = 10$, that is 2 more than Newton's method.

Comparisons between the convergence of both the Newton and secant methods can be further made, using the inequalities (2.31) and (2.42), as $e_n = C|r - r_n|$ satisfies respectively:

1. $e_n \leq \delta^{2^n}$ in Newton's method and

2. $e_n \leq \delta^{f_n}$ in the secant method.

with $\delta = e_0 = C|r - r_0|$. Thus

$$\frac{|r - r_n|}{|r - r_0|} = \frac{e_n}{e_0} = \leq \delta^{2^n - 1}$$

for Newton's method and

$$\frac{|r - r_n|}{|r - r_0|} = \frac{e_n}{e_0} = \leq \delta^{f_n - 1}$$

for the secant method.

In the same way that this was done for the preceding methods (Corollaries 2.1 and 2.2), one can also derive the minimum number of iterations needed theoretically to reach requested precisions using the secant method. However, in this chapter, in order to confirm that Newton's method is faster, we will only consider for example the specific case of $\delta = \frac{1}{2}$, seeking the minimum n_0 for which $\frac{|r - r_n|}{|r - r_0|} \leq 2^{-p}$, (i.e a precision p in a floating-point system $\mathbb{F}(2, p, E_{\min}, E_{\max})$). Straightforwardly, it can be shown that such n_0 satisfies:

$$2^{n_0^{(1)}} \geq 1 + p > 2^{n_0^{(1)} - 1}$$

p	$n_0^{(1)}$	$n_0^{(2)}$
10	4	6
24 (IEEE-single)	5	8
53 (IEEE-double)	6	9

TABLE 2.9: Comparing Newton's and secant methods for precisions $p = $ 10, 24, 53

for Newton's method and

$$f_{n_0^{(2)}} \geq 1 + p > f_{n_0^{(2)}-1}$$

for the secant method. Comparisons of these estimates to $p = 10, 24, 53$ are shown in Table 2.9. Thus although Newton's method is faster, it takes at most about 2 to 3 more iterations for the secant method to reach the same precision.

2.6 Exercises

The Bisection Method

1. Locate all the roots of f, then approximate each one of them up to 3 decimal figures using the bisection method.

   (a) $f(x) = x - 2\sin x$

   (b) $f(x) = x^3 - 2\sin x$

   (c) $f(x) = e^x - x^2 + 4x + 3$

   (d) $f(x) = x^3 - 5x - x^2$

2. Show that the following equations have infinitely many roots by graphical methods. Use the bisection method to determine the smallest positive value up to 4 decimal figures.

   (a) $\tan x = x$

   (b) $\sin x = e^{-x}$

   (c) $\cos x = e^{-x}$

   (d) $\ln(x + 1) = \tan(2x)$

3. The following functions have a unique root in the interval $[1, 2]$. Use the Bisection method to approximate that root up to 4 decimal figures. Compare the number of iterations used with the "theoretical" estimate.

   (a) $f(x) = x^3 - e^x$

   (b) $f(x) = x^2 - 4x + 4 - \ln x$

   (c) $f(x) = x^3 + 4x^2 - 10$

   (d) $f(x) = x^4 - x^3 - x - 1$

   (e) $f(x) = x^5 - x^3 + 3$

   (f) $f(x) = e^{-x} - \cos x$

   (g) $f(x) = \ln(1 + x) - \frac{1}{x+1}$

4. The following functions have a unique root in the interval $[0, 1]$. Use the bisection method to approximate that root up to 5 decimal figures. Compare the number of iterations needed to reach that precision with the predictable "theoretical" value.

   (a) $f(x) = e^{-x} - 3x$

   (b) $f(x) = e^x - 2$

   (c) $f(x) = e^{-x} - x^2$

(d) $f(x) = \cos x - x$

(e) $f(x) = \cos x - \sqrt{x}$

(f) $f(x) = e^x - 3x$

(g) $f(x) = x - 2^{-x}$

(h) $f(x) = 2x + 3\cos x - e^x$

5. Prove that the function $f(x) = \ln(1-x) - e^x$ has a unique negative root. Use the bisection method to calculate the first four iterations.

6. Prove that the function $f(x) = e^x - 3x$ has a unique positive root. Use the bisection method to calculate the first four iterations.

7. The bisection method generates a sequence of intervals $\{[a_0, b_0], [a_1, b_1], ...\}$. Prove or disprove the following estimates.

   (a) $|r - a_n| \le 2|r - b_n|$

   (b) $|r - b_n| \le 2^{-n}(b_0 - a_0)$

   (c) $r_{n+1} = \frac{a_n + r_n}{2}$

   (d) $r_{n+1} = \frac{b_n + r_n}{2}$

Newton's and the Secant Methods

8. Use three iterations of Newton's method to compute the root of the function $f(x) = e^{-x} - \cos x$ that is nearest to $\pi/2$.

9. Use three iterations of Newton's method to compute the root of the function $f(x) = x^5 - x^3 - 3$ that is nearest to 1.

10. The polynomial $p(x) = x^4 + 2x^3 - 7x^2 + 3$ has 2 positive roots. Find them by Newton's method, correct to four significant figures.

11. Use Newton's method to compute $\ln 3$ up to five decimal figures.

12. Approximate $\pm\sqrt{e}$ up to 7 decimal figures using Newton's method.

13. Compute the first four iterations using Newton's method to find the negative root of the function $f(x) = x - e/x$.

14. Use Newton's method to approximate the root of the following functions up to 5 decimal figures, located in the interval $[0, 1]$. Compare the number of iterations used to reach that precision with the predictable "theoretical" value.

    (a) $f(x) = e^x - 3x$

    (b) $f(x) = x - 2^{-x}$

15. To approximate the reciprocal of 3, i.e., $r = \frac{1}{3}$, using Newton's method:

(a) Define some appropriate non-polynomial function that leads to an iterative formula not dividing by the iterate. Specify the restrictions on the initial condition if there are any.

(b) Choose two different values for the initial condition to illustrate the local character of convergence of the method.

16. Based on Newton's method, approximate the reciprocal of the square root of a positive number R, i.e., $\frac{1}{\sqrt{R}}$, using first a polynomial function, and secondly a non polynomial function. Determine the necessary restrictions on the initial conditions, if there are any.

17. To approximate the negative reciprocal of the square root of 7, i.e., $r = \frac{-1}{\sqrt{7}}$, using Newton's method:

(a) Define some appropriate non-polynomial function that leads to an iterative formula not dividing by the iterate. Specify the restrictions on the initial condition if there are any.

(b) Use Newton's method to approximate $r = \frac{-1}{\sqrt{7}}$ up to 4 decimal figures.

18. Approximate $\sqrt{2}$ up to 7 decimal figures using Newton's method.

19. The number $\sqrt{R}$ $(R > 0)$ is a zero of the functions listed below. Based on Newton's method, determine the iterative formulae for each of the functions that compute $\sqrt{R}$. Specify any necessary restriction on the choice of the initial condition, if there is any.

(a) $a(x) = x^2 - R$

(b) $b(x) = 1/x^2 - 1/R$

(c) $c(x) = x - R/x$

(d) $d(x) = 1 - R/x^2$

(e) $e(x) = 1/x - x/R$

(f) $f(x) = 1 - x^2/R$

20. Based on Newton's method , determine an iterative sequence that converges to π. Compute π up to 3 decimal figures.

21. Let $f(x) = x^3 - 5x + 3$.

(a) Locate all the roots of f.

(b) Use successively the bisection and Newton's methods to approximate the largest root of f correct to 3 decimal places.

22. To approximate the cubic root of a positive number a, i.e., $r = a^{\frac{1}{3}}$, where $1 < a \leq 2$, using Newton's method:

(a) Define some appropriate polynomial function $f(x)$ with unique root $r = a^{\frac{1}{3}}$, then write the formula of Newton's iterative sequence $\{r_n\}$.

(b) Assume that, for $r_0 = 2$, the sequence $\{r_n\}$ is decreasing and satisfies: $a^{\frac{1}{3}} = r < ... < r_{n+1} < r_n < r_{n-1} < ... < r_1 < r_0 = 2$. Prove then that: $|r_{n+1} - r| \le (r_n - r)^2$ for all $n \ge 0$.

(c) Prove by recurrence that: $|r - r_n| \le |r - r_0|^{2^n}$, for all $n \ge 0$

(d) Assuming $|r_0 - r| \le \frac{1}{2}$. Estimate the least integer n_0 such that $|r_{n_0} - r| \le (\frac{1}{2})^{32}$.

23. Let $p(x) = c_2 x^2 + c_1 x + c_0$ be a **quadratic polynomial** with one of its roots r located in an interval (a, b), with:

$$\min_{a \le x \le b} |p'(x)| \ge d > 0 \text{ and } \frac{2d}{|c_2|} \le (b - a).$$

Using Newton's method with r_0 sufficiently close to r:

(a) Show that if $r_n \in (a, b)$ then $|r_{n+1} - r| \le C|r_n - r|^2$, where $C = \frac{|c_2|}{d}$.

(b) Let $e_n = C|r - r_n|$. Show that if $r_n \in (a, b)$ then $e_{n+1} \le e_n^2$. Give also the condition on $|r_0 - r|$ that makes $e_0 < 1$, and therefore $e_n < 1$ for all n.

(c) Assume $|r_0 - r| = \frac{1}{2C}$. Show by recurrence that $e_n \le (e_0)^{(2^n)}$, then estimate the smallest value n_p of n, so that:

$$\frac{|r_{n_p} - r|}{|r_0 - r|} \le 2^{-p}.$$

24. Calculate an approximate value for $4^{3/4}$ using 3 steps of the secant method.

25. Use three iterations of the secant method to approximate the unique root of $f(x) = x^3 - 2x + 2$.

26. Show that the iterative formula for the secant method can also be written

$$x_{n+1} = \frac{x_{n-1} f(x_n) - x_n f(x_{n-1})}{f(x_n) - f(x_{n-1})}$$

Compare it with the standard formula. Which one is more appropriate to use in the algorithm of the secant method?

27. Use the secant method to approximate the root of the following functions up to 5 decimal figures, located in the interval $[0, 1]$. Compare the number of iterations used to reach that precision with the number of iterations obtained in exercise 14.

(a) $f(x) = e^x - 3x$

(b) $f(x) = x - 2^{-x}$

(c) $f(x) = -3x + 2\cos x - e^x$

2.7   Computer Projects

Exercise 1: Newton's Method

The aim of this exercise is to approximate π by computing the root of $f(x) = \sin(x)$ in the interval $(3, 4)$, based on Newton's method. For that purpose:

1. Write two MATLAB functions:
   ```
   function[sinx]=mysin(x,p)
   function[cosx]=mycos(x,p)
   ```
   With inputs:

   - a variable x representing some angle in **radians**
   - a precision p

   Then using Taylor's series expansion, these functions should compute respectively the sin and cos of x, up to a tolerance $Tol = 0.5*10^{(-p+1)}$, and output respectively the values of $\sin(x)$ or $\cos(x)$ in $\mathbb{F}(10, p, -20, +20)$.
   Hint: First compute $\sin(x)$ and $\cos(x)$, then use **num2str(. , p)**
   Note that:
   $$\sin(x) = x - \frac{x^3}{3!} + \frac{x^5}{5!} - \frac{x^7}{7!} \ldots$$
   $$\cos(x) = 1 - \frac{x^2}{2!} + \frac{x^4}{4!} - \frac{x^6}{6!} + \ldots$$

   PS. Do not use the MATLAB built-in function for the factorials.
   Test each of the functions above for $x = \pi/3, \pi/4$ and $\pi/6$ with $p = 14$ and save your results in a Word document.

2. Write a MATLAB
   ```
   function [root,k]= myNewton(f,df,a,b,p,kmax)
   ```
   That takes as inputs:

   - a function f and its derivative df (as function handles)
   - 2 real numbers a and b, where (a, b) is the interval locating the root of f,
   - a precision p
   - a maximum number of iterations $kmax$

   Then, based on Newton's method, this function should output:

   - root: the approximation to the root of f up to p decimal figures
     Hint: first compute *root*, then use num2str(root , p)
   - k: the number of iterations used to reach the required precision p, whereas:
     $Tol = 0.5 * 10^{(-p+1)}$ is the relative error to be reached when computing the root.

Test your function for 2 different functions f and save your results in a word document.

3. Write a `MATLAB`
   `function [mypi, errpi, k]= mypiNewton(p, kmax)`
   That takes as inputs p and `kmax` as defined in the previous question. Applying **Newton's method** on the interval $(3, 4)$ and using the functions `mysin` and `mycos` programmed in part 1, this function should output:

   - `mypi`: the approximation to π up to p decimal figures
   - `errpi`: the absolute error $|mypi - \pi|$ where π is considered in double precision
   - `k`: the number of iterations used in Newton's method to reach the precision p

   Hint: Note that after calling the functions `myNewton`, `mysin`, `mycos`, their outputs should be converted back to numbers using the command `str2num(.)`

   Test this function for $kmax = 20$ and successively for $p = 2, 3, 7, 10, 15$. Save your numerical results in a Word document.

Exercise 2 : Newton's Method on Polynomials
Let $p = p_n(x) = a_{n+1}x^n + a_n x^{n-1} + ... + a_2 x + a_1$, be a polynomial of degree n and let $a = [a_1, a_2, ..., a_{n+1}]$ be the corresponding coefficients row vector. The objective of this exercise is to approximate the roots of p included in some interval $[-int, +int]$, using Newton's method. For this purpose, starting with a set of equally-spaced points on $[-int, +int]$, given by the `MATLAB` instruction:

$$x = -\text{int} : \text{incr} : +\text{int};$$

and selecting `incr` appropriately (`incr=0.5` in this exercise), use the following function:

```
function [S] = SignPoly(a,x)
N=length(a); m=length(x); S=[ ];
for j=1:m
     p=a(N);
     for i=N-1:-1:1
          p=p*x(j) + a(i);
     end
S(j)=sign(p);
end
```

This function takes as inputs the vector $a = [a_1, a_2, ..., a_{n+1}]$ of coefficients of p, and the vector x, then computes $p(x_i)$ at all components of the vector x, using nested polynomial evaluation. The required output is a vector S whose components represent the signs of $p(x_i) \, \forall \, i = 1 : length(x)$

1. Write a MATLAB
   function [A,B]=LocateRoots(a, x)
   With the same inputs as SignPoly and outputs 2 vectors A and B of
   equal length $m \leq length(x)$, such that for each k, $1 \leq k \leq m$, there
   exists a root r of p, with $A(k) < r < B(k)$.
   Hint: LocateRoots should call the function SignPoly.
   A pseudo-code for LocateRoots is as follows.

   ```
   A=[];B=[];
   for k=1: length(x)-1
       if S(k) * S(k+1) <0
       %then there exists a root r with   a=x(k) < r < x(k+1)=b,
       % where a and b are components of A and  B respectively
       end
   end
   ```

2. Write a MATLAB
   function R=PolyEvaluate(a,r)
   That takes as inputs the vector a of coefficients of p and a real number
   r and computes the ratio $R = \frac{p(r)}{p'(r)}$.
   Hint: Use one "for loop" only and nested polynomial evaluation to compute first the ratio $R1 = \frac{p(r)}{r.p'(r)}$,

3. Write a MATLAB
   function [roots]=PolyNewton(a,A,B,pr,kmax)
   That takes as inputs the vectors a, A and B as introduced in parts 1
   and 2, an integer pr representing some precision and $kmax$ a maximum
   number of iterations as a safeguard. Based on Newton's method, this
   function should output the vector "roots," whose components are the
   roots of p computed up to "pr" decimal figures.
   Hint: Compute first the vector "roots," then use the MATLAB function
   num2str(roots, pr) to round all roots of $p(.)$ up to pr decimals. ($tol =$
   $0.5 * 10^{1-pr}$).

4. Test each one of the 3 functions above on the following Hermite poly-
   nomials of degree $n \geq 3$ and save the results in a separate document.
   (Compare the computed roots with those listed in the table.)
   Note the following properties of all Hermite polynomials:

   - The roots are irrational numbers that are symmetric with respect
     to the origin.
   - The value "zero" is a root of all odd orders Hermite polynomials.
   - To obtain higher order Hermite polynomials, use the relation:

   $$H_{n+1}(x) = 2xH_n(x) - 2nH_{n-1}(x)$$

n	$H_n(x)$	Approximate non-negative roots
0	1	...
1	$2x$	0
2	$4x^2 - 2$	0.707
3	$8x^3 - 12x$	0 ; 1.224
4	$16x^4 - 48x^2 + 12$	0.524 ; 1.650
5	$32x^5 - 160x^3 + 120x$	0; 0.958 ; 2.020
6	$64x^6 - 480x^4 + 720x^2 - 120$	0.436 ; 1.335 ; 2.350
7	$128x^7 - 1344x^5 + 3360x^3$ $-1680x$	0; 0.816 ; 1.673; 2.651
8	$256x^8 - 3584x^6 + 13440x^4$ $-13440x^2 + 1680$	0.381; 1.157 ; 1.981 2.930
9	$512x^9 - 9216x^7 + 48384x^5$ $-80640x^3 + 30240x$	0 ; 0.723 ; 1.468 ; 2.266 ; 3.190
10	$1024x^{10} - 23040x^8 + 161280x^6$ $-403200x^4 + 302400x^2 - 30240$	0.342 ; 1.036 ; 1.756 ; 2.532 ; 3.436

Exercise 3: Testing the Order of Convergence of Root-Finding methods

The order of convergence of a "root finding" method is α, if:

$$|r_n - r| \approx C_n |r_{n-1} - r|^\alpha \text{ for } n \geq 2$$

the constants C_n being bounded. To experiment numerically the value of α, the above equality is transformed to

$$\ln |r_n - r| \approx \ln C_n + \alpha \ln |r_{n-1} - r|$$

Letting $A_n = \ln |r_n - r|$ and $C_{1,n} = \ln C_n$, then

$$A_n \approx C_{1,n} + \alpha A_{n-1}$$

Equivalently:

$$\frac{A_n}{A_{n-1}} \approx \frac{C_{1,n}}{A_{n-1}} + \alpha$$

Since $\lim_{n\to\infty} |r_n - r| = 0$, then $\lim_{n\to\infty} A_n = -\infty$, implying that for "relatively large" values of n

$$\alpha_n = \frac{A_n}{A_{n-1}} \approx \alpha$$

meaning that the sequence $\{\alpha_n = \frac{A_n}{A_{n-1}}\}$ provides an approximation to α, with the values of α_n obtained through this numerical process **oscillating** about the theroetical value of α.

(Note that respectively for the secant, bisection and Newton's methods, $\alpha = 1.6, 1$ and 2.)

1. Write a `MATLAB`
   `function [k,R, AbsErr,alphan]=SecantConverge(f,a,b,Tol,Kmax,r)`
   That takes as inputs:

   - a function **f** having a <u>known root **r**</u>, located in some interval **(a,b)**
   - a tolerance **Tol** $= \frac{1}{2}10^{1-p}$ that is the relative error bound, with $p \geq 5$
   - the maximum number `kmax` of iterations to be used

   and returns:

   - the number **k** of iterations needed to reach the precision **p**
   - a column vector **R** whose components are the successive iterates $\{r_n\}$ approximating the root of $f(.)$ by the secant method.
   - a column vector `AbsErr`$= |R - r|$ whose components are the absolute errors $|r_n - r|$
   - a column vector alphan, whose components are $\alpha_n = \frac{\ln |(r_n - r)|}{\ln |(r_{n-1} - r|}$

2. Write a `MATLAB`
   `function[k,R,AbsErr,alphan]=BisectConverge(f,a,b,Tol,Kmax,r)`
   Which inputs and outputs are similar to those defined for the secant method in part 1.

3. Write a `MATLAB`
   `function [k,R,AbsErr,alphan]=NewtonConverge(f,df,a,b,Tol,Kmax,r)`
   Which inputs and outputs are similar to those defined in part 1. (`df` is the derivative function of **f**.)

4. Test each of the 3 functions above for 2 different test cases and display your results in a table of the form:

R(n)	AbsErr(n)	alphan(n)

5. Use the `MATLAB` `plot` function for plotting `alphan(n)` as a function n (the nth iteration). Plot also on the same graph (with a different color), a horizontal line representing the theoretical value α.
   Save your numerical results obtained in part 3 and your graphs in a Word document.
   Suggested interesting functions:

   (a) $f(x) = \ln x - 1$, whose root is the irrational number e
   (b) $f(x)$: Polynomial functions of degree $>= 4$
   (c) Approximate square or cubic roots

(d) Approximate reciprocals of square or cubic roots

(e) $f(x) = x^2 - x - 1$, whose positive root is the "golden number"

Chapter 3

Solving Systems of Linear Equations by Gaussian Elimination

3.1   Mathematical Preliminaries

This chapter assumes basic knowledge of linear algebra, in particular *Elementary Matrix Algebra* as one can find these notions in a multitude of textbooks such as [32]. Thus, we consider the problem of computing the **solution of a system** of n linear equations in n unknowns. The scalar form of that system is as follows:

$$(S) \begin{cases} a_{11}x_1 & +a_{12}x_2 & +... & +... & +a_{1n}x_n & = b_1 \\ a_{21}x_1 & +a_{22}x_2 & +... & +... & +a_{2n}x_n & = b_2 \\ ... & ... & ... & ... & ... \\ a_{n1}x_1 & +a_{n2}x_2 & +... & +... & +a_{nn}x_n & = b_n \end{cases}$$

Written in matrix form, (S) is equivalent to:

$$Ax = b, \tag{3.1}$$

where the coefficient square matrix $A \in \mathbb{R}^{n,n}$, and the column vectors $x, b \in \mathbb{R}^{n,1} \cong \mathbb{R}^n$. Specifically,

$$A = \begin{pmatrix} a_{11} & a_{12} & ... & ... & a_{1n} \\ a_{21} & a_{22} & ... & ... & a_{2n} \\ ... & ... & ... & ... & ... \\ a_{n1} & a_{n2} & ... & ... & a_{nn} \end{pmatrix}$$

$$x = \begin{pmatrix} x_1 \\ x_2 \\ \cdots \\ x_n \end{pmatrix} \text{ and } b = \begin{pmatrix} b_1 \\ b_2 \\ \cdots \\ b_n \end{pmatrix}.$$

We assume that the basic linear algebra property for systems of linear equations like (3.1) are satisfied. Specifically:

Proposition 3.1 *The following statements are equivalent:*

1. *System (3.1) has a unique solution.*

2. $\det(A) \neq 0$.

3. *A is invertible.*

Our objective is to present the basic ideas of a **linear system solver** which consists of two main procedures allowing to solve (3.1) with the **least number of floating point arithmetic operations (flops)**.

1. The first, referred to as **Gauss elimination (or reduction)** reduces (3.1) into an equivalent system of linear equations, which matrix is **upper triangular**. Specifically one shows in section 4 that

$$Ax = b \iff Ux = c,$$

where $c \in \mathbb{R}^n$ and $U \in \mathbb{R}^{n,n}$ is given by:

$$U = \begin{pmatrix} u_{11} & u_{12} & \cdots & \cdots & u_{1n} \\ 0 & u_{22} & \cdots & \cdots & u_{2n} \\ \cdots & \cdots & \cdots & \cdots & \cdots \\ 0 & 0 & \cdots & u_{n-1,n-1} & u_{n-1,n} \\ 0 & 0 & \cdots & 0 & u_{nn} \end{pmatrix}.$$

Thus, $u_{ij} = 0$ for $i > j$. Consequently, one observes that A is invertible if and only if

$$\Pi_{i=1}^{n} u_{ii} = u_{11} u_{22} \dots u_{nn} \neq 0, \text{ i.e., } u_{ii} \neq 0 \ \forall i.$$

2. The second procedure consists in solving by **back substitution** the upper triangular system

$$Ux = c. \tag{3.2}$$

A picture that describes the two steps of the linear solver is:

$$A, b \rightarrow \boxed{\textbf{Gauss Reduction}} \rightarrow U, c \rightarrow \boxed{\textbf{Back Substitution}} \rightarrow x$$

k	$n = 2^k$	$N = n \times (n+1)$	$\approx$in Megabytes	
			IEEE single precision	IEEE double precision
3	8	72	2.7×10^{-4}	5.5×10^{-4}
6	64	4160	1.6×10^{-2}	3.2×10^{-2}
8	256	65792	0.25	0.5
10	1024	1049600	4	8

TABLE 3.1: Computer memory requirements for matrix storage

Our plan in this chapter is as follows. We start in Section 3.2 by discussing issues related to computer storage. This is followed in Section 3.3 by discussing the back substitution procedure that solves upper triangular systems, such as (3.2). Finally in Section 3.4 we present various versions of Gauss reduction, the simplest of which is **Naive Gaussian elimination**.
Extensive details regarding *Numerical Solutions of Linear Equations* can be found at a basic level in [31] and at a higher level in [16].

3.2 Computer Storage and Data Structures for Matrices

The data storage for A and b is through one data structure: the **augmented matrix** $AG \in \mathbb{R}^{n,n+1}$, given by:

$$AG = \begin{pmatrix} a_{11} & a_{12} & \ldots & \ldots & a_{1n} & b_1 \\ a_{21} & a_{22} & \ldots & \ldots & a_{2n} & b_2 \\ \ldots & \ldots & \ldots & \ldots & \ldots & \ldots \\ a_{n1} & a_{n2} & \ldots & \ldots & a_{nn} & b_n \end{pmatrix}$$

We generally assume that the matrix A is a **full** matrix, that is, "most of its elements are non-zero." Storing the augmented matrix AG for a full matrix in its standard form, would then require $N = n \times (n+1)$ words of computer memory. If one uses single precision, $4N$ bytes would be necessary, while using double precision would necessitate $8N$ bytes for that storage.
For instance, when the matrix size is $n = 2^k$, the computer memory for double precision computation should exceed $N = 8 \times 2^k(2^k + 1) \approx O(2^{2k+3})$ bytes. Table 3.1 illustrates some magnitudes of memory requirements.

Practically, computer storage is usually **one-dimensional**. As a result, matrix elements are either stored **column-wise** (as in MATLAB), or **row-wise**. In the case where the elements of the augmented matrix AG are contiguously stored by columns, this storage would obey the following sequential pattern:

$$\underbrace{| a_{11}\, a_{21}\, ...\, a_{n1}}_{\text{column 1}}\, \underbrace{| a_{12}\, ...\, a_{n2}}_{\text{column 2}}\, | ... | \underbrace{a_{1n}\, ...\, a_{nn}}_{\text{column n}}\, | \underbrace{b_1\, b_2\, ...,\, b_n}_{\text{column n+1}}\, |$$

while if stored by rows, the storage pattern for the augmented matrix elements becomes:

$$\underbrace{| a_{11}\, a_{12}\, ...\, a_{1n}\, b_1}_{\text{line 1}}\, | \underbrace{a_{21}\, ...\, a_{2n}\, b_2}_{\text{line 2}}\, | ... | \underbrace{a_{n1}\, ...\, a_{nn}\, b_n}_{\text{line n}}\, |$$

Once Gauss reduction has been applied to the original system $Ax = b$, the resulting upper triangular system $Ux = c$ would necessitate the storage of the upper triangular matrix U and the right hand side vector c. Obviously, the augmented matrix for this system is given by:

$$UG = \begin{pmatrix} u_{11} & u_{12} & ... & ... & u_{1n} & c_1 \\ 0 & u_{22} & ... & ... & u_{2n} & c_2 \\ ... & ... & ... & ... & ... & ... \\ 0 & ... & ... & 0 & u_{nn} & c_n \end{pmatrix}$$

Since by default, the lower part of the matrix U consists of zeros, this part of the storage shall not be waisted but used for other purposes, particularly that of storing the **multiplying factors**, which are essential parameters to carry out Gauss elimination procedure. Hence, at this stage we may consider the data structure UG whether stored by rows or by columns as consisting of the elements of U and c and unused storage space:

$$UG = \begin{pmatrix} u_{11} & u_{12} & ... & ... & u_{1n} & c_1 \\ \boxed{\text{unused}} & u_{22} & ... & ... & u_{2n} & c_2 \\ ... & ... & ... & ... & ... & ... \\ \boxed{\text{unused}} & ... & ... & \boxed{\text{unused}} & u_{nn} & c_n \end{pmatrix}$$

We turn now to the back substitution procedure.

3.3   Back Substitution for Upper Triangular Systems

Although this procedure comes after the completion of the Gauss reduction step, we shall deal with it from the start. It indeed provides the importance of this global approach.

Considering (3.2) in its scalar form, with all diagonal elements $u_{ii} \neq 0$, gives:

$$\begin{pmatrix} u_{11} & u_{12} & \cdots & \cdots & u_{1n} \\ 0 & u_{22} & \cdots & \cdots & u_{2n} \\ \cdots & \cdots & \cdots & \cdots & \cdots \\ 0 & 0 & \cdots & u_{n-1,n-1} & u_{n-1,n} \\ 0 & 0 & \cdots & 0 & u_{nn} \end{pmatrix} \begin{pmatrix} x_1 \\ x_2 \\ \cdots \\ x_{n-1} \\ x_n \end{pmatrix} = \begin{pmatrix} c_1 \\ c_2 \\ \cdots \\ c_{n-1} \\ c_n \end{pmatrix}$$

Solving this system by the **back substitution** procedure reduces such procedure to solving n equations, each one in one unknown only.

We give two versions of the back substitution process: the first one is **column oriented**, while the second one is **row oriented**. We then evaluate and compare the computational complexity of each version.

1. **Column-version:** The two main steps are as follows:

   (a) Starting with $j = n : -1 : 1$, solve the last equation for x_j, where $x_j = c_j/u_{j,j}$.

   (b) In all rows above, that is from row $i = 1 : (j-1)$, compute the new right hand side vector that results by "shifting" the last column of the matrix (terms in x_j) to the right hand side. For example when $j = n$, the new system to solve at this step is as follows:

$$\begin{pmatrix} u_{11} & u_{12} & \cdots & \cdots & u_{1,n-1} \\ 0 & u_{22} & \cdots & \cdots & u_{2,n-1} \\ \cdots & \cdots & \cdots & \cdots & \cdots \\ 0 & 0 & \cdots & 0 & u_{n-1,n-1} \end{pmatrix} \begin{pmatrix} x_1 \\ x_2 \\ \cdots \\ x_{n-1} \end{pmatrix} = \begin{pmatrix} c_1 - u_{1n}x_n \\ c_2 - u_{2n}x_n \\ \cdots \\ c_{n-1} - u_{n-1,n}x_n \end{pmatrix}$$

This process is repeated till the first row is reached, where:

$$u_{11}x_1 = c_1 - u_{1,n}x_n - u_{1,n-1}x_{n-1} - \ldots - u_{12}x_2$$

leading thus to $x_1 = (c_1 - u_{1,n}x_n - u_{1,n-1}x_{n-1} - \ldots - u_{12}x_2)/u_{11}$.

The corresponding algorithm is implemented as follows:

Algorithm 3.1 Column Back Substitution

```
function x = ColBackSubstitution(U,c)
% Input: U  an upper-triangular invertible matrix, and
%            c a column vector
% Output: solution vector x of system Ux = c
% Storage is column oriented
n=length(c) ;
for j=n:-1:1
```

```
            x(j)=c(j)/U(j,j);
            for i=1: j-1
                  c(i)=c(i) - U(i,j) * x(j);
            end
      end
```

The number of floating point operations used in this algorithm is n^2, and is computed as follows:

- For every $j, (j = 1 : n)$: 1 division is needed to compute $x(j)$ adding up therefore to a total of n flops.

- For every $j, (j = 1 : n)$ and for every $i, (i = 1 : j - 1)$, to compute each modified right hand side term $c(i)$: 1 addition + 1 multiplication are used, that sum up to a total of:

$$\sum_{j=1}^{n}\sum_{i=1}^{j-1} 2 = \sum_{j=1}^{n} 2[(j-1) - 1 + 1] = 2(1+2+...+(n-1)) = n(n-1)$$

As for the $2nd$ version, the rows are successively and completely solved for one unknown, starting with the last one $(i = n)$.

2. **Row-version:**

Algorithm 3.2 Row Back Substitution

```
% Input and Output as in "ColBackSubstitution" above
% Storage is row oriented
function x = RowBackSubstitution(U,c)
n=length(c);
x(n)=c(n)/U(n,n);
for i=n-1:-1:1
      for j=i+1:n
            c(i)=c(i)-U(i,j) * x(j);
      end
x(i)=c(i)/U(i,i);
end
```

It is easy to verify in that case that the total number of flops used remains equal to n^2 .

3.4 Gauss Reduction

Our starting point is to assume "ideal mathematical conditions" allowing one to carry the **reduction** without any safeguard. Before setting formally these assumptions, we work out the following example:

Example 3.1 *Consider the reduction of the following system into upper triangular form :*

$$\begin{cases} x_1 & -x_2 & +2x_3 & +x_4 & = 1 \\ 3x_1 & +2x_2 & +x_3 & +4x_4 & = 1 \\ 5x_1 & 8x_2 & +6x_3 & +3x_4 & = 1 \\ 4x_1 & +2x_2 & +5x_3 & +3x_4 & = -1 \end{cases} \tag{3.3}$$

The corresponding augmented matrix being:

$$\begin{pmatrix} 1 & -1 & 2 & 1 & 1 \\ 3 & 2 & 1 & 4 & 1 \\ 5 & 8 & 6 & 3 & 1 \\ 4 & 2 & 5 & 3 & -1 \end{pmatrix}$$

We proceed by applying successively 3 Gauss reductions. In each one of these, the following **linear algebra elementary operation** is being used: at the kth reduction, $k = 1, 2, 3$, and for $i = k + 1, ..., 4$

$$\text{(New) Equ } i \leftarrow \text{ (Previous) Equ } i - \text{(multiplier)} \times \text{ Pivot Equ } k \tag{3.4}$$

More explicitly:

1. **Reduction 1.** The **pivot equation** is the first equation ($k = 1$), the **pivot element** is $a_{11} = 1$. The respective **multipliers** for i successively $2, 3, 4$ are $\{\frac{a_{1i}}{a_{11}} = 3, 5, 4\}$. Thus, performing (3.4) repeatedly:

$$\text{Equation } 2 \leftarrow \text{ Equation } 2 - 3 \times \text{ Pivot Equation } 1,$$

$$\text{Equation } 3 \leftarrow \text{ Equation } 3 - 5 \times \text{ Pivot Equation } 1,$$

$$\text{Equation } 4 \leftarrow \text{ Equation } 4 - 4 \times \text{ Pivot Equation } 1,$$

At this stage, the modified augmented matrix is:

$$\begin{pmatrix} 1 & -1 & 2 & 1 & 1 \\ 0 & 5 & -5 & 1 & -2 \\ 0 & 13 & -4 & -2 & -4 \\ 0 & 6 & -3 & -1 & -5 \end{pmatrix}.$$

In order not to waste the implicitly zero storage locations, we use them to place the multipliers of the first reduction. Hence, at the accomplishment of reduction 1, the augmented matrix takes the form:

$$\begin{pmatrix} 1 & -1 & 2 & 1 & 1 \\ \boxed{3} & 5 & -5 & 1 & -2 \\ \boxed{5} & 13 & -4 & -2 & -4 \\ \boxed{4} & 6 & -3 & -1 & -5 \end{pmatrix}.$$

with the understanding that "boxed" elements are the corresponding multipliers.

2. **Reduction 2.** Perform repeatedly operation (3.4) with the **second pivot equation** ($k = 2$), the pivot element being here $a_{22} = 5$, and i successively 3,4. The multipliers are respectively $\{\frac{a_{2i}}{a_{22}} = \frac{13}{5}, \frac{6}{5}\}$.

$$\text{Equation } 3 \leftarrow \text{Equation } 3 - \frac{13}{5} \times \text{Equation } 2,$$

$$\text{Equation } 4 \leftarrow \text{Equation } 4 - \frac{6}{5} \times \text{Equation } 2,$$

The second reduction yields the following augmented matrix:

$$\begin{pmatrix} 1 & -1 & 2 & 1 & 1 \\ 0 & 5 & -5 & 1 & -2 \\ 0 & 0 & 9 & -23/5 & 6/5 \\ 0 & 0 & 3 & -11/5 & -13/5 \end{pmatrix}.$$

Adding the multipliers of the second reduction, the contents of the augmented matrix updated data structure are as follows:

$$\begin{pmatrix} 1 & -1 & 2 & 1 & 1 \\ \boxed{3} & 5 & -5 & 1 & -2 \\ \boxed{5} & \boxed{13/5} & 9 & -23/5 & 6/5 \\ \boxed{4} & \boxed{6/5} & 3 & -11/5 & -13/5 \end{pmatrix}.$$

Finally, we come to the last reduction.

3. **Reduction 3.** Perform operation (3.4) with the **third pivot equation** ($k = 3$), the pivot element being $a_{33} = 9$, and the sole multiplier being $\{\frac{a_{3i}}{a_{33}} = \frac{1}{3}\}$, for $i = 4$. Specifically:

$$\text{Equation } 4 \leftarrow \text{Equation } 4 - \frac{1}{3} \times \text{Equation } 3,$$

yields the augmented matrix:

$$\begin{pmatrix} 1 & -1 & 2 & 1 & 1 \\ 0 & 5 & -5 & 1 & -2 \\ 0 & 0 & 9 & -23/5 & 6/5 \\ 0 & 0 & 0 & -2/3 & -3 \end{pmatrix}.$$

Placing the multipliers, the updated augmented matrix is then:

$$\begin{pmatrix} 1 & -1 & 2 & 1 & 1 \\ \boxed{3} & 5 & -5 & 1 & -2 \\ \boxed{5} & \boxed{13/5} & 9 & -23/5 & 6/5 \\ \boxed{4} & \boxed{6/5} & \boxed{1/3} & -2/3 & -3 \end{pmatrix}.$$

The back substitution applied on the upper triangular system yields:

$$(x_1 = -217/30, x_2 = 17/15, x_3 = 73/30, x_4 = 9/2)$$

We may now discuss the assumptions leading to the naive Gauss elimination.

3.4.1 Naive Gauss Elimination

The adjective **naive** applies because this form is the simplest form of Gaussian elimination. It is not usually suitable for automatic computation unless essential modifications are made. We give first the condition that allows theoretically the procedure to work out successfully.

Definition 3.1 *A square matrix A_n has the* **principal minor property**, *if all its principal sub-matrices A_i, $i = 1, ..., n$ are invertible, where*

$$A_i = \begin{pmatrix} a_{11} & a_{12} & ... & ... & a_{1i} \\ a_{21} & a_{22} & ... & ... & a_{2i} \\ ... & ... & ... & ... & ... \\ a_{i1} & a_{i2} & ... & ... & a_{ii} \end{pmatrix}$$

If a matrix A verifies Definition 3.1, the pivot element at each reduction is well defined and is located on the main diagonal. Thus, $\forall b \in \mathbb{R}^{n,1}$, the following algorithms can be applied on the augmented matrix $[A|b]$. The first one assumes that the matrix A is stored column-wise.

Algorithm 3.3 Column Naive Gauss

```
% The algorithm is column oriented
% The matrix A is assumed to have the principal minor property
% At reduction k, the kth equation is the pivot equation,  A(k,k)
% is the pivot element, and equations 1,..,k remained unchanged
function[U, c]=NaiveGauss(A,b)
n=length(b) ;
for k=1:n-1
% Get the pivot element and the multipliers proceeding by columns
    piv=A(k,k);
```

```
    for i=k+1:n
        A(i,k)=A(i,k)/piv;
    end
% Modify the body of matrix A  proceeding by columns
    for j=k+1:n
        for i=k+1:n
            A(i,j)=A(i,j)-A(i,k)*A(k,j);
        end
    end
% Modify the right hand side b
    for i=k+1:n
        b(i)=b(i)-A(i,k)*b(k);
    end
end
% Extract c and U proceeding by columns
c=b;
U=triu(A);
```

The flop count for this algorithm can be easily evaluated:

1. To find the multipliers:

$$\sum_{k=1}^{n-1}\sum_{i=k+1}^{n} 1 = \sum_{k=1}^{n-1} n - k = 1 + 2 + ... + (n-1) = \frac{n(n-1)}{2} \text{ divisions}$$

2. To modify the body of the matrix:

$$\sum_{k=1}^{n-1}\sum_{j=k+1}^{n}\sum_{i=k+1}^{n} 2 = \sum_{k=1}^{n-1}\sum_{j=k+1}^{n} 2(n-k) = 2\sum_{k=1}^{n-1}(n-k)^2$$

$$= 2[1^2 + 2^2 + ... + (n-1)^2] = 2\left[\frac{n(n-1)(2n-1)}{6}\right] \text{ operations.}$$

3. To modify the right hand side vector:

$$\sum_{k=1}^{n-1}\sum_{i=k+1}^{n} 2 = 2\sum_{k=1}^{n-1} n - k = 2[1+2+...+(n-1)] = 2\left[\frac{n(n-1)}{2}\right] \text{ operations}$$

In terms of **flops**, these would total to:

$$\frac{n(n-1)}{2} + \frac{n(n-1)(2n-1)}{3} + n(n-1) = \frac{n(n-1)}{6}(7+4n) = O(\frac{2n^3}{3}).$$

The next version requires the same number of flops but is row oriented.

Algorithm 3.4 Row Naive Gauss

```
% The algorithm is row oriented
% The matrix A is assumed to have the principal minor property
% At reduction k, the kth equation is the pivot equation and A(k,k)
% is the pivot element, and equations 1,..,k remained unchanged
function[c,U]=naiveGauss(A,b)
n=length(b) ;
for k=1:n-1
% Get the pivot element
    piv=A(k,k);
% Proceed by row: get the multiplier for equation i
    for i=k+1:n
        A(i,k)=A(i,k)/piv;
% and modify its remaining coefficients, then its right hand side
        for j=k+1:n
            A(i,j)=A(i,j)-A(i,k)*A(k,j);
        end
    b(i)=b(i)-A(i,k)*b(k);
    end
end
% Extract c and U
c=b;
U=triu(A);
```

The above 2 versions, that are the simplest expressions of Gaussian elimination, do not take into account the eventual sensitivity of the system to **propagate round-off errors**.

3.4.2 Partial Pivoting: Unscaled and Scaled Partial Pivoting

When computing in floating point systems $\mathbb{F}$, there are several situations where the application of the naive Gaussian elimination algorithms fails although the matrix A may verify the principal minor property.

As an illustration consider first the case where the pivot element is relatively small in $\mathbb{F}$. This would lead to large multipliers that worsen the round-off errors, as shown in the following example.

Example 3.2 *Consider the following 2×2 system of equations, where ϵ is a small non-zero number:*

$$\begin{cases} \epsilon x_1 + x_2 = 2 \\ 3x_1 + x_2 = 1 \end{cases} \tag{3.5}$$

The exact solution to this problem in $\mathbb{R}$ is $x_1 \approx \frac{-1}{3}$ and $x_2 \approx 2$.
Naive Gauss elimination where the pivot is ϵ leads to:

$$\begin{cases} \epsilon x_1 + x_2 = 1 \\ (1 - \frac{3}{\epsilon})x_2 = 1 - \frac{6}{\epsilon} \end{cases}$$

and the back substitution procedure would give:

$$\begin{cases} x_2 = \frac{1-6/\epsilon}{1-3/\epsilon} \\ x_1 = \frac{2-x_2}{\epsilon} \end{cases}$$

If these calculations are performed in a floating point system $\mathbb{F}$, as $1/\epsilon$ is large, then

$$\begin{cases} 1 - \frac{6}{\epsilon} \approx -\frac{6}{\epsilon} \\ 1 - \frac{3}{\epsilon} \approx -\frac{3}{\epsilon} \end{cases}$$

The computed solutions in that case are incorrect, with:

$$x_2 \approx 2 \text{ and } x_1 \approx 0.$$

However, if we perform a permutation of the equations before the reduction process, then the equivalent system becomes:

$$\begin{cases} 3x_1 + x_2 = 1 \\ \epsilon x_1 + x_2 = 2 \end{cases}$$

Carried out, naive Gauss reduction would lead to:

$$\begin{cases} 3x_1 + x_2 = 1 \\ (1 - \frac{\epsilon}{3})x_2 = 2 - \frac{\epsilon}{3} \end{cases}$$

Back substitution in this case would clearly give: $x_2 \approx 2$ and $x_1 \approx -1/3$. ∎

This example leads us to conclude that some type of strategy is essential for selecting new pivot equations and new pivots at each Gaussian reduction. Theoretically **complete pivoting** would be the best approach. This process requires at each stage, first searching over all entries of adequate submatrices - in all rows and all columns - for the largest entry in absolute value and then permuting rows and columns to move that entry into the required pivot position. This would be quite expensive as a great amount of searching and data movement would be involved. However, scanning just the first column in the submatrix at each reduction and selecting as pivot the greatest absolute value entry accomplishes our goal, thus avoiding too small or zero pivots. This is **unscaled (or simple) partial pivoting**. It would solve the posed problem, but compared to Complete Pivoting strategy, it does not involve an examination of the entries in the rows of the matrix.

Moreover, rather than interchanging rows through the partial pivoting procedure, that is to avoid the data movement, we use an **indexing array**. Thus, the order in which the equations are used is denoted by the row vector IV called the **index vector**. At first, IV is set to $[1, 2, ..., n]$, then at each reduction, if there would be a permutation in the rows, it is performed only on IV which acts as a vector of pointers to the memory location of the rows. In fact, at each reduction, $IV=[i_1, i_2, ..., i_n]$ which is a permutation of the initial vector

IV. This definitely eliminates the time consuming and unnecessary process of moving around the coefficients of equations in the computer memory.

We formalize now the unscaled partial pivoting procedure.

1. **Gaussian Elimination with Unscaled Partial Pivoting**
   This strategy consists in first finding at reduction k, the "best" pivot equation. This is achieved by identifying the maximum absolute value element in the kth column, located in some row ranging from the kth row to the last. More explicitly:
   - At reduction $k = 1$, seek i_1 in the set $\{1, 2, ..., n\}$ such that:

   $$|a_{i_1,1}| = \max_{1 \leq i \leq n} |a_{i1}| = \max\{|a_{11}|, |a_{21}|, ..., |a_{n1}|\}$$

   then perform a permutation of row 1 and row i_1 in IV only. Row i_1 is the first pivot equation, and $a_{i1,1}$ is the pivot element. We write $IV([1, i_1]) = IV([i_1, 1])$, meaning that at this stage,

   $$IV = [i_1, ..., 1, ..., n] = [i_1, i_2, ..., i_n]$$

   - At reduction k, seek i_k in $\{IV(k), ..., IV(n)\}$, such that:

   $$|a_{i_k,k}| = \max_{IV(k) \leq i \leq IV(n)} |a_{ik}| = \max\{|a_{IV(k),k}|, |a_{IV(k+1),k}|, ..., |a_{IV(n),k}|\}$$

   repositioning i_k in IV will set $i_k = i_k + (k-1)$, so that row $IV(i_k)$ is the pivot equation and $a_{IV(ik),k}$ is the pivot element. Perform then a permutation of rows $IV(k)$ and $IV(i_k))$ in the last IV. Therefore one writes:

   $$IV([k, i_k]) = IV([i_k, k])$$

   As such, in case of effective row permutation, the Naive Gauss Elimination algorithm is modified as follows:

```
% The algorithm is column oriented
% At reduction k, find in the kth column (from row k to row n)
% the element with maximum absolute value
n=length(b);
for k=1:n-1
          [p,ik]=max(abs(A(k:n,k)));
          % Permutation of rows k and ik is then performed
          A([k ik])=A([ik k]);
          piv=A(k,k);
          . . . . . . . . . . . . . . . . . .
```

If an index vector is referred to, the algorithm proceeds as follows.

Algorithm 3.5 Column Unscaled Partial Pivoting Gauss

```
function[U,c]=PartialPivotingGauss(A,b)
n=length(b);
% Initialize IV
IV=1:n
%At reduction k, find the absolute value maximum column element
%and its position in IV starting from kth component
for k=1:n-1
      [p, ik]=max(abs(A(IV(k:n),k)));
% find the position of ik in last IV
      ik=ik + k - 1 ;
      % Permutation of rows k and ik is then performed through IV
      IV([k ik])=IV([ik k]);
      % Identify the pivots
       piv=A(IV(k),k);
% Find the multipliers
for i=k+1:n
        A(IV(i),k)=A(IV(i),k)/piv;
end
% Modify the body of matrix A and right hand side b
for j=k+1:n
        for i=k+1:n
               A(IV(i),j)=A(IV(i),j)-A(IV(i),k)*A(IV(k),j);
        end
end
for i=k+1:n
        b(IV(i))=b(IV(i))-A(IV(i),k)*b(IV(k));
end
%Extract U,c
c=b(IV);
U=triu(A(IV,:));
```

Example 3.3 *Solve the following system using unscaled partial pivoting Gaussian reduction.*

$$\begin{cases} 3x_1 & -13x_2 & +9x_3 & +3x_4 & = -19 \\ -6x_1 & +4x_2 & +x_3 & -18x_4 & = -34 \\ 6x_1 & -2x_2 & +2x_3 & +4x_4 & = 16 \\ 12x_1 & -8x_2 & +6x_3 & +10x_4 & = 26 \end{cases} \qquad (3.6)$$

We first initialize the index vector of the system:

IV	1	2	3	4

The augmented matrix for the system above is:

$$\begin{pmatrix} 3 & -13 & 9 & 3 & -19 \\ -6 & 4 & 1 & -18 & -34 \\ 6 & -2 & 2 & 4 & 16 \\ 12 & -8 & 6 & 10 & 26 \end{pmatrix}$$

(a) **Reduction 1** Seek the pivot equation:

$$\max\{3, |-6|, 6, 12\} = 12.$$

First occurrence of the maximum is at the fourth position, i.e., at $IV(4){=}4$ (meaning that at this stage, the fourth component of IV is equation 4). So, one needs to perform the permutation of rows 1 and 4 through the index vector IV, the pivot equation becoming effectively equation 4 and the pivot element being 12. Updating the index vector, computing the multipliers $a_{IV(i),1}/12$, $i = 2, 3, 4$ and simultaneouslly modifying the body of matrix and right hand side leads to:

IV	4	2	3	1

$$\begin{pmatrix} \boxed{1/4} & -11 & 15/2 & 1/2 & -51/2 \\ \boxed{-1/2} & 0 & 4 & -13 & -21 \\ \boxed{1/2} & 2 & -1 & -1 & 3 \\ 12 & -8 & 6 & 10 & 26 \end{pmatrix}$$

(b) **Reduction 2** Similarly, one starts with a search for the pivot equation:

$$\max_{IV(2),IV(3),IV(4)} \{|a_{IV(2),2}|, |a_{IV(3),2}, |a_{IV(4),2}|\}$$
$$= \max\{|-11|, 0, 2\} = 11$$

The maximum 11 occurs at $IV(4) = 1$. Hence we perform the permutation of Equations $IV(2) = 2$ and $IV(4) = 1$. Thus, the pivot equation is row 1 and the pivot element is -11. Computing the multipliers and proceeding into the modifications of the remaining part of the augmented matrix leads to the following profile of the index vector and of the matrix data:

IV	4	1	3	2

$$\begin{pmatrix} \boxed{1/4} & -11 & 15/2 & 1/2 & -51/2 \\ \boxed{-1/2} & \boxed{0} & 4 & -13 & -21 \\ \boxed{1/2} & \boxed{-2/11} & 4/11 & -10/11 & -18/11 \\ 12 & -8 & 6 & 10 & 26 \end{pmatrix}$$

(c) **Reduction 3** In this last stage, seek the pivot equation:

$$\max_{IV(3),IV(4)} \{|a_{IV(3),3}|, |a_{IV(4),3}|\} = \max\{4, 4/11\} = 4.$$

The maximum 4 occurs at $IV(4) = 2$. Hence we perform the permutation of Equations $IV(4) = 2$ and $IV(3) = 3$. It is easily verified at the end of the process the contents of the data structure are as follows:

$$\begin{array}{|c|cccc|}\hline IV & 4 & 1 & 2 & 3 \\\hline\end{array}$$

$$\begin{pmatrix} \boxed{1/4} & -11 & 15/2 & 1/2 & -51/2 \\ \boxed{-1/2} & \boxed{0} & 4 & -13 & -21 \\ \boxed{1/2} & \boxed{-2/11} & \boxed{1/11} & 3/11 & 3/11 \\ 12 & -8 & 6 & 10 & 26 \end{pmatrix}$$

Obviously, back substitution yields:

$$x_4 = 1, x_3 = -2, x_2 = 1, x_1 = 3$$

Consider now the special case of a system of equations where the coefficients in a same row have a relatively large variation in magnitude. Gaussian elimination with simple partial pivoting is not sufficient and could lead to incorrect solutions as shown in the following example.

Example 3.4 *Consider the following 2×2 system of equations, where C is a large positive number.*

$$\begin{cases} 3x_1 + Cx_2 = C \\ x_1 + x_2 = 3 \end{cases} \tag{3.7}$$

The exact solution to this problem in $\mathbb{R}$ is $x_1 \approx 2$ and $x_2 \approx 1$. Applying the simple partial pivoting Gauss elimination, and since

$$\max\{3, 1\} = 3$$

the first row is the pivot equation, the pivot is 3 and the sole multiplier is $\frac{1}{3}$. This leads to:

$$\begin{cases} 3x_1 + Cx_2 = C \\ (1 - \frac{1}{3}C)x_2 = 3 - \frac{1}{3}C \end{cases}$$

where the back substitution procedure gives:

$$\begin{cases} x_2 = \frac{3 - \frac{1}{3}C}{1 - \frac{1}{3}C} \\ x_1 = \frac{C(1 - x_2)}{3} \end{cases}$$

If these calculation are performed in a floating point system $\mathbb{F}$ with finite fixed precision, and since C is large, then

$$\begin{cases} 3 - \frac{1}{3}C \approx -\frac{1}{3}C \\ 1 - \frac{1}{3}C \approx -\frac{1}{3}C \end{cases}$$

Therefore, the computed solutions would be:

$$x_2 \approx 1 \text{ and } x_1 \approx 0.$$

However scaling the rows first then selecting as pivot the scaled absolute value entry, improves the situation. The row scales vector being $S = [C, 1]$, to select the pivot equation, one would compute

$$\max\{\frac{3}{C}, \frac{1}{1}\} = 1$$

Consequently, in this example, the second row is selected as pivot equation. Now the pivot is 1 and the multiplier is 3. Carried out, the scaled partial pivoting Gauss reduction would lead to:

$$\begin{cases} (C - 3)x_2 = (C - 9) \\ x_1 + x_2 = 3 \end{cases}$$

Back substitution in this case would clearly give: $x_2 \approx 1$ and $x_1 \approx 2$. ∎

In view of this example, a more elaborate version than the simple partial pivoting would be the **scaled partial pivoting**, where we set up a strategy that simulates a scaling of the row vectors and then selects as a pivot element the relatively largest scaled absolute value entry in a column. This process would, in some way, load balance the entries of the matrix.

We formalize now this variation of simple pivoting strategies.

2. **Gaussian Elimination with Scaled Partial Pivoting**

In this strategy, scaled values are used to determine the best partial pivoting possible, particularly if there are large variations in magnitude of the elements within a row. Besides the index vector IV that is created to keep track of the equation-permutations of the system, a **scale factor** must be computed for each equation. We define the absolute value maximum element of each row s_i by:

$$s_i = \max_{1 \le j \le n} \{|a_{ij}|\} ; 1 \le i \le n$$

The column **scale vector** is therefore: $s = [s_1, s_2, ..., s_n]'$.

For example in starting the forward elimination process, we do not arbitrarily use the first equation as the pivot equation as in the naive Gauss elimination, nor do we select the row with maximum absolute value in the entries of the first column, as in the simple partial pivoting strategy. Instead we scan first in column 1 the ratios

$$\left\{ \frac{|a_{i,1}|}{s_i} \,, i = 1, ..., n \right\}$$

and select the equation (or row) for which this ratio is greatest. Let i_1 be the first index for which the ratio is greatest, then:

$$\frac{|a_{i_1,1}|}{s_{i_1}} = \max_{1 \le i \le n} \left\{ \frac{|a_{i,1}|}{s_i} \right\}$$

Interchange i_1 and 1 in the index vector only, which is now $IV = [i_1, i_2, ... i_n]$. In a similar way, proceed next to further reduction steps. Notice that through this procedure, the scale factors are computed once. They are not changed after each pivot step as the additional amount of computations are not worthwhile.

We give now a version of the newly devised algorithm.

Algorithm 3.6 Column Scaled Partial Pivoting Gauss

```
% Initialize IV and seek the scales
IV=1:n ;
for i=1:n
    s(i)=max(abs(A(i,1:n))
end
% Alternatively: s=(max(abs(A')))'
% At reduction k, find maximum scaled column element
for k=1:n-1
    [p, ik]=max(abs(A(IV(k:n),k) ./ s(IV(k:n)) ) ;
    ik=ik+k-1;
    IV([k  ik])= IV([ik  k]) ;
.........Same as  Partial Pivoting.............
```

As an illustration to the method, let us apply the scaled partial pivoting Gaussian reduction on the system of equations of the preceding example.

Example 3.5

We first set the index vector and evaluate the scales of the system:

IV	1	2	3	4

Augmented matrix					Scales
3	−13	9	3	−19	13
−6	4	1	−18	−34	18
6	−2	2	4	16	6
12	−8	6	10	26	12

(a) **Reduction 1** Seek the pivot equation:

$$\max\{3/13, 6/18, 1, 1\}.$$

First occurrence of the maximum is the 3rd one, i.e., at $IV(3)=3$ (meaning that the third component of IV is equation 3). So, one needs to perform the permutation of rows 1 and 3 through the index vector IV, the pivot equation becoming equation 3 and the pivot element being 6. Updating the index vector and computing the multipliers $a_{IV(i),1}/6, i = 2, 3, 4$ would yield:

IV	3	2	1	4

	Augmented matrix				Scales
1/2	−13	9	3	−19	13
-1	4	1	−18	−34	18
6	−2	2	4	16	6
2	−8	6	10	26	12

Modifying the body of matrix and right hand side leads to:

	Augmented matrix				Scales
1/2	−12	8	1	−27	13
-1	2	3	−14	−18	18
6	−2	2	4	16	6
2	−4	2	2	−6	12

(b) **Reduction 2** Similarly to reduction 1, one starts with a search for the pivot equation:

$$\max_{IV(2),IV(3),IV(4)}\{\frac{|a_{IV(2),2}|}{s_{IV(2)}}, \frac{|a_{IV(3),2}|}{s_{IV(3)}}, \frac{|a_{IV(4),2}|}{s_{IV(4)}}\}$$

$$= \max\{2/18, 12/13, 4/12\}.$$

The maximum 12/13 occurs at $IV(3) = 1$. Hence we perform the permutation of Equations $IV(2) = 2$ and $IV(3) = 1$. Thus, the pivot equation is row 1 and the pivot element is -12. Computing the multipliers and proceeding into the modifications of the remaining part of the augmented matrix leads to the following profile of the index vector and of the matrix data:

IV	3	1	2	4

	Augmented matrix				Scales
1/2	−12	8	1	−27	13
-1	-1/6	13/3	−83/6	−45/2	18
6	−2	2	4	16	6
2	1/3	−2/3	5/3	3	12

(c) **Reduction 3** This last step keeps the index vector unchanged since $\max\left\{|\frac{13}{3\times18}|;|\frac{2}{3\times12}|\right\} = \frac{13}{3\times18}$. It is easily verified at the end of the process the contents of the data structure are as follows:

IV	3	1	2	4

Augmented matrix					Scales
1/2	−12	8	1	−27	13
-1	-1/6	13/3	−83/6	−45/2	18
6	−2	2	4	16	6
2	1/3	-2/13	−6/13	−6/13	12

Obviously, back substitution yields:

$$x_4 = 1, \; x_3 = -2, \; x_2 = 1, \; x_1 = 3.$$

3.5   LU Decomposition

A major by-product of Gauss elimination is the **decomposition** or **factorization** of a matrix A into the product of a **unit lower triangular matrix L** by an upper triangular one U. We will base our arguments on the systems of equations (3.3) and (3.6).

1. **First case : Naive Gauss**

   Going back to (3.3) and on the basis of the multipliers of naive Gauss elimination, let L and U be respectively the unit lower and the upper triangular matrices of the process:

$$L = \begin{pmatrix} 1 & 0 & 0 & 0 \\ 3 & 1 & 0 & 0 \\ 5 & \frac{13}{5} & 1 & 0 \\ 4 & \frac{6}{5} & \frac{1}{3} & 1 \end{pmatrix} \quad ; \quad U = \begin{pmatrix} 1 & -1 & 2 & 1 \\ 0 & 5 & -5 & 1 \\ 0 & 0 & 9 & -23/5 \\ 0 & 0 & 0 & -2/3 \end{pmatrix}$$

Note that the product LU verifies:

$$\begin{pmatrix} 1 & 0 & 0 & 0 \\ 3 & 1 & 0 & 0 \\ 5 & \frac{13}{5} & 1 & 0 \\ 4 & \frac{6}{5} & \frac{1}{3} & 1 \end{pmatrix} \begin{pmatrix} 1 & -1 & 2 & 1 \\ 0 & 5 & -5 & 1 \\ 0 & 0 & 9 & -\frac{23}{5} \\ 0 & 0 & 0 & -\frac{2}{3} \end{pmatrix} = \begin{pmatrix} 1 & -1 & 2 & 1 \\ 12 & -8 & 6 & 10 \\ 3 & 2 & 1 & 4 \\ 4 & 2 & 5 & 3 \end{pmatrix}$$

$$(3.8)$$

which is precisely:
$$LU = A.$$

This identity obeys the following theorem ([10], [16]):

Theorem 3.1 *Let $A \in \mathbb{R}^{n,n}$ be a square matrix verifying the principal minor property. If A is processed through naive Gauss reduction, then A is factorized uniquely into the product of a unit lower triangular matrix L and an upper triangular matrix U associated to the reduction process, with*
$$A = LU$$

2. **Second case: Partial Pivoting**
   Consider now the scaled partial pivoting reduction applied on (3.6). Based on the last status of $IV = [3, 1, 2, 4]$, we extract successively the unit lower and the upper triangular matrices of the process:

$$L = \begin{pmatrix} 1 & 0 & 0 & 0 \\ 1/2 & 1 & 0 & 0 \\ -1 & -1/6 & 1 & 0 \\ 2 & 1/3 & -2/13 & 1 \end{pmatrix} \quad ; \quad U = \begin{pmatrix} 6 & -2 & 2 & 4 \\ 0 & -12 & 8 & 1 \\ 0 & 0 & 13/3 & -83/6 \\ 0 & 0 & 0 & -6/13 \end{pmatrix}$$

Computing the product LU gives:

$$\begin{pmatrix} 1 & 0 & 0 & 0 \\ 1/2 & 1 & 0 & 0 \\ -1 & -1/6 & 1 & 0 \\ 2 & 1/3 & -2/13 & 1 \end{pmatrix} \begin{pmatrix} 6 & -2 & 2 & 4 \\ 0 & -12 & 8 & 1 \\ 0 & 0 & 13/3 & -83/6 \\ 0 & 0 & 0 & -6/13 \end{pmatrix}$$

$$= \begin{pmatrix} 6 & -2 & 2 & 4 \\ 3 & -13 & 9 & 3 \\ -6 & 4 & 1 & -18 \\ 12 & -8 & 6 & 10 \end{pmatrix}$$

The product matrix is the matrix A up to a **permutation matrix** $P = P(IV)$, associated to the final status of the index vector. We write then
$$LU = P(IV)A$$

where P is defined as follows:

Definition 3.2 *Let $I \in \mathbb{R}^{n,n}$, be the identity matrix defined by its **rows**, i.e.,*

$$I = \begin{pmatrix} e_1 \\ e_2 \\ \dots \\ e_n \end{pmatrix}$$

Let $IV = [i_1, i_2, ..., i_n]$ be the last status of the index vector through *the partial pivoting procedures. The permutation matrix P associated to IV is a permutation of the identity matrix I, and is given by the* **row matrix:**

$$P = P(IV) = \begin{pmatrix} e_{i_1} \\ e_{i_2} \\ \cdots \\ e_{i_n} \end{pmatrix}$$

In example 3.5, the final status of $IV = [3, 1, 2, 4]$. Thus,

$$P = P(IV) = \begin{pmatrix} e_3 \\ e_1 \\ e_2 \\ e_4 \end{pmatrix} = \begin{pmatrix} 0 & 0 & 1 & 0 \\ 1 & 0 & 0 & 0 \\ 0 & 1 & 0 & 0 \\ 0 & 0 & 0 & 1 \end{pmatrix}$$

Note then that the product:

$$PA = \begin{pmatrix} 0 & 0 & 1 & 0 \\ 1 & 0 & 0 & 0 \\ 0 & 1 & 0 & 0 \\ 0 & 0 & 0 & 1 \end{pmatrix} \begin{pmatrix} 3 & -13 & 9 & 3 \\ -6 & 4 & 1 & -18 \\ 6 & -2 & 2 & 4 \\ 12 & -8 & 6 & 10 \end{pmatrix}$$

is precisely the product LU found above. Hence the LU decomposition theorem which generalizes Theorem 3.1 stands as follows:

Theorem 3.2 *Let a square matrix $A \in \mathbb{R}^{n,n}$ be processed through partial pivoting Gauss reduction. If the unit lower triangular matrix L , the upper triangular matrix U and the index vector IV are extracted from the final status of the process then:*

$$P(IV)A = LU$$

where $P(IV)$ is the permutation matrix associated to the reduction process.

Note also that this decomposition of A is unique.

The LU decomposition or factorization of A is particularly helpful in computing the **determinant** of A, in solving different systems of equations $Ax = b$, where the coefficient matrix A is held constant, or also in computing the **inverse of** A.

3.5.1   Computing the Determinant of a Matrix

Clearly from Theorems 3.1 and 3.2, we conclude respectively that in the first case

$$\det(A) = \det(L) \times \det(U)$$

while in the second case

$$\det(A) = (-1)^s \times \det(L) \times \det(U)$$

as $\det(P) = (-1)^s$, s being the number of permutations performed on IV through the partial pivoting procedures.

These results are stated hereafter:

Theorem 3.3 *(a) Under the hypothesis of Theorem 3.1,*

$$\det(A) = \prod_{i=1}^{n} u_{ii},$$

(b) Under the hypothesis of Theorem 3.2,

$$\det(A) = (-1)^s \prod_{i=1}^{n} u_{ii},$$

where u_{ii} , $i = 1, ..., n$ are the diagonal elements of the upper triangular matrix U associated to the reduction process.

One easily verifies that in example 3.4

$$\det(A) = 1 \times 5 \times 9 \times \frac{2}{3} = 30$$

while in example 3.5:

$$det(A) = 6 \times (-12) \times 13/3 \times (-6/13) = 144$$

since $s = 2$.

3.5.2 Computing the Inverse of A

The LU decomposition of a matrix A is also useful in computing its inverse denoted by A^{-1} and verifying the property

$$AA^{-1} = I$$

where I is the identity matrix. Let c_j and e_j represent respectively the jth column of A^{-1} and that of I, then one writes:

$$A[c_1 \, c_2 \, ... \, c_n] = [e_1 \, e_2 \, ... \, e_n] \tag{3.9}$$

1. **First case : Naive Gauss**

   Under the hypothesis of Theorem 1 and since $LU = A$, then (3.9) is equivalent to:
   $$LU[c_1 \, c_2 \, ... \, c_n] = [e_1 \, e_2 \, ... \, e_n]$$

   To obtain A^{-1} it is therefore enough to solve for c_j, in turn:

   $$LUc_j = e_j, \text{ for } j = 1, ..., n$$

By letting $Uc_j = y$, one has then to solve successively the following 2 triangular systems:

(i) The lower triangular system $Ly = e_j$, and get the vector y by **forward substitution**.

(ii) The upper triangular system $Uc_j = y$, and get the jth column c_j by **backward substitution**.

Example 3.6 *Use the LU decomposition of A based on the naive Gauss reduction applied to (3.3), to find the first column of A^{-1}*

Referring to Example 3.1, solving:

(i) The lower triangular system $Ly = e_1$, gives $y = [1, -3, 14/5, 4/3]'$ by forward substitution.

(ii) The upper triangular system $Uc_1 = y$, gives by backward substitution:

$$c_1 = [158/45, -41/45, -32/45, -2]'.$$

2. **Second case : Partial Pivoting**

Under the hypothesis of Theorem 2 and since $LU = PA$, then (3.9) is equivalent to:

$$PAA^{-1} = P$$

or equivalently:

$$LU[c_1 \, c_2 \, ... \, c_n] = [p_1 \, p_2 \, ... \, p_n]$$

where p_j is the jth column of P.

To obtain A^{-1} it is therefore enough to solve for c_j, in turn:

$$LUc_j = p_j, \text{ for } j = 1, ..., n$$

using the same 2 steps as in the first case above.

Remark 3.1 *Note that in Definition 2, the permutation matrix P is defined in terms of its rows, while in the process of computing A^{-1}, one has first to identify the columns of P.*

Example 3.7 *Use the LU decomposition of A based on the scaled partial pivoting reduction applied to (3.6), to find the last column of A^{-1}*

Referring to Example 3.3, solving:

(i) The lower triangular system $Ly = p_4$, gives $y = [0, 0, 0, 1]'$ by forward substitution.

(ii) The upper triangular system $Uc_1 = y$, gives by backward substitution:

$$c_4 = [155/72, -115/24, -83/12, -13/6]'.$$

Remark 3.2 Forward Substitution
In finding the inverse of the matrix A using its LU decomposition, one has to solve lower triangular systems of the form Ly = v, for some well defined vector v. For that purpose the forward substitution algorithm is needed prior to back substitution. We give successively the column then the row oriented version.

Algorithm 3.7 Column Forward Substitution

```
function x = ColForwardSubstitution(L,c)
% Input: L  a lower-triangular invertible matrix, and
%            c a column vector
% Output: solution vector x of system Lx = c
% Storage is column oriented
n=length(c) ;
for j=n:-1:1
        x(j)=c(j)/U(j,j);
        for i=1: j-1
             c(i)=c(i) - U(i,j) * x(j);
        end
end
```

As for the row version, it is implemented as follows:

Algorithm 3.8 Row Forward Substitution

```
function x = RowForwardSubstitution(L,c)
% Input: L  a lower-triangular invertible matrix, and
%            c a column vector
% Output: solution vector x of system Lx = c
% Storage is row oriented
n=length(c) ;
for j=n:-1:1
        x(j)=c(j)/U(j,j);
        for i=1: j-1
             c(i)=c(i) - U(i,j) * x(j);
        end
end
```

3.5.3   Solving Linear Systems Using *LU* Factorization

Generalizing the method above, if the *LU* factorization of *A* is available, one can as well solve systems $Ax = v$ varying the right hand side vector v only. That is, one solves consecutively 2 triangular systems:
(i) A lower triangular system by forward substitution.

(ii) An upper triangular system by backward substitution.

Straightforward illustrations to this procedure are left as exercises.

3.6 Exercises

1. Solve each of the following systems using naive Gaussian elimination and back substitution. Show the multipliers at each stage. Carry five significant figures and round to the closest.

(a) $\begin{cases} 3x_1 + 4x_2 + 3x_3 = 5 \\ x_1 + 5x_2 - x_3 = 0 \\ 6x_1 + 3x_3 + 7x_3 = 3 \end{cases}$

(b) $\begin{cases} 3x_1 + 2x_2 - 5x_3 = 0 \\ 4x_1 - 6x_2 + 2x_3 = 0 \\ x_1 + 4x_2 - x_3 = 4 \end{cases}$

(c) $\begin{cases} 9x_1 + x_2 + 7x_3 = 1 \\ 4x_1 + 4x_2 + 9x_3 = 0 \\ 8x_1 + 9x_2 + 6x_3 = 1 \end{cases}$

2. Apply the naive Gaussian elimination on the following matrices, showing the multipliers at each stage.

(a) $\begin{bmatrix} 4 & 2 & 1 & 2 \\ 1 & 3 & 2 & 1 \\ 1 & 2 & 4 & 1 \\ 2 & 1 & 2 & 3 \end{bmatrix}$

(b) $\begin{bmatrix} 1 & -1 & 2 & 1 \\ 4 & 2 & 5 & 3 \\ 3 & 2 & 1 & 4 \\ 5 & 8 & 6 & 3 \end{bmatrix}$

3. Solve each of the following systems using Gaussian elimination with unscaled partial pivoting and back substitution. Write the index vector and the multipliers at each step. Carry five significant figures and round to the closest.

(a) $\begin{cases} 8x_1 + 9x_2 + 2x_3 = 99 \\ 9x_1 + 6x_2 - 5x_3 = 132 \\ 1x_1 + 0x_2 + 9x_3 = 90 \end{cases}$

(b) $\begin{cases} 8x_1 + x_2 + 8x_3 = 49 \\ 9x_1 + x_2 + 2x_3 = 33 \\ 5x_1 + 2x_2 + 8x_3 = 52 \end{cases}$

(c) $\begin{cases} 3x_1 + 2x_2 - x_3 = 7 \\ 5x_1 + 3x_2 + 2x_3 = 4 \\ -x_1 + x_2 - 3x_3 = -1 \end{cases}$

4. Apply the unscaled partial pivoting Gaussian elimination on the following matrices, showing the multipliers and the index vector at each stage.

(a)
$$\begin{bmatrix} 6 & 6 & 2 & 6 \\ 7 & 1 & 0 & 3 \\ 7 & 7 & 0 & 9 \\ 3 & 0 & 8 & 0 \end{bmatrix}$$

(b)
$$\begin{bmatrix} 1 & -1 & 2 & 1 \\ 3 & 2 & 1 & 4 \\ 5 & 8 & 6 & 3 \\ 4 & 2 & 5 & 3 \end{bmatrix}$$

5. Solve each of the following systems using Gaussian elimination with scaled partial pivoting and Back substitution. Write the scales vector, the index vector, and the multipliers at each step. Carry five significant figures and round to the closest.

(a)
$$\begin{cases} x_1 & +x_2 & +6x_3 & +2x_4 & = 2 \\ 7x_1 & +6x_2 & +7x_3 & +9x_4 & = 0 \\ 3x_1 & +2x_2 & +4x_3 & +x_4 & = -1 \\ 5x_1 & +6x_2 & + & +8x_4 & = 0 \end{cases}$$

(b)
$$\begin{cases} 3x_1 & + -9x_3 = 3 \\ 5x_1 + 5x_2 + x_3 = -20 \\ 7x_2 + 5x_3 = 0 \end{cases}$$

(c)
$$\begin{cases} x_1 + 8x_2 + 2x_3 + x_4 = 5 \\ 9x_1 + 8x_2 + 8x_3 + 2x_4 = 4 \\ +4x_3 + x_4 = 0 \\ 7x_1 + 3x_2, +9x_3 + x_4 = -1 \end{cases}$$

6. Apply the scaled partial pivoting Gauss elimination on the following matrices, showing the scales vector, the index vector, and the multipliers at each stage.

(a)
$$\begin{bmatrix} 1 & 4 & 5 \\ 2 & 3 & 5 \\ 6 & 8 & 9 \end{bmatrix}$$

(b)
$$\begin{bmatrix} 1 & 7 & 6 & 9 \\ 4 & 7 & 1 & 3 \\ 4 & 2 & 1 & 5 \\ 6 & 6 & 4 & 2 \end{bmatrix}$$

(c)
$$\begin{bmatrix} 1 & 0 & 3 & 0 \\ 0 & 1 & 3 & -1 \\ 3 & -3 & 0 & 6 \\ 0 & 2 & 4 & -6 \end{bmatrix}$$

7. Consider the following system of 2 equations in 2 unknowns:

$$(S) \begin{cases} 10^{-5}x + y = 7 \\ x + y = 1 \end{cases}$$

(a) Find the exact solution of (S) in $\mathbb{R}$.

(b) Use the naive Gaussian reduction to solve (S) in $F(10, 4, -25, +26)$ and compare the result with the exact solution.

(c) Use the partial pivoting Gaussian reduction to solve (S) in $F(10, 4, -25, +26)$ and compare the result with the exact solution.

8. Consider the following system of 2 equations in 2 unknowns:

$$(S) \begin{cases} 2x + 10^5 y = 10^5 \\ x + y = 3 \end{cases}$$

(a) Find the exact solution of (S) in $\mathbb{R}$.

(b) Use the simple partial pivoting Gaussian reduction to solve (S) up to 4 significant figures and compare the result with the exact solution.

(c) Use the scaled partial pivoting Gaussian reduction to solve (S) up to 4 significant figures and compare the result with the exact solution.

9. Based on the naive Gaussian reduction applied to each coefficient matrix A of Exercise 2:

(a) Determine the unit lower triangular matrix L and the upper triangular matrix U.

(b) Use the LU decomposition of A to compute the determinant of A.

(c) Use the LU decomposition of A to determine the inverse of A.

10. Based on the unscaled partial pivoting Gaussian reduction applied to to the first matrix A of Exercise 4:

(a) Determine the unit lower triangular matrix L, the upper unit triangular matrix U and the permutation matrix P.

(b) Use the LU decomposition of A to compute the determinant of A.

(c) Use the LU decomposition of A to determine the first column of the inverse of A.

11. Based on the scaled partial pivoting Gaussian reduction applied to each coefficient matrix A of Exercise 6:

(a) Determine the unit lower triangular matrix L, the upper unit triangular matrix U and the permutation matrix P.

(b) Use the *LU* decomposition of A to compute the determinant of A.

(c) Use the *LU* decomposition of A to determine the second column of the inverse of A.

12. Use the *LU* decomposition of the matrix A in exercises 2(a), 4(a) and 6(a) to solve:

(a) $Ax = [1, 1, 1, 1]'$

(b) $Ax = [-1, 2, 0, -1]'$

13. Based on the following definition:

Definition 3.3 *A square matrix A of size $n \times n$ is **strictly diagonally dominant** if for every row, the magnitude of the diagonal entry is larger then the sum of the magnitude of all the other non- diagonal entries, in that row. i.e.,*

$$|A(i, i)| > \sum_{j=1}^{n} |A(i, j)| \;; \forall \, i = 1, 2, ..., n$$

Determine which of the following matrices is strictly diagonally dominant? Satisfies the principal minor property?

(a) $A = \begin{bmatrix} 8 & -1 & 4 \\ 1 & -10 & 3 \\ -5 & 0 & 15 \end{bmatrix}$

(b) $B = \begin{bmatrix} 8 & -1 & 4 & 9 \\ 1 & 7 & 3 & 0 \\ -5 & 0 & -11 & 3 \\ 4 & 3 & 2 & 12 \end{bmatrix}$

(c) $C = \begin{bmatrix} 28 & -1 & 4 & 9 & 2 \\ 1 & 30 & 3 & 9 & 7 \\ 0 & 0 & 7 & 3 & 0 \\ 4 & 3 & 2 & 20 & 7 \\ 3 & 0 & 0 & 0 & 9 \end{bmatrix}$

14. Find a set of values for a, b and c for which the following matrix is strictly diagonally dominant.

$$\begin{bmatrix} a & 1 & 0 & b & 0 \\ a & 9 & 1 & 3 & -1 \\ c & a & 10 & 5 & -1 \\ a & b & c & -6 & 1 \\ 1 & c & 0 & 0 & a \end{bmatrix}$$

15. Apply the naive Gauss reduction on the following strictly diagonally dominant band matrices. (As such, the naive Gauss reduction is successfully applicable on the matrix).

   (a) For each matrix below, determine at each reduction, the multipliers and the elements of the matrix that are modified.

   (b) Extract the upper triangular matrix U and the lower unit triangular matrix L obtained at the end of this process.

   (c) Determine the total number of operations needed for the LU decomposition of the given matrix.

   - Let T_n be a triangular matrix, with

$$T_n = \begin{bmatrix} a1 & b1 & 0 & 0 & \cdots & 0 \\ c1 & a2 & b2 & 0 & \cdots & 0 \\ 0 & c2 & a3 & b3 & \cdots & 0 \\ \cdots & \cdots & \cdots & \cdots & \cdots & \cdots \\ \cdots & \cdots & \cdots & \cdots & \cdots & \cdots \\ \cdots & \cdots & \cdots & c_{n-2} & a_{n-1} & b_{n-1} \\ 0 & \cdots & 0 & 0 & c_{n-1} & a_n \end{bmatrix}$$

   - Let UQ_n be an upper quadri-diagonal matrix, with

$$UQ_n = \begin{bmatrix} a1 & b1 & d1 & 0 & \cdots & 0 \\ c1 & a2 & b2 & d2 & \cdots & 0 \\ 0 & c2 & a3 & b3 & \cdots & 0 \\ \cdots & \cdots & \cdots & \cdots & \cdots & \cdots \\ \cdots & \cdots & c_{n-3} & a_{n-2} & b_{n-2} & d_{n-2} \\ \cdots & \cdots & \cdots & c_{n-2} & a_{n-1} & b_{n-1} \\ 0 & \cdots & 0 & 0 & c_{n-1} & a_n \end{bmatrix}$$

   - Let LQ_n be a lower quadri-diagonal matrix, with

$$LQ_n = \begin{bmatrix} a1 & b1 & 0 & 0 & \cdots & 0 \\ c1 & a2 & b2 & 0 & \cdots & 0 \\ d1 & c2 & a3 & b3 & \cdots & 0 \\ \cdots & \cdots & \cdots & \cdots & \cdots & \cdots \\ \cdots & \cdots & \cdots & \cdots & \cdots & \cdots \\ \cdots & \cdots & d_{n-3} & c_{n-2} & a_{n-1} & b_{n-1} \\ 0 & \cdots & 0 & d_{n-2} & c_{n-1} & a_n \end{bmatrix}$$

16. Consider the following 5×5 strictly diagonally dominant lower Hessenberg matrix

$$A = \begin{pmatrix} 4 & 1 & 0 & 0 & 0 \\ 1 & 4 & 1 & 0 & 0 \\ 1 & 1 & 4 & 1 & 0 \\ 1 & 1 & 1 & 4 & 1 \\ 1 & 1 & 1 & 1 & 4 \end{pmatrix}$$

1- Apply the naive Gauss reduction on the matrix A showing the status of that matrix after each elimination, then extract out of this process, the upper triangular matrix U and the unit lower triangular matrix P.
2- Check that at each reduction, the multipliers reduce to one value, and at each reduction except the last, the modified elements reduce to two values, in addition to the diagonal element at last reduction. Compute the total number of flops needed for the LU-decomposition of the matrix A.
3- Deduce the total number of flops needed for the LU-decomposition of the $(n \times n)$ diagonally dominant lower Hessenberg matrix B where c is a constant and

$$B = \begin{pmatrix} c & 1 & 0 & 0 & . & . & . & 0 \\ 1 & c & 1 & 0 & . & . & . & 0 \\ 1 & 1 & c & 1 & 0 & . & . & 0 \\ . & . & . & . & . & . & . & . \\ . & . & . & . & . & . & . & . \\ 1 & 1 & 1 & . & . & 1 & c & 1 \\ 1 & 1 & 1 & 1 & . & . & 1 & c \end{pmatrix}$$

Express your answer in terms of n.

17. Evaluate the complexity of the following algorithms used in this chapter

- The Column-Backward Substitution algorithm.
- The Row-Forward Substitution algorithm.

3.7 Computer Projects

Exercise 1: Naive Gauss for Special Pentadiagonal Matrices

Definition 3.4 *A pentadiagonal matrix A is a square matrix with 5 non-zero diagonals: the main diagonal d, 2 upper subdiagonals u and v, then 2 lower subdiagonals l and m:*

$$A = \begin{bmatrix} d(1) & u(1) & v(1) & 0 & . & . & 0 \\ l(1) & d(2) & u(2) & v(2) & . & . & 0 \\ m(1) & l(2) & d(3) & u(3) & v(3) & . & 0 \\ 0 & m(2) & l(3) & d(4) & u(4) & . & 0 \\ 0 & . & . & . & . & . & . \\ . & . & . & . & . & . & . \\ . & 0 & m(n-4) & l(n-3) & d(n-2) & u(n-2) & v(n-2) \\ 0 & . & 0 & m(n-3) & l(n-2) & d(n-1) & u(n-1) \\ 0 & 0 & . & 0 & m(n-2) & l(n-1) & d(n) \end{bmatrix}$$

Definition 3.5 *A penta-diagonal matrix A is* **strictly diagonally dominant** *if for every row, the magnitude of the diagonal entry is larger than the sum of the magnitude of all the other non diagonal entries in that row.*

$$|d(i)| > |u(i)| + |v(i)| + |l(i-1)| + |m(i-2)| \; ; \forall \, i = 1, 2, ..., n$$

As such, the matrix A will satisfy the principal minor property, and therefore naive Gauss reduction is successfully applicable on A.

Let A be a **strictly diagonally dominant penta-diagonal** matrix.

1. Write a MATLAB
   function [m1,l1,d1,u1,v1]=NaiveGaussPenta(m,l,d,u,v,tol)
   that takes as input 5 column vectors m, l, d, u and v representing the 5 diagonals of A, and some tolerance tol. At each reduction, if the absolute value of the pivot element is less then tol, an error message should be displayed, otherwise this function performs naive Gauss reduction on the matrix A and returns through the process, the 5 modified diagonals m1, l1, d1, u1 and v1.
   Your function should neither use the built-in MATLAB function that factorizes A into L and U nor use the general code for naive Gauss reduction. Your code should be specifically designed for **pentadiagonal matrices only** and should use the least number of flops.

2. Write a MATLAB
   function x = RowForwardPenta(d, l, m, c)
   that takes as input 3 column vectors representing the main diagonal d and 2 lower diagonals l and m of an invertible **lower triangular matrix** L and a column vector c. This function performs row-oriented forward

substitution to solve the system $Lx = c$, using the least number of flops. Your code should be designed for **pentadiagonal matrices only**.

3. Write a MATLAB
   `function x = RowBackwardPenta(d, u, v, c)`
   which takes as input 3 column vectors representing the main diagonal d and 2 upper diagonals u and v of an invertible upper triangular matrix U and a column vector c. This function performs row-oriented backward substitution to solve the system $Ux = c$, using the least number of flops. Your code should be designed for **pentadiagonal matrices only**.

4. Write a MATLAB
   `function B = InversePenta(m, l, d, u, v, tol)`
   that takes as input the 5 diagonals of the **pentadiagonal** matrix A and outputs B, the inverse of the matrix A.
   Your function should call for the previous functions programmed in parts 1, 2 and 3.

5. Write a MATLAB
   `function T =InverseTransposePenta(m, l, d, u, v, tol)`
   that takes as input the 5 diagonals of the **pentadiagonal** matrix A and outputs $T = (A^T)^{-1}$, the inverse of the transpose of A. Your function should be based on the LU-decomposition of A, and call for the functions programmed in parts 1, 2 and 3. **Hint:** If $A = LU$, then:

   - $A^T = (LU)^T = U^T L^T$.
   - Since $A^T T = I$, then

   $$A^T[c_1, c_2, ..., c_n] = U^T L^T[c_1, c_2, ..., c_n] = [e_1, e_2, ..., e_n]$$

   $$\Leftrightarrow U^T L^T[c_i] = [e_i], \text{ for } i = 1, 2, ..., n,$$

   where c_i is the ith column of T and e_i is the ith column of the identity matrix I.

6. Test each of your functions on 3 different **strictly diagonally dominant pentadiagonal matrices** with $n \geq 5$. (In one of the test cases, choose one of the Pivot elements smaller than tol). Save your inputs and outputs in a separate document.

Exercise 2: Naive Gauss Reduction on Upper Hessenberg Matrices
A Hessenberg matrix is a special kind of square matrix, one that is "almost" triangular. To be exact, an upper Hessenberg matrix has zero entries below the first sub-diagonal.

$$H = \begin{bmatrix} H(1,1) & H(1,2) & & \cdot & & \cdot & H(1,n) \\ H(2,1) & H(2,2) & & H(2,4) & & \cdot & H(2,n) \\ & \cdot & & \cdot & & \cdot & \cdot \\ & \cdot & & \cdot & & \cdot & \cdot \\ & \cdot & & \cdot & & \cdot & \cdot \\ 0 & \cdot & & H(n-1,n-2) & H(n-1,n-1) & & H(n-1,n) \\ 0 & \cdot & & 0 & H(n,n-1) & & H(n,n) \end{bmatrix}$$

Definition: An upper Hessenberg matrix H is strictly diagonally dominant if for every row, the magnitude of the diagonal entry is larger than the sum of the magnitude of all the other non-diagonal entries in that row.

$$|H(i,i)| > |H(i,i-1)| + |H(i,i+1)| + ... + |H(i,n)| \, \forall \, i = 1,2,...,n$$

(As such, the matrix H will satisfy the principal minor property, and the naive Gauss reduction is successfully applicable on H.)

Let H be a strictly diagonally dominant upper Hessenberg matrix.

1. Write a MATLAB
   function [L, U] = NaiveGaussUHessenberg(H)
   that takes as input an $n \times n$ strictly diagonally dominant upper Hessenberg matrix H. This function performs naive Gauss reduction on H and returns at the end of the process, the upper and unit lower triangular matrices U and L.

   Your function should neither use the built-in MATLAB function that factorizes A into the product of L and U, nor use the general code for naive Gauss reduction. Your code should be designed for upper Hessenberg matrices only, and should use the least number of flops.

2. Write a MATLAB
   function [x] = RowForwardUHessenberg(L, c)
   that takes as input an invertible bi-diagonal lower triangular square matrix L of size n (displayed below) and a column vector c of length n. This function performs row-oriented forward substitution to solve the system $Lx = c$, using the least number of flops. Your code should be designed for bi-diagonal lower triangular matrices only and should use the least number of flops.

$$L = \begin{bmatrix} L(1,1) & 0 & \cdot & & \cdot & & \cdot & & 0 \\ L(2,1) & L(2,2) & \cdot & & \cdot & & \cdot & & 0 \\ 0 & L(3,2) & \cdot\cdot & & \cdot & & \cdot & & 0 \\ 0 & 0 & 0 & & \cdot & & \cdot & & 0 \\ \cdot & \cdot & \cdot & & \cdot & & \cdot & & \cdot \\ 0 & \cdot & 0 & L(n-1,n-2) & L(n-1,n-1) & & 0 \\ 0 & 0 & \cdot & & 0 & L(n,n-1) & & L(n,n) \end{bmatrix}$$

3. Write a MATLAB
   function [B] = InverseUHessenberg(H)

that takes as input an invertible upper Hessenberg matrix H, and outputs B, the inverse of H, using the *LU*-decomposition of H. Your function should call for the previous functions programmed in parts 1 and 2.

4. Test each of your functions above for 2 different upper Hessenberg strictly diagonally dominant matrices, with $n \geq 5$, and save the results in a separate document.
Call for previous functions when needed.
Do not check validity of the inputs.
Hint: To construct an $n \times n$ upper Hessenberg strictly diagonally dominant matrix H, proceed using the following MATLAB instructions:

```
A = rand(n);
m=max(sum(A));
m1=max(sum(A'));
s=max(m, m1);
B=A + s*eye(n);
H=triu(B, -1);
```

Exercise 3: Naive Gauss on Arrow Matrices
An arrow matrix is a special kind of square sparse matrix, in which there is a tri-diagonal banded portion, with a column at one side and a row at the bottom.

$$
A = \begin{bmatrix}
d(1) & u(1) & 0 & . & . & . & . & 0 & c(1) \\
l(1) & d(2) & u(2) & 0 & . & . & . & 0 & c(2) \\
0 & l(2) & d(3) & u(3) & 0. & . & . & 0 & c(3) \\
0 & 0 & l(3) & d(4) & u(4) & 0 & . & 0 & c(4) \\
. & . & . & . & . & . & . & . & . \\
0 & . & . & 0 & l(n-3) & d(n-2) & u(n-2) & & c(n-2) \\
0 & . & . & 0 & 0 & l(n-2) & d(n-1) & & c(n-1) \\
r(1) & r(2) & r(3) & r(4) & . & . & r(n-1) & & d(n)
\end{bmatrix}
$$

Definition: An arrow matrix A is strictly diagonally dominant if for every row, the magnitude of the diagonal entry is larger than the sum of the magnitude of all the other non-diagonal entries in that row, i.e.,

$$|d(n-1) > |l(i-1)| + |u(i+1)| + |c(i)| \, \forall \, i = 1, 2, ..., n-2$$

$$|d(n-1)| > |l(n-2)| + |c(n-1)| \text{ and } |d(n)| > |r(1)| + |r(2)| + ... + |r(n-1)|$$

(As such, the matrix A will satisfy the principal minor property, and the naive Gauss reduction is successfully applicable on A, without need for pivoting.)
Let A be a strictly diagonally dominant arrow matrix where:
- $d = [d(1), ..., d(n)]$ is a vector of length (n) representing the main diagonal of A.

- $u = [u(1), ..., u(n-2)]$ is a vector of length (n-2), and $[u(1), ..., u(n-2), c(n-1)]$ represents the first upper diagonal of A.
- $l = [l(1), ..., l(n-2)]$ is a vector of length (n-2), and $[l(1), ..., l(n-2), r(n-1)]$ represents the first lower diagonal of A.
- $c = [c(1), ..., c(n-1)]$ is a vector of length (n-1), and $c = [c(1), ..., c(n-1), d(n)]$ represents the last column of A.
- $r = [r(1), ..., r(n-1)]$ is a vector of length (n-1), and $r = [r(1), ..., r(n-1), d(n)]$ represents the last row of A.

1. Write a `MATLAB`
   function `[d1,u1,l1,c1,r1]=NaiveGaussArrow(d,u,l,c,r)`
   that takes as input the 5 vectors defined above representing A. This function performs <u>naive Gauss reduction</u> on the matrix A and returns at the end of the process, the modified vectors: `d1`, `u1`, `l1`, `c1`, `r1` (including the multipliers.)
   Your function should neither use the built-in `MATLAB` function that factorizes A into the product of L and U, nor use the general code for naive Gauss reduction. Your code should be designed for arrow matrices only, and should use the least number of flops.

2. Write a `MATLAB`
   function`[x]=RowBackwardArrow(d,u,c,b)`
   that takes as input 3 vectors as defined above, representing an invertible nearly bi-diagonal upper triangular square matrix U of size n (displayed below) and a column vector b of length n. This function performs row-oriented backward substitution to solve the system $Ux = b$, using the least number of flops. Your code should be designed for nearly bi-diagonal upper triangular matrices only and should use the least number of flops.

$$U = \begin{bmatrix} d(1) & u(1) & 0 & . & . & . & 0 & c(1) \\ 0 & d(2) & u(2) & 0 & . & . & 0 & c(2) \\ 0 & 0 & d(3) & u(3) & 0 & . & 0 & c(3) \\ . & . & . & . & . & . & . & . \\ . & . & . & . & . & . & . & . \\ 0 & . & . & 0 & d(n-2) & u(n-2) & c(n-2) \\ 0 & . & . & . & 0 & d(n-1) & c(n-1) \\ 0 & . & . & . & . & 0 & d(n) \end{bmatrix}$$

3. Write a `MATLAB`
   function `[x] = RowForwardArrow(d, l, r, b)`
   that takes as input 3 vectors as defined above, representing an invertible nearly bi-diagonal lower triangular square matrix L of size n (displayed below) and a column vector b of length n. This function performs row-oriented forward substitution to solve the system $Lx = b$, using the least

number of flops. Your code should be designed for nearly bi-diagonal lower triangular matrices only and should use the least number of flops.

$$
L = \begin{bmatrix}
d(1) & 0 & . & . & . & . & & . & 0 \\
l(1) & d(2) & 0 & 0 & . & . & & . & 0 \\
0 & l(2) & d(3) & 0 & . & . & & . & 0 \\
. & . & . & . & . & . & & . & . \\
. & . & . & . & . & . & & . & . \\
. & . & . & . & . & . & . & & . \\
0 & . & . & . & 0 & l(n-2) & d(n-1) & & 0 \\
r(1) & r(2) & . & . & . & . & & r(n-1) & d(n)
\end{bmatrix}
$$

4. Write a MATLAB
   function [B] = InverseArrow((d, u, l, c, r)
   that takes as input the 5 vectors defined above representing an invertible arrow matrix A, and outputs B, the inverse of A, using the LU-decomposition of A. Your function should call for the previous functions programmed in parts 1, 2 and 3.

5. Test each of your functions above for 2 different arrow strictly diagonally dominant matrices A, with $n \geq 6$, and save the results in a separate document.
   Call for previous functions when needed.
   Do not check validity of the inputs.

Chapter 4

Polynomial Interpolation and Splines Fitting

4.1 Definition of Interpolation

Consider a set D_n of $n+1$ data points in the (x, y) plane:

$$D_n = \{(x_i, y_i) |\ i = 0, 1 ..., n;\ n \in \mathbb{N} \text{ with } x_i \neq x_j \text{ for } i \neq j\}. \qquad (4.1)$$

We assume that D_n represents partially the values of a function $y = f(x)$, i.e.,

$$f(x_i) = y_i\ \forall\ i = 0, 1 ..., n \qquad (4.2)$$

Our basic objective in this chapter is to construct a continuous function $r(x)$ that "represents" $f(x)$ (or the empirical law $f(x)$ behind the set of data D_n). Thus $r(x)$ would represent $f(x)$ for all x, in particular for $x \notin$ the set of **nodes** $\{x_0, x_1, ..., x_n\}$. Such a function $r(x)$ is said to **interpolate** the set of data D_n if it satisfies the **interpolation conditions**:

$$r(x_i) = y_i\ \forall\ i = 0, 1 ..., n \qquad (4.3)$$

i.e., $r(x)$ fits the function $f(x)$ at the nodes $\{x_i\}$, .

Several kinds of interpolation may be considered by choosing $r(x)$ to be a polynomial function, a rational function or even a trigonometric one. The most natural is to consider polynomial or piecewise polynomial interpolation (spline functions), as polynomial functions are the simplest in reproducing the basic arithmetic (floating-point) operations of addition, subtraction, multiplication and division $\{+, -, \times, \div\}$. Consistently, we only analyze in this chapter **polynomial** and **spline interpolations**. Such type of interpolation is referred to as **Lagrange interpolation**.

4.2   General Lagrange Polynomial Interpolation

For simplicity, we assume that the set of data D_n is given as a natural increasing sequence of x-values, i.e.,

$$x_0 < x_1 < ... < x_n.$$

Let also $h_i = x_i - x_{i-1}$, $\forall i \geq 1$. We state now the basic Lagrange interpolation theorem.

Theorem 4.1 *There exists a unique polynomial of degree less than or equal to n:*

$$p_n(x) = p_{01...n}(x)$$

interpolating D_n, i.e., such that $p_n(x_i) = y_i$, $\forall i = 0, 1, ..., n$.

Proof. The proof of this theorem is based on the **Lagrangian cardinal basis** associated with D_n that is given by:

$$L_n = \{l_i(x) : 0 \leq i \leq n\}$$

where the **cardinal functions** l_i are special polynomials of degree exactly n in $\mathbb{P}_n$ ($\mathbb{P}_n$ being the set of all polynomials of degree less than or equal to n). They are defined as follows, $\forall i = 0, ..., n$:

$$l_i(x) = \frac{\prod_{0 \leq j \neq i \leq n}(x - x_j)}{\prod_{0 \leq j \neq i \leq n}(x_i - x_j)} = \frac{(x - x_0)...(x - x_{i-1})(x - x_{i+1})...(x - x_n)}{(x_i - x_0)...(x_i - x_{i-1})(x_i - x_{i+1})...(x_i - x_n)}$$

$$(4.4)$$

Once the cardinal functions (4.4) are available, we can interpolate any function f using **Lagrange form of the interpolation polynomial**:

$$p_{01...n}(x) = \sum_{i=0}^{n} l_i(x) f(x_i). \tag{4.5}$$

Obviously, the following properties are satisfied by a Lagrangian basis function, $\forall i, j = 0, 1, \ldots, n$:

$$- \; l_i(x_j) = \delta_{ij} = \begin{cases} 0 \text{ if } i \neq j \\ 1 \text{ if } i = j \end{cases}$$

$$- \; p_{01\ldots n}(x_i) = y_i$$

The definition of the Lagrange polynomial above is enough to establish the existence part of Theorem 4.1.

As for obtaining uniqueness of such a polynomial $p_{01\ldots n}$, we proceed by contradiction by supposing the existence of another polynomial $q(x) \in \mathbb{P}_n$, claiming to accomplish what $p(x)$ does; that is $q(x)$ satisfies as well the interpolation conditions $q(x_i) = y_i$ for $0 \leq i \leq n$. The polynomial:

$$(p_{01\ldots n}(x) - q(x))$$

is then of degree at most n, and takes on the value 0 at all nodes $x_0, x_1, \ldots, x_n$. Recall however that a non-zero polynomial of degree n can have at most n roots, implying that $(p_{01\ldots n}(x) - q(x)) = 0$. One concludes therefore that $p_{01\ldots n}(x) = q(x) \, \forall x$, which establishes the uniqueness of $p_{01\ldots n}(x)$. ∎

Remark 4.1 *It is obvious from equation (4.5) that:*

$$p_{01\ldots n}(x) = p_{i_0 i_1 \ldots i_n}(x)$$

for any permutation $\{i_0, i_1, \ldots, i_n\}$ of the set of indices $\{0, 1, \ldots, n\}$.

Example 4.1 *Write out the cardinal functions and the corresponding Lagrange interpolating polynomial based on the following data:*

$$D_2 = \{\ (1/4, -1),\ (1/3, 2),\ (1, 7)\}$$

Using equation (4.4), we have:

$$l_0(x) = \frac{(x - \frac{1}{3})(x - 1)}{(\frac{1}{4} - \frac{1}{3})(\frac{1}{4} - 1)} = 16(x - \frac{1}{3})(x - 1)$$

$$l_1(x) = \frac{(x - \frac{1}{4})(x - 1)}{(\frac{1}{3} - \frac{1}{4})(\frac{1}{3} - 1)} = -18(x - \frac{1}{4})(x - 1)$$

$$l_2(x) = \frac{(x - \frac{1}{3})(x - \frac{1}{4})}{(1 - \frac{1}{3})(1 - \frac{1}{4})} = 2(x - \frac{1}{3})(x - \frac{1}{4})$$

The interpolating polynomial in Lagrange's form is therefore given by:

$$p_{012}(x) = -36(x - \frac{1}{4})(x-1) - 16(x - \frac{1}{3})(x-1) + 14(x - \frac{1}{3})(x - \frac{1}{4}) = -38x^2 + \frac{349}{6}x - \frac{79}{6}$$

This form of the polynomial might be useful in computing $f(x)$ in the vicinity of the nodes $1/3, 1/4, 1$.

Example 4.2 *Consider the following table of data associated with the function $f(x) = \ln(x)$.*

i	x_i	y_i
0	1.0	0
1	1.5	0.17609
2	2.0	0.30103
3	3.0	0.47712
4	3.5	0.54407
5	4.0	0.60206

Use Lagrange polynomials of orders 1 then 2 to approximate $f(1.2)$, noting that the exact value is $\ln(1.2) = 0.0791812460476480$.

- **Linear interpolation** based on the points $\{x_0,\ x_1\} = \{1.0,\ 1.5\}$, where $l_0(.)$ and $l_1(.) \in \mathbb{P}_1$. Using (4.5), one has:

$$p_{01}(x) = y_0 l_0(x) + y_1 l_1(x) = 0\,\frac{x - 1.5}{1.0 - 1.5} + 0.17609\,\frac{x - 1.0}{1.5 - 1.0}$$

Thus $p_{01}(1.2) = 0.070436$, and the relative error in this approximation is 6.136716×10^{-1}

- **Quadratic interpolation** based on the points $\{1.0,\ 1.5,\ 2.0\}$, where $l_0(.),\ l_1(.)$ and $l_2(.) \in \mathbb{P}_2$.

$$p_{012}(x) = y_0 l_0(x) + y_1 l_1(x) + y_2 l_2(x) = 0\,\frac{(x - 1.5)(x - 2)}{(1.0 - 1.5)(1.0 - 2)} +$$

$$+0.17609\,\frac{(x - 1.0)(x - 2)}{(1.5 - 1.0)(1.5 - 2)} + 0.30103\,\frac{(x - 1.0)(x - 1.5)}{(2 - 1.0)(2 - 1.5)}$$

Thus $p_{012}(1.2) = 0.076574$, and the relative error is now 3.292757×10^{-2}.

Remark 4.2 *Note that Lagrange's formula is not computationally practical in the sense that computing $p_{01...k}(x)$, with $k < n$, cannot be obtained from $p_{01...k-1}(x)$. The cardinal functions of the latter are polynomials of degree exactly $k - 1$ in $\mathbb{P}_{k-1}$, while those of the former are polynomials of degree exactly k in $\mathbb{P}_k$. Thus, after computing the Lagrange cardinal functions for $p_{01...k-1}(x)$, one has to compute a totally distinct set of cardinal functions for $p_{01...k}(x)$.*

This motivates one to look for recurrence formulae to the Lagrange interpolating polynomial.

4.3 Recurrence Formulae

These recurrence formulae are obtained through relations between two consecutive-order interpolation polynomials, specifically and for $k \geq 1$:

- Consider first $p_{012..k} \in \mathbb{P}_k$ and $p_{012..k-1} \in \mathbb{P}_{k-1}$. As

$$p_{012..k}(x_i) - p_{012..k-1}(x_i) = 0, \ \forall \, i = 0, 1, ..., k - 1$$

This implies that:

$$p_{012..k}(x) - p_{012..k-1}(x) = C(x - x_0)(x - x_1)...(x - x_{k-1}) \qquad (4.6)$$

Note that C is a constant as the right hand side polynomial is exactly of degree k.

- In a similar way considering now $p_{012..k} \in \mathbb{P}_k$ and $p_{12..k} \in \mathbb{P}_{k-1}$, one obtains:

$$p_{012..k}(x) - p_{12..k}(x) = C'(x - x_1)...(x - x_{k-1})(x - x_k). \qquad (4.7)$$

It is clear that $C = C'$ as both constants are the coefficient of x^k in $p_{012..k}$.

(4.6) and (4.7) constitute the basis for Neville's and Newton's recurrence formulae, as shown hereafter.

4.3.1 Neville's Formula

Given that $C = C'$, the algebraic operation:

$$(x - x_k) \times (4.6) \ -(x - x_0) \times (4.7)$$

yields:

$$(x_0 - x_k)p_{01...k-1\,k}(x) = (x - x_k)p_{01...k-1}(x) - (x - x_0)p_{12...k-1\,k}(x).$$

Hence one reaches Neville's formula (also called Aitken-Neville's), given by:

$$p_{01...k-1\,k}(x) = \frac{(x - x_0)p_{12...k-1\,k}(x) - (x - x_k)p_{012...k-1}(x)}{x_k - x_0}, \ k \geq 1. \qquad (4.8)$$

i	x_i	$p_i(x)$	$p_{i,i+1}(x)$	$p_{i,i+1,i+2}(x)$	$\cdots$	$p_{i,i+1,\ldots,i+n}(x)$
0	x_0	$p_0(x)$				
1	x_1	$p_1(x)$	$p_{0,1}(x)$			
2	x_2	$p_2(x)$	$p_{1,2}(x)$	$p_{0,1,2}(x)$		
3	x_3	$p_3(x)$	$p_{2,3}(x)$	$p_{1,2,3}(x)$		
4	x_4	$p_4(x)$	$p_{3,4}(x)$	$p_{2,3,4}(x)$	$\cdots$	
$\vdots$	$\cdots$	$\cdots$	$\cdots$	$\cdots$	$\cdots$	$\cdots$
n	x_n	$p_n(x)$	$p_{n-1,n}(x)$	$p_{n-2,n-1,n}(x)$	$\cdots$	$p_{0,1,\ldots,n}(x)$

TABLE 4.1: Neville's array constructing Lagrange interpolation polynomials

A more general **Neville's recurrence formulae** can be concluded. Specifically, for any $i \in \{0, 1, \ldots, n\}$:

- **Base statement**: $p_i(x) = y_i$, $i = 0, 1, \ldots, n$

- **Recurrence statement**:

$$p_{i\,i+1\ldots i+k}(x) = \frac{(x - x_i)p_{i+1\,i+2\ldots i+k}(x) - (x - x_{i+k})p_{i\,i+1\ldots i+k-1}(x)}{x_{i+k} - x_i},$$

(4.9)

  with

$$0 \leq i < i + k \leq n.$$

Based on the set of data D_n in (4.1) and using the formulas above repeatedly, we can create an array of interpolating polynomials $\in \mathbb{P}_n$, where each successive polynomial can be determined from 2 adjacent polynomials in the previous column, as is shown in Table 4.1. For example,

$$p_{01}(x) = \frac{(x - x_0)p_1(x) - (x - x_1)p_0(x)}{x_1 - x_0}$$

$$p_{123} = \frac{(x - x_1)p_{23}(x) - (x - x_3)p_{12}(x)}{x_3 - x_1}$$

Remark 4.3 *Neville's recurrence expressions of the interpolating polynomial can be easily programmed. The consequent algorithms can be written either in a recursive or iterative form.*

In what follows, we write a recursive algorithm for Neville's formula leaving it as an exercise to transform it into an iterative one.

Algorithm 4.1 Algorithm for Neville's Formula(Recursive Version)

```
function [ z ]= Neville(x, y, s)
% Input data vectors x=[x1,x2,...,xk]  and y=[y1,y2,...,yk]
% s : value (or vector) at which we seek the interpolation
```

```
% Output z=p_{12...k}(s)=p(s)
 k = length(x);
if k=1
    z=y;
% z1=p_{12...(k-1)}(s) ;  z2=p_{2...k}(s)
% z= [(s-x1)*z2 - (s-xk)*z1] / (xk-x1)
else
   z1= Neville(x(1:k-1), y(1:k-1) , s);
   z2= Neville(x(2:k), y(2:k) , s);
   z= ((s-x(1))*z2 - (s-x(k))*z1)/(x(k)-x(1));
end
```

4.3.2 Newton's Form for the Interpolation Polynomial

As for Neville's formula, we proceed with (4.6) by rewriting it in a more general recurrence form as follows:

$$p_{i\,i+1...,i+k}(x) = p_{i\,i+1...i+k-1}(x) + C(x - x_i)...(x - x_{i+k-1}), \qquad (4.10)$$

with

$$0 \le i < i + k \le n.$$

Newton's formula is obtained by determining a proper expression for the constant C as a function of the data $D_k = \{(x_i, y_i)|i = 0, 1, ..., k\}$. Note that such constant can be computed by setting $x = x_{i+k}$ in (4.10), so that:

$$y_{i+k} = p_{i\,i+1...i+k-1}(x_{i+k}) + C(x_{i+k} - x_i)...(x_{i+k} - x_{i+k-1}),$$

and therefore:

$$C = C(x_i, x_{i+1}, ..., x_{i+k}; y_i, y_{i+1}, ..., y_{i+k}) = \frac{y_{i+k} - p_{i\,i+1...i+k-1}(x_{i+k})}{(x_{i+k} - x_i)...(x_{i+k} - x_{i+k-1})}.$$

For $k = 1$, this gives:

$$C = C(x_i, x_{i+1}; y_i, y_{i+1}) = \frac{y_{i+1} - y_i}{x_{i+1} - x_i}. \qquad (4.11)$$

Define then

$$[x_i, x_{i+1}] = \frac{y_{i+1} - y_i}{x_{i+1} - x_i}$$

as the first order divided difference associated with $\{x_i, x_{i+1}\}$, so that (4.10) is expressed as follows:

$$p_{i,i+1}(x) = p_i(x) + [x_i, x_{i+1}](x - x_i) \qquad (4.12)$$

which is Newton's formula of order 1. More generally, we may define divided differences of any order $k \ge 1$, through a recurrence process as follows:

Definition 1 *Given the set of data*

$$D_n = \{(x_i, y_i) | i = 0, 1, ..., n\}, \ x_i \neq x_j \ for \ i \neq j.$$

Let $[x_i] = y_i$, $i = 0, 1, ..., n$. *Then, for* $0 \leq i < i + k \leq n$, *the* k^{th} *order divided difference is given through the recurrence formula:*

$$[x_i, x_{i+1}, ..., x_{i+k}] = \frac{[x_{i+1}, ..., x_{i+k}] - [x_i, x_{i+1}, ..., x_{i+k-1}]}{x_{i+k} - x_i}. \qquad (4.13)$$

Consequently, we prove that the constant C in (4.10) is a kth order divided difference. This is done in the following proposition.

Theorem 4.2 *Let* $0 \leq i < i + k \leq n$. *Let*

$$p_{i\,i+1...i+k}(x) = p_{i\,i+1...i+k-1}(x) + C(x - x_i)...(x - x_{i+k-1}),$$

is the interpolating polynomial based on the nodes $\{x_i, ..., x_{i+k}\}$, *as defined in (4.6). Then, the constant C is the kth order divided difference*

$$C = [x_i, x_{i+1}, ..., x_{i+k}] = \frac{[x_{i+1}, ..., x_{i+k}] - [x_i, x_{i+1}, ..., x_{i+k-1}]}{x_{i+k} - x_i}$$

Proof. To obtain this result we use a mathematical induction process on k. Clearly, (4.12) indicates that the result is true for $k = 1$.
Assuming now that the proposition is correct for all $j \leq k - 1$ with $i + j < n$, then, one writes on the basis of the induction hypothesis for $j = k - 1$, successively:

$$p_{i...i+k-1}(x) = p_{i...i+k-2}(x) + [x_i, x_{i+1}, ...x_{i+k-1}](x - x_i)...(x - x_{i+k-2})$$

where $[x_i, x_{i+1}, ...x_{i+k-1}]$ is the coefficient of x^{k-1} in the polynomial $p_{i...i+k-1}(x)$ and

$$p_{i+1...i+k}(x) = p_{i+1, ..., i+k-1}(x) + [x_{i+1}, ..., x_{i+k}](x - x_{i+1})...(x - x_{i+k-1})$$

where $[x_{i+1}, ..., x_{i+k}]$ is the coefficient of x^{k-1} in the polynomial $p_{i+1...i+k}(x)$. Using now Neville's formula, one has:

$$p_{i...i+k}(x) = \frac{(x - x_i)p_{i+1...i+k}(x) - (x - x_{i+k})p_{i...i+k-1}(x)}{x_{i+k} - x_i}.$$

By equating the coefficients of x^k on both sides of this identity one has:

$$C = \frac{[x_{i+1}, ..., x_{i+k-1}, x_{i+k}] - [x_i, x_{i+1}, ..., x_{i+k-1}]}{x_{i+k} - x_i},$$

which is the targeted result of the theorem.   ∎

As a consequence of this theorem, we may write now Newton's formula for Lagrange interpolating polynomial as follows: for $i < i + k \leq n$:

$$p_{i\,i+1...i+k-1\,i+k}(x) = y_i + ... + [x_i, x_{i+1}, ..., x_{i+k}](x - x_i)...(x - x_{i+k-1}). \quad (4.14)$$

or equivalently as:

$$p_{i\,i+1...i+k-1\,i+k}(x) = y_i + \sum_{j=1}^{k} [x_i, ..., x_{i+j}](x - x_i)...(x - x_{i+j-1}) \quad (4.15)$$

$$= \sum_{j=0}^{k} [x_i, ..., x_{i+j}] \prod_{j=0}^{i-1}(x - x_{i+j})$$

More specifically:

$$p_{01...n}(x) = y_0 + [x_0, x_1](x - x_0) + [x_0, x_1, x_2](x - x_0)(x - x_1) + ...$$

$$... + [x_0, x_1, x_2, ..., x_n](x - x_0)(x - x_1)...(x - x_{n-1})$$

Remark 4.4 *Note that, as expressed in (4.10), Newton's formula of the interpolating polynomial is built up in steps, in the sense that once $p_{i\,i+1...i+k-1}(x)$ is found reproducing part of the data, determining $p_{i\,i+1...i+k}(x)$, necessitates the computation of one new divide difference coefficient only.*

4.3.3   Divided Differences Table and Newton's Formula

Let $\{i_0, i_1, ..., i_k\}$ be any permutation of the set of integers $\{i, i+1, ..., i+k\}$. Based on the uniqueness property of the interpolating polynomials:

$$p_{i\,i+1...i+k-1\,i+k}(x) = p_{i_0\,i_1...i_{k-1}\,i_k}(x)$$

and consequently the kth order divided differences $[x_i, x_{i+1}, ..., x_{i+k}]$ and $[x_{i_0}, x_{i_1}..., x_{i_{k-1}}, x_{i_k}]$ representing respectively the (same) coefficient of x^k in the two polynomials above, are equal. This leads to the following invariance property satisfied by divided differences:

Theorem 4.3 *Let $\{i_0, i_1, ..., i_k\}$ of be any permutation of the set of integers $\{i, i+1, ..., i+k\}$. Then:*

$$[x_i, x_{i+1}, ..., x_{i+k}] = [x_{i_0}, x_{i_1}, ..., x_{i_k}]$$

Obviously, use of Newton's formula necessitates the computation of divided differences. As such, constructing divided differences tables associated with a set of data $D_n = \{(x_i, y_i)|i = 0, 1, ..., n\}$ is a preliminary step to any implementation of Newton's formula. The construction of divided differences is shown in Table 4.2 for the case $n = 5$.   The following `MATLAB` code takes as input 2 vectors x and y of equal length and returns the divided difference table of the first (n-1)-order divided differences, as a lower triangular matrix, using the `MATLAB` `diff` operator.

i	x_i	y_i	$[\cdot,\cdot]$	$[\cdot,\cdot,\cdot]$	$[\cdot,\cdot,\cdot,\cdot]$	$[\cdot,\cdot,\cdot,\cdot,\cdot]$	$[\cdot,\cdot,\cdot,\cdot,\cdot,\cdot]$
0	x_0	y_0					
			$[x_0,x_1]$				
1	x_1	y_1		$[x_0,x_1,x_2]$			
			$[x_1,x_2]$		$[x_0,x_1,x_2,x_3]$		
2	x_2	y_2		$[x_1,x_2,x_3]$		$[x_0,...,x_4]$	
			$[x_2,x_3]$		$[x_1,x_2,x_3,x_4]$		$[x_0,...,x_5]$
3	x_3	y_3		$[x_2,x_3,x_4]$		$[x_1,...,x_5]$	
			$[x_3,x_4]$		$[x_2,x_3,x_4,x_5]$		
4	x_4	y_4		$[x_3,x_4,x_5]$			
			$[x_4,x_5]$				
5	x_5	y_5					

TABLE 4.2: Divided difference table for $n = 5$

Algorithm 4.2 Constructing a Divided Difference Table

```
function D = DivDiffTable(x,y)
% D is a lower Triangular matrix
% If   x=[x(1),x(2),...,x(n)]  is a vector of length n, then
%  diff(x)=[(x(2)-x(1)), (x(3)-x(2)),....,(x(n)-x(n-1))]
% is a vector of length (n-1)
n=length(x) ;
m=length(y) ;
if m==n
   D=zeros(n,n) ;
   D(1:n, 1) = y(1:n) ;
   Y= D(1:n, 1)   ;
     for j=2: n
        V1=x(1:n-j+1) ; V2=x(j:n) ;
        D(j:n, j)= (diff(Y) ./ (V2-V1)' ) ;
        Y=D(j:n, j) ;
     end
end
```

Example 4.3 *Create the divided difference table based on the set of data of Example 4.2 representing the function* $f(x) = \ln(x)$ *where:*

$$D_5 = \{(1,0)\,,\,(1.5, 0.17609)\,,\,(2.0, 0.30103)\,,\,(3, 0.47712)\,,\,(3.5, 0.54407)\,,\,(4, 0.60206)\}$$

Let us consider now approximations of $f(x)$ for values of x first at the top of Table 4.3, for example $x = 1.2$, then at the middle of the table, as $x = 2.5$. (Note that, in general, one can prove that the approximation-error is smaller when x is centered with respect to the nodes).

i	x_i	y_i	$[\cdot,\cdot]$	$[\cdot,\cdot,\cdot]$	$[\cdot,\cdot,\cdot,\cdot]$	$[\cdot,\cdot,\cdot,\cdot,\cdot]$	$[\cdot,\cdot,\cdot,\cdot,\cdot,\cdot]$
0	1.0	0					
			0.35218				
1	1.5	0.17609		−0.1023			
			0.24988		0.02655		
2	2.0	0.30103		−0.0492		−0.006404	
			0.17609		0.01054		0.001411
3	3.0	0.47712		−0.02813		−0.002172	
			0.13390		0.00511		
4	3.5	0.54407		−0.01792			
			0.11598				
5	4.0	0.60206					

TABLE 4.3: A divided difference table for $f(x) = \ln(x)$ for unequally sized data $x = \{1.0, 1.5, 2.0, 3.0, 3.5, 4.0\}$

$p_{...}(1.2)$	Value	Relative error
$p_{01}(1.2)$	0.070436	1.10446×10^{-1}
$p_{012}(1.2)$	0.076574	3.2928×10^{-2}
$p_{0123}(1.2)$	0.0778484	1.6833×10^{-2}
$p_{01234}(1.2)$	0.07840171	9.845×10^{-3}
$p_{012345}(1.2)$	0.0786821	6.30384×10^{-3}

TABLE 4.4: Errors in polynomial interpolation for $f(x) = \ln(1.2)$

1. The first interpolating polynomials of degrees 1, 2, 3, 4 and 5 are successively as follows:

   - $p_{01}(x) = 0.35218(x - 1)$,
   - $p_{012}(x) = p_{01}(x) - 0.1023(x - 1)(x - 1.5)$,
   - $p_{0123}(x) = p_{012}(x) + 0.02655(x - 1)(x - 1.5)(x - 2)$,
   - $p_{01234}(x) = p_{0123}(x) - 0.006404(x - 1)(x - 1.5)(x - 2)(x - 3)$,
   - $p_{012345}(x) = p_{01234}(x) + 0.001411(x - 1)(x - 1.5)(x - 2)(x - 3)(x - 3.5)$.

   As a result, approximations to $\ln(1.2) = 0.0791812460476248$ using:

   $$p_{01}(1.2), \ p_{012}(1.2), \ p_{0123}(1.2), \ p_{01234}(1.2) \text{ and } p_{012345}(1.2)$$

   are displayed in Table 4.4.

2. To get approximations to $f(2.5)$, using Theorem 4.3), we obtain successively linear, quadratic and cubic polynomials as follows:

$p_{...}(2.5)$	Value	Relative error
$p_{23}(2.5)$	0.389075	2.227725×10^{-2}
$p_{234}(2.5)$	0.3961067	4.6070814×10^{-3}
$p_{231}(2.5)$	0.4013733	$8.62774435 \times 10^{-3}$
$p_{2345}(2.5)$	0.3973825	1.4009867×10^{-3}
$p_{2341}(2.5)$	0.39874	2.0103315×10^{-3}

TABLE 4.5: Errors in polynomial interpolation for $f(x) = \ln(2.5)$

- $p_{23}(x) = y_2 + [x_2, x_3](x - x_2) = 0.30103 + 0.17609(x - 2)$
- $p_{231}(x) = p_{23}(x) + [x_2, x_3, x_1](x - x_2)(x - x_3)$
  $= p_{23}(x) + [x_1, x_2, x_3](x - x_2)(x - x_3) = p_{23}(x) - 0.0492(x - 2)(x - 3)$
- $p_{2314}(x) = p_{231}(x) + [x_2, x_3, x_1, x_4](x - x_2)(x - x_3)(x - x_1)$
  $= p_{231}(x) + [x_1, x_2, x_3, x_4](x - x_2)(x - x_3)(x - x_1) = p_{231}(x) +$
  $0.01054(x - 2)(x - 3)(x - 1.5)$
- $p_{2310}(x) = p_{231}(x) + [x_2, x_3, x_1, x_0](x - x_2)(x - x_3)(x - x_1)$
  $= p_{231}(x) + [x_0, x_1, x_2, x_3](x - x_2)(x - x_3)(x - x_1) = p_{231}(x) +$
  $0.02655(x - 2)(x - 3)(x - 1.5)$

Another alternative, starting with $p_{23}(x)$, would be:

- $p_{234}(x) = p_{23}(x) + [x_2, x_3, x_4](x - x_2)(x - x_3) = p_{23}(x) -$
  $0.02813(x - 2)(x - 3)$
- $p_{2345}(x) = p_{234}(x) + [x_2, x_3, x_4, x_5](x - x_2)(x - x_3)(x - x_4)p_{234}(x) +$
  $0.00511(x - 2)(x - 3)(x - 3.5)$
- $p_{2341}(x) = p_{234}(x) + [x_2, x_3, x_4, x_1](x - x_2)(x - x_3)(x - x_4)$
  $= p_{234}(x) + [x_1, x_2, x_3, x_4](x - x_2)(x - x_3)(x - x_4) = p_{234}(x) -$
  $0.002172(x - 2)(x - 3)(x - 3.5)$

This process can be carried through to obtain higher order interpolation polynomials. Table 4.5 gives the results obtained for the approximation of $\ln(2.5) = 0.39794001$.

Hence, it appears clear that increasing the degree of the interpolation polynomial does not improve much the approximation of the exact value of $f(x)$. Using **Algorithm 4.2**, we may now write an algorithm that implements Newton's formula.

Algorithm 4.3 Program for Newton's Formula

```
function p=NewtonForm(x,y,X)
%Input: two equally sized vectors x and y of length k
%          One vector X of length n
```

```
%Output: p(X) based on Newton interpolation formula on the data (x,y)
D=DivDiffTable(x,y);
k=length(x);%(equal to length of y)
n=length(X);X=X(:);
term=ones(n,1);
p=zeros(n,1);
for i=1:k
    p=p+D(i,i)*term;
    term=term.*(X-x(i));
end
```

To conclude on recurrence formulae for the Lagrange interpolation polynomial, a rule of thumb would be to use Neville's formula in case of computer implementation as it takes only one algorithm to program (Algorithm 4.1). On the other hand, Newton's formula requires writing 2 programs: one for divided differences (Algorithm 4.2) before developing Algorithm 4.3 for a straightforward evaluation of the interpolation polynomial.

4.4 Equally Spaced Data: Difference Operators

Consider now the set of data D_n with equidistant x nodes, i.e.,

$$x_{i+1} - x_i = h, \ \forall i = 0, 1, ..., n - 1.$$

In this case, we can compute divided differences associated with D_n by using the "**difference functions**" or "**difference operators**," based on the y data only. Specifically, we make the following definitions:

Definition 4.1 *Let* $Y = [y_0, y_1, ..., y_n]$, *then:*

1. $\Delta^1 Y = [\Delta y_0, \Delta y_1, ..., \Delta y_{n-1}]$ *is the vector of* n *first order differences associated with* Y, *where* $\Delta y_i = y_{i+1} - y_i$ *for* $i = 0, 1..., n - 1$.

2. *By recurrence, for* $k = 2, 3, ...n$, *we may then define the vector of* k^{th} *order differences* $\Delta^k Y = [\Delta^k y_0, \Delta^k y_1, ..., \Delta^k y_{n-k}]$, *where* $\Delta^k y_i = \Delta^{k-1} y_{i+1} - \Delta^{k-1} y_i$ *for* $i = 0, 1, ..., n - k$.

Difference operators are linear in the sense that:

$$\Delta^k (Y + Z) = \Delta^k Y + \Delta^k Z \text{ and } \Delta^k (aY) = a\Delta^k Y, \ a \in \mathbb{R}, \ k = 2, 3, ...n.$$

Besides, one easily obtains a relation between divided differences and differences of all orders as shown below.

Theorem 4.4 *Let D_n be a set of data as defined in (4.1), where the x-nodes are equally spaced with $x_{i+1} - x_i = h$, $\forall i = 0, 1, ..., n-1$. Then for all k where $1 \leq k \leq$ with $i + k \leq n$:*

$$[x_i, x_{i+1}, ..., x_{i+k}] = \frac{\Delta^k y_i}{h^k k!} \tag{4.16}$$

Proof. The proof is done by induction on k. After verifying the result for $k = 1$, assume that it is true for $1, ..., k-1$, i.e.,

$$[x_i, x_{i+1}, ..., x_{i+k-1}] = \frac{\Delta^{k-1} y_i}{h^{k-1}(k-1)!}.$$

Since,

$$[x_i, x_{i+1}, ..., x_{i+k}] = \frac{[x_{i+1}, ..., x_{i+k}] - [x_i, x_{i+1}, ..., x_{i+k-1}]}{(x_{i+k} - x_i)},$$

then:

$$[x_i, x_{i+1}, ..., x_{i+k}] = \frac{\Delta^{k-1} y_{i+1} - \Delta^{k-1} y_i}{h^{k-1}(k-1)!(x_{i+k} - x_i)} = \frac{\Delta^{k-1} y_{i+1} - \Delta^{k-1} y_i}{h^{k-1}(k-1)!\, kh}.$$

that reaches the required result.   ∎

Based on the theorem above and in case of equally spaced data, Newton's interpolating polynomial is expressed as follows::

$$p(x) = y_0 + \frac{\Delta y_0}{1!h}(x-x_0) + \frac{\Delta^2 y_0}{2!h^2}(x-x_0)(x-x_1) + \cdots + \frac{\Delta^n y_0}{n!h^n}(x-x_0)\ldots(x-x_{n-1}) \tag{4.17}$$

where it is understood that $p(x) = p_{012...n}(x)$.

Remark 4.5 *Note the resemblance of this formula with that of Taylor's formula for a function $f(x)$ where the nth degree polynomial representing $f(x)$ is given by:*

$$q(x) = f(x_0) + f'(x_0)(x - x_0) + \cdots + \frac{f^{(n)}(x_0)}{n!}(x - x_0)^n$$

This remark will be exploited in Chapter 5 when approximating derivatives such as $f^{(k)}(x_0)$ by k^{th} order differences $\frac{\Delta^k f(x_0)}{h^n}$.

The result of the above theorem allows us therefore to compute divided difference tables by simply first computing differences as displayed in Table 4.6. The algorithm of the difference table for the first differences up to order $(n-1)$ is implemented as follows:

i	x_i	y_i	Δ	Δ^2	$\cdots$	Δ^n
0	x_0	y_0			$\vdots$	$\vdots$
			$\Delta y_0 = y_1 - y_0$			
1	x_1	y_1		$\Delta^2 y_0$		
			$\Delta y_1 = y_2 - y_1$			
2	x_2	y_2		$\Delta^2 y_1$		
			$\Delta y_2 = y_3 - y_2$			
3	x_3	y_3		$\vdots$	$\vdots$	$\Delta^n y_0$
$\vdots$			$\vdots$	$\vdots$	$\vdots$	
$n-2$	x_{n-2}	y_{n-2}				
			$\Delta y_{n-2} = y_{n-1} - y_{n-2}$			
$n-1$	x_{n-1}	y_{n-1}		$\Delta^2 y_{n-2}$		
			$\Delta y_{n-1} = y_n - y_{n-1}$			
n	x_n	y_n				
					$\vdots$	$\vdots$

TABLE 4.6: A difference table for equally spaced x data

Algorithm 4.4 Constructing a Difference Table

```
function D = DiffTable(x,y)
% D is a lower Triangular matrix
n=length(x) ;
m=length(y) ;
if m==n
   D=zeros(n,n) ;
   D(:,1) = y ;
    for j=2:n
        D(j:n, j)= diff(D(j-1:n, j-1 ) );
     end
end
```

Example 4.4 *The following set of data D_4 is associated with 0-th order Bessel's function of the first kind.*

i	x_i	y_i
0	1.0	0.7651977
1	1.3	0.6200860
2	1.6	0.4554022
3	1.9	0.2818186
4	2.2	0.1103623

i	x_i	y_i	Δ	Δ^2	Δ^3	Δ^4
0	1.0	0.7651977				
			−0.1451117			
1	1.3	0.6200860		−0.0195721		
			−0.1646838		0.0106723	
2	1.6	0.4554022		−0.0088998		0.0003548
			−0.1735836		0.0110271	
3	1.9	0.2818186		0.0021273		
			−0.1714563			
4	2.2	0.1103623				

TABLE 4.7: An example of a difference table for equally spaced x data

Since the x-data are equally space with $h = 0.3$, the differences Table 4.7 can therefore be easily constructed out of this data. Using Table 4.7, we may subsequently write any of the interpolation polynomials based on D_4. For example:

$p_{234}(x) = y_2 + \frac{\Delta y_2}{0.3}(x - x_2) + \frac{\Delta^2 y_2}{(0.3)^2 2!}(x - x_2)(x - x_3)$

$= 0.4554022 - \frac{0.1735836}{0.3}(x - 1.6) + \frac{0.0021273}{(0.3)^2}(x - 1.6)(x - 1.9)$

$p_{231}(x) = p_{23}(x) + [x_2, x_3, x_1](x - x_2)(x - x_3) = p_{23}(x) + [x_1, x_2, x_3](x - x_2)(x - x_3)$

$= p_{23}(x) + \frac{\Delta^2 y_1}{(0.3)^2 2!}(x - x_2)(x - x_3) = p_{23}(x) - \frac{0.0088998}{(0.3)^2 2!}(x - 1.6)(1 - 1.9)$

4.5   Errors in Polynomial Interpolation

When a function f is approximated on an interval $[a, b] = [x_0, x_n]$ by means of an interpolating polynomial p_n, it is naturally expected that the function be well approximated at all intermediate points between the nodes, and that as the number of nodes increases, this agreement will become more and more accurate. Nevertheless, this expectation is incorrect.

A theoretical estimate of the error is derived in ([21], page 189) and leads to the following result:

Theorem 4.5 *Let f be a function in $C^{n+1}[a, b]$, and p_n the Lagrange polynomial of degree at most n, that interpolates f based on the set of data D_n. There exists some point $c \in (a, b)$ such that the error function:*

$$E_n(f(x)) = f(x) - p_n(x) = w_n(x)\frac{f^{(n+1)}(c)}{(n + 1)!},$$

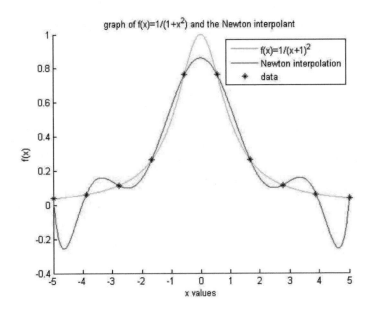

FIGURE 4.1: Runge counter example for non convergence of the interpolation polynomial

where $w_n(x) = (x - x_0)(x - x_1)....(x - x_n)$, and $x \in (a, b)$.

However, such result does **not** lead to a convergence result in the sense of:

$$\lim_{n \to \infty} |f(x) - p_n(x)| = 0, \ \forall x \in (a, b),$$

even if the function f possesses continuous derivatives of all orders in that interval.

Example 4.5 *A well-known counter example of this phenomenon is provided by the* **Runge function**

$$f(x) = \frac{1}{1 + x^2}$$

Let $p_{01...n}(x)$ be the polynomial that interpolates this function at $n+1$ equally spaced nodes on the interval $[-5, +5]$ for example, including the endpoints. It is easy to verify the following contradictory results in Figure 4.1.

1. The curve representing $p_{01...n}(x)$ assumes negative values, which obviously $f(x)$ does not have.

2. Adding more equally spaced nodes, leading to higher degree polynomials worsens the situation. The graphs of the resulting polynomials have

wilder oscillations, especially near the endpoints of the interval, and the error increases beyond all bounds as confirmed in the graph.

Thus, in this case it can be shown that:

$$\lim_{n\to\infty} \max_{-5\le x\le+5} |f(x) - p_n(x)| = \infty.$$

This behavior is called the "Runge's phenomenon."
In a more advanced study of this topic [26], it is proved that the divergence of the polynomials is often due to the fact that the nodes of interpolation are equally spaced, which contrary to intuition, is usually a very poor and inappropriate choice. Specifically, one can show that:

$$|w_n(x)| \le n!\frac{h^{n+1}}{4}$$

and therefore

$$\max_x |f(x) - p_n(x)| \le \frac{\max_x |f^{(n+1)}(x)|}{4(n+1)} h^{n+1}$$

If $n \to \infty$, the order of magnitude of $\max_x |f^{(n+1)}(x)|$ could outweigh the nearly-zero order of $h^{n+1}/4(n+1)$.
In [26], numerical results are conducted in the case of the Runge function confirming this hypothesis. More specifically, it is verified that

$$\max_{-5\le x\le+5} |f^{(22)}(x)| = O(10^{19})$$

while the corresponding value of $\max \frac{w_n(x)}{(n+1)!} = O(10^{-10})$
A much better and more adequate choice of nodes leading to more accurate results that help minimizing Runge's phenomenon is obtained for example with the set of Chebyshev nodes defined over the unit interval $[-1, +1]$ by:

$$x_i = \cos[\frac{2i-1}{2n}\pi], \ 1 \le i \le n$$

(Note that these values are graphically obtained by projecting equally spaced points on the unit circle, down on the unit interval $[-1, +1]$). More generally over arbitrary interval $[a, b]$ the coordinates of Chebyshev nodes are:

$$x_i = \frac{1}{2}(a + b) + \frac{1}{2}(b - a)\cos[\frac{2i-1}{2n}\pi]$$

It is possible then to prove that

$$\lim_{n\to\infty} |f(x) - p_n(x)| = 0$$

This problem motivates the use of local piecewise polynomial interpolation.

4.6 Local Interpolation: Spline Functions

As the **global** approach of interpolating polynomials does not provide in general a systematic and efficient way to approximate a function $f(x)$ on the basis of the data

$$D_n = \{(x_i, y_i)|i = 0, 1, ...n, , x_0 = a, < x_1 < ... < x_{n-1} < x_n = b\},$$

we consider hereafter a **local** approach that considers approximating a function $f(x)$ by **spline functions**. Such functions are piecewise polynomials joined together with certain imposed continuity conditions, to which we will refer as the **imposed "smoothness conditions"** of interpolation.
In the theory of splines, the interior points $\{x_i\}_{i=0}^n$ at which the function changes its expression are called the "nodes" or "knots" of the partition.
In this chapter, we analyze successively linear quadratic and cubic spline functions interpolating D_n.

4.6.1 Linear Spline Interpolation

The simplest connection between two points is a line segment. A **spline of degree one** or **linear spline**, is therefore a function that consists of **linear** polynomial pieces joined together to achieve continuity of the polygonal curve representing it. Its formal definition is given as follows:

Definition 4.2 *A linear spline interpolating the data D_n is a function $s(x)$ such that:*

1. *$s_i(x) = \{s(x)|_{x \in [x_i, x_{i+1}]}\}$ is a polynomial of degree at most 1, i.e.,*

$$s_i \in \mathbb{P}^1, \forall i = 0, 1, ..., n - 1.$$

2. *$s(x)$ is **globally continuous** on $[a, b]$, (i.e., $s \in C([a, b])$, the set of all continuous functions on $[a, b]$).*

3. *$s(x)$ satisfies the interpolation conditions: $s(x_i) = y_i, \forall i = 0, 1, ..., n$.*

Note that there exists a unique function $s(x)$ verifying these three criteria:
- To determine $s(x)$, a total of $2n$ unknowns have to be evaluated as each of the linear polynomials $s_i(x)$ defined by the first criterion over the subinterval $[x_i, x_{i+1}]$ is determined by 2 parameters.
- The second and third criteria impose respectively continuity at the $n - 1$ interior nodes in addition to the $n + 1$ interpolation conditions, that add up to a total of $(n - 1) + (n + 1) = 2n$ restrictions or "imposed smoothness conditions."
The number of unknown parameters being equal to the number of imposed

conditions, the equations of the linear spline are uniquely determined.
We proceed directly to write them using Newton's linear interpolating polynomial form on each subinterval $[x_i, x_{i+1}]$. This gives in a straightforward way, $\forall i = 0, 1, ..., n - 1$:

$$s_i(x) = [x_i] + [x_i, x_{i+1}](x - x_i) = y_i + \frac{y_{i+1} - y_i}{x_{i+1} - x_i}(x - x_i) \qquad (4.18)$$

Clearly, by joining the linear pieces $\{s_i(x) \, | i = 0, 1, ..., -1n\}$, one obtains the unique linear spline satisfying the definition above.

The implementation of the linear spline algorithm is as follows:

Algorithm 4.5 Linear Splines

```
function l = LinearSpline(x, y, r)
% Input: 2 vectors x and y of equal length, and a real number r
% Output: l= s(r): s=linear spline
n=length(x);
% seek i : x(i) < r < x(i+1)
i=max(find(x<r));
% compute l=s(r)=s_i(r)
l=y(i) + (y(i+1)-y(i)) / (x(i+1)-x(i)) * (r-x(i));
```

Example 4.6 *Consider the set of data $D_4 = \{(x_i, y_i = f(x_i)) \ i = 0, 1, 2, 3, 4\}$ where*

i	x_i	y_i
0	1.0	7.6
1	1.3	2.0
2	1.6	4.5
3	1.9	2.8
4	2.2	11

Determine the linear spline function interpolating D_4, then interpolate $f(1.4)$

1. Given that $s_i(x) = y_i + \frac{y_{i+1}-y_i}{x_{i+1}-x_i}(x - x_i)$:

$s_0(x) = y_0 + \frac{y_1-y_0}{x_1-x_0}(x - x_0) = 7.6 - 18.6(x - 1)$; $1.0 \le x \le 1.3$
$s_1(x) = y_1 + \frac{y_2-y_1}{x_2-x_1}(x - x_1) = 2 + 8.3(x - 1.3)$; $1.3 \le x \le 1.6$
$s_2(x) = y_2 + \frac{y_3-y_2}{x_3-x_2}(x - x_2) = 4.5 - 5.6(x - 1.6)$; $1.6 \le x \le 1.9$
$s_3(x) = y_3 + \frac{y_4-y_3}{x_4-x_3}(x - x_3) = 2.8 + 27.3(x - 1.9)$; , $1.9 \le x \le 2.2$

2. Since $x_1 < x = 1.4 < x_2 \Rightarrow f(1.4) \approx s_1(1.4) = 2 + 8.3(1.4 - 1.3) = 2.83$

As confirmed by the graph, first order splines are not smooth functions, the first derivative being discontinuous at each interior node. This deficiency is overcome by looking at higher order degree splines.

4.6.2 Quadratic Spline Interpolation

We start by providing a definition for interpolating quadratic splines based on the data D_n.

Definition 4.3 *A quadratic spline interpolating the data D_n is a function $s(x)$ such that:*

1. $s_i(x) = \{s(x)|_{x \in [x_i, x_{i+1}]}\}$, *is a polynomial of degree at most 2, i.e.,*

$$s_i \in \mathbb{P}^2, \forall i = 0, 1, ..., n - 1.$$

2. $s(x)$ *is globally of class C^1, that is:*

   (a) $s \in C([a, b])$,
   (b) $s' \in C([a, b])$.

3. $s(x)$ *satisfies the interpolation conditions:* $s(x_i) = y_i$, $\forall i = 0, 1, ..., n$.

In order to determine the equations of the interpolating quadratic spline, we start by counting the number of unknown parameters and imposed smoothness conditions from the definition.

- From the first criterion, each of the $s_i(x)$ - being a quadratic polynomial - is determined by 3 parameters. Hence, complete obtainment of $s(x)$ requires $3n$ unknowns.

- The second and third criteria impose respectively continuity of s and s' at the $n - 1$ interior nodes in addition to the $n + 1$ interpolation conditions, that add up to a total of $2(n - 1) + (n + 1) = 3n - 1$ restrictions.

Obviously, to obtain a unique determination of the interpolating quadratic spline, there appears to be a deficit of one further constraint!

There is a variety of ways of providing an additional condition. For example, one may impose specific values on $s'(x_0)$, such as:

$$s'(x_0) = 0, \quad \text{(``natural quadratic spline'')} \tag{4.19}$$

or use the **forward difference approximation** formula to the derivative

$$s'(x_0) = [x_0, x_1] = \frac{y_1 - y_0}{x_1 - x_0} \approx f'(x_0) \tag{4.20}$$

Instead of deriving the quadratic spline through a system of $3n$ equations in $3n$ unknowns, a shorter way to proceed is by noting first that $s'(x)$ is a linear interpolating spline on the set of data $D'_n = \{(x_i, s'(x_i)) | i = 0, 1, ..., n\}$. In this view, introduce first the set of unknowns:

$$\{z_i = s'(x_i), \text{ for } i = 0, 1, ...n\}$$

Obviously, it is enough to start first by writing the equation of $s'_i(x)$ followed by an integration process.

- On the subinterval $[x_i, x_{i+1}]$:

$$s_i'(t) = z_i + \left(\frac{z_{i+1} - z_i}{x_{i+1} - x_i}\right)(t - x_i), \forall t \in [x_i, x_{i+1}], \forall i = 0, 1, ..n - 1.$$

- Integration of this last equation from x_i to x : $x_i \leq x \leq x_{i+1}$, yields:

$$s_i(x) = y_i + z_i(x - x_i) + \frac{1}{2}\left(\frac{z_{i+1} - z_i}{x_{i+1} - x_i}\right)(x - x_i)^2. \tag{4.21}$$

- Imposing then the interpolation conditions $s_i(x_{i+1}) = y_{i+1}, i = 0, 1, ...n - 1$, provides a new set of n equations:

$$y_{i+1} = y_i + z_i(x_{i+1} - x_i) + \left(\frac{z_{i+1} - z_i}{x_{i+1} - x_i}\right)\frac{(x_{i+1} - x_i)^2}{2}$$

- Through algebraic simplification and letting $h_{i+1} = x_{i+1} - x_i$, one obtains:

$$\frac{y_{i+1} - y_i}{h_{i+1}} = \frac{z_{i+1} + z_i}{2}, \ i = 0, 1, ..., n - 1.$$

Obviously, to determine $s_i(x)$, the values of the sequence $\{z_i\}$ should be computed first. Given an arbitrary z_0 chosen as suggested in equations (4.19) or (4.20), the sequence $\{z_i\}_{i=1}^n$ can be found from the recurrence relation:

$$z_{i+1} = -z_i + 2[x_i, x_{i+1}], \ i = 0, 1, ..., n - 1, \tag{4.22}$$

where $[x_i, x_{i+1}]$, is the set of first order divided differences associated with the data D_n. It suffices then to determine the equations of the quadratic spline over the interval $[a, b]$, from equation (4.21).

Algorithm 4.6 Algorithm for Quadratic Spline

```
function q = QuadraticSpline(x, y, r)
% Input: 2 vectors x and y of equal length, and a real number r
% Output: q= s(r): s= quadratic spline based on data x and y
n=length(x); z=zeros(1, n);
% compute z(1) (or set z(1)=0) then compute z(i), i=2,...,n
z(1)=(y(2)-y(1)) / (x(2)-x(1));
for i=1:n-1
  z(i+1)=-z(i)+2*(y(i+1)-y(i))/(x(i+1)-x(i));
end
% seek i : x(i) < r < x(i+1) and compute q=s(r)=s_i(r)
i=max(find(x<r));
q=y(i)+z(i)*(r-x(i))+(z(i+1)-z(i))/(x(i+1)-x(i))*((r-x(i))^2/2));
```

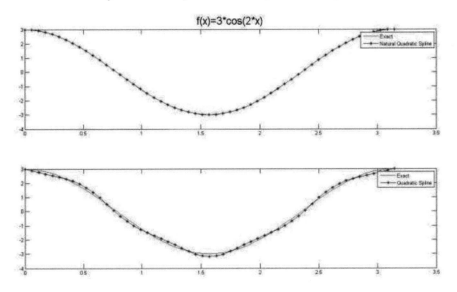

FIGURE 4.2: Natural quadratic spline approximation for $f(x) = 3\cos(2x)$

Example 4.7 *Find the natural quadratic spline interpolant for the following data:*

$$D_5 = \{(-1, 2); (0, 1); (0.5, 0); (1, 1); (2, 2); (2.5, 3)\}$$

Let $z_0 = 0$. Using equation (4.21) recursively:

$$\{z_i\}_{i=0}^5 = \{0, -2, -2, 6, -4, 8\}$$

From equation (4.22) the natural quadratic spline is given by:
$$s_0(x) = -(x+1)^2 + 2 \; ; -1.0 \le x \le 0$$
$$s_1(x) = -2x + 1 \; ; 0 \le x \le 0.5.$$
$$s_2(x) = 8(x - \tfrac{1}{2})^2 - 2(x - \tfrac{1}{2}) \; ; 0.5 \le x \le 1.0$$
$$s_3(x) = -5(x-1)^2 + 6(x-1) + 1 \; ; 1.0 \le x \le 2.0$$
$$s_4(x) = 12(x-2)^2 - 4(x-2) + 2 \; ; 2.0 \le x \le 2.5$$

Remark 4.6 *On the choice of the additional condition on z_0.*

When conducting numerical tests regarding the use of the natural spline condition $z_0 = 0$, it was found that such constraint provides a good quadratic spline approximation results **only** in the case where the data $\{x_i, y_i\}$ correspond to a function $f(x)$ for which $f'(x_0) = 0$. This is shown in Figure 4.2 for $f(x) = 3\cos(2x)$. Otherwise, changing the additional condition from $z_0 = 0$ to $z_0 = [x_0, x_1] = \frac{y_1 - y_0}{x_1 - x_0}$ proved to be an $O(h)$ approximation to $f'(x_0)$, appears to provide more accurate results. The next 2 figures attest to such facts for the functions:

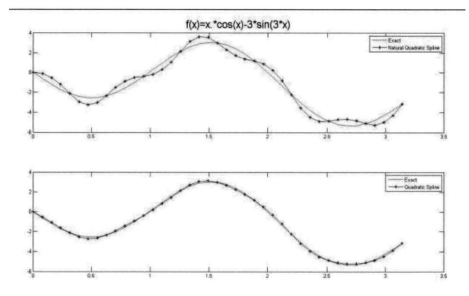

FIGURE 4.3: Comparison of approximations between natural quadratic spline and quadratic spline using $z_0 = \frac{y_1 - y_0}{x_1 - x_0}$ for $f(x) = x\cos(x) - 3\sin(3x)$

- $f(x) = x\cos(x) - 3\sin(3x)$, in Figure 4.3

- $f(x) = 3\sin(2x)$, in Figure 4.4.

4.6.3   Cubic Spline Interpolation

In the previous two cases one notes the following:
- The polygonal lines representing linear splines lack smoothness as the slope of the spline may change abruptly through each node.
- As for quadratic splines, the discontinuity is in the second derivative, and is therefore not so evident; nevertheless, the curvature of the spline changes abruptly through each node, and the curve may not be visually smooth enough.
Cubic splines allow for smoother data fitting and they are most frequently used in applications. It can be proved that cubic spline functions are among the best interpolation functions that are available at an acceptable computational cost. In this case, we join cubic polynomials together in such a way that the resulting spline function has its first and second derivatives continuous everywhere in the interval $[a, b]$. At each interior node, 3 conditions will be imposed, so that the graph of the function will look smoother than in the case of linear and quadratic splines. Discontinuities of course may occur in the

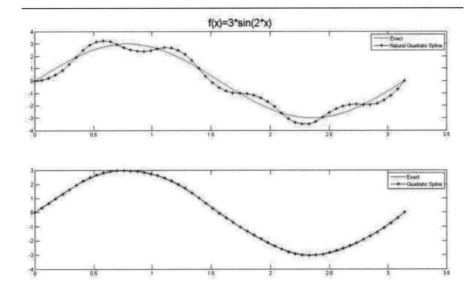

FIGURE 4.4: Comparison of approximations between natural quadratic spline and quadratic spline using $z_0 = \frac{y_1 - y_0}{x_1 - x_0}$ for $f(x) = 3\sin(2x)$

third derivative, but these cannot be easily detected visually. Cubic splines are formally defined as follows.

Definition 4.4 *A cubic spline that interpolates the data D_n, is a function $s(x)$ such that:*

1. *$s_i(x) = \{s(x)|_{x \in [x_i, x_{i+1}]}\}$ is a polynomial of degree at most 3, i.e.,*

$$s_i \in \mathbb{P}^3 \, \forall i = 0, 1, ..., n - 1.$$

2. *$s(x)$ is globally of class C^2, that is:*

   (a) $s \in C([a, b])$.

   (b) $s' \in C([a, b])$.

   (c) $s'' \in C([a, b])$.

3. *$s(x)$ satisfies the interpolation conditions: $s(x_i) = y_i$, $\forall i = 0, 1, ..., n$.*

Following the same pattern as previously, and counting the number of unknown parameters and imposed conditions from the definition, we note the following:
- From the first criterion, each of the $s_i(x)$ is determined by 4 parameters. Hence, complete obtainment of $s(x)$ requires $4n$ unknowns.

- The second and third criteria impose now respectively $3(n-1)$ continuity conditions for s, s' and s'' at the interior nodes, in addition to the $n+1$ interpolation conditions.

Hence for a total of $4n$ unknowns, one has a total of $3(n-1)+n+1 = 4n-2$ constraints. Obviously, to allow <u>unique</u> determination of the interpolating cubic spline, there appears to be a deficit of two constraints!

These two supplementary conditions may be for example supplied as follows:

1. Letting $s''(x_0) = s''(x_n) = 0$, the spline is called a **natural spline** (or free boundary.)

2. An alternative to the natural spline is to use:

   $$s''(x_0) = 2[x_0, x_1, x_2] \approx f''(x_0) \text{ and } s''(x_n) = 2[x_{n-2}, x_{n-1}, x_n] \approx f''(x_n)$$

3. Letting $s'(x_0) = f_0$ and $s'(x_n) = f_n$, the spline is called a **clamped spline**. However, for this type of boundary condition to hold, it is necessary to have the values of $f'(x_0)$ and $f'(x_n)$ (or at least an accurate approximation.)

In this chapter, we will restrict our analysis to **natural cubic splines** only. Instead of determining the solution of the problem through a system of $4n$ equations in $4n$ unknowns, we note that $s'(x)$ and $s''(x)$ are quadratic and linear splines based respectively on the data sets $D'_n = \{(x_i, s'(x_i)\}$ and $D''_n = \{(x_i, s''(x_i)\}$, where the unknowns:

$$\{z_i = s'(x_i) | i = 0, 1, ..., n\}$$

and

$$\{w_i = s''(x_i) | i = 0, 1, ..., n\}$$

represent respectively the sets of **slopes** and **moments** at the nodes. Obviously, we should proceed by first writing $s''_i(x)$ on the interval $[x_i, x_{i+1}]$ followed by 2 successive integrations.

- On the subinterval $[x_i, x_{i+1}]$:

$$s''_i(t) = w_i + \frac{w_{i+1} - w_i}{x_{i+1} - x_i}(t - x_i), \forall t \in [x_i, x_{i+1}], \forall i = 0, 1, ..., n-1. \quad (4.23)$$

- Integration of (4.23) from x_i to x, $x_i \leq x \leq x_{i+1}$ yields:

$$s'_i(x) - z_i = w_i(x - x_i) + \frac{w_{i+1} - w_i}{x_{i+1} - x_i} \frac{(x - x_i)^2}{2}, \forall x \in [x_i, x_{i+1}], \forall i = 0, 1, ..., n-1$$

$$(4.24)$$

- Imposing in (4.24) the conditions $s'_i(x_{i+1}) = z_{i+1}$ at internal nodes, provide a new set of $n-1$ equations. Specifically:

$$\frac{z_{i+1} - z_i}{h_{i+1}} = \frac{w_{i+1} + w_i}{2}, \ i = 0, 1, ..., n-1. \tag{4.25}$$

which is equivalent to:

$$z_{i+1} = z_i + h_{i+1} \frac{w_i + w_{i+1}}{2}, \ i = 1, ..., n-1. \tag{4.26}$$

- A second integration of equation (4.24) from x_i to x yields the cubic polynomials $s_i(x)$:

$$s_i(x) = y_i + z_i(x - x_i) + w_i \frac{(x - x_i)^2}{2} + \frac{w_{i+1} - w_i}{6h_{i+1}}(x - x_i)^3, \quad (4.27)$$

$\forall x \in [x_i, x_{i+1}], \ \forall i = 0, 1, ..., n-1.$

- Imposing then the interpolation conditions $s_i(x_{i+1}) = y_{i+1}$ provides a new set of $n-1$ equations given by:

$$y_{i+1} = y_i + z_i h_{i+1} + w_i \frac{h_{i+1}^2}{2} + \frac{(w_{i+1} - w_i) h_{i+1}^2}{6}, \ \forall i = 0, 1, ..., n-1 \tag{4.28}$$

- This last equation leads to 2 simultaneous equations satisfied at all internal nodes of the spline, i.e., for all $i = 1, .., n-1$:

$$\begin{cases} \frac{y_{i+1} - y_i}{h_{i+1}} = z_i + (w_{i+1} + 2w_i) \frac{h_{i+1}}{6} \\ \frac{y_i - y_{i-1}}{h_i} = z_{i-1} + (w_i + 2w_{i-1}) \frac{h_i}{6} \end{cases}$$

Subtracting these last 2 equations and using (4.25) gives:

$$[x_i, x_{i+1}] - [x_{i-1}, x_i] = h_i \frac{w_{i-1} + w_i}{2} + h_{i+1} \frac{(w_{i+1} + 2w_i)}{6} - h_i \frac{(w_i + 2w_{i-1})}{6}$$

Equivalently:

$$\frac{h_i}{6} w_{i-1} + \frac{h_i + h_{i+1}}{3} w_i + \frac{h_{i+1}}{6} w_{i+1} = (h_i + h_{i+1})[x_{i-1}, x_i, x_{i+1}], i = 1, 2, ..., n-1 \tag{4.29}$$

Since the sought spline is "natural" ($w_0 = w_n = 0$), equation (4.29) provides therefore a system of $n-1$ equations in $n-1$ unknowns given by:

$$Aw = r, \tag{4.30}$$

where the coefficient matrix A of size $n-1 \times n-1$ is:

$$\begin{pmatrix} (h_1 + h_2)/3 & h_2/6 & 0...0 & 0 \\ h_2/6 & (h_2 + h_3)/3 & 0...0 & 0 \\ 0 & h_3/6 & 0...0 & 0 \\ 0 & 0 & ... & h_{n-1}/6 \\ 0 & & ... & h_{n-1}/6 \quad (h_{n-1} + h_n)/3 \end{pmatrix} \tag{4.31}$$

and the vectors w and r are respectively:

$$w = (w_1, w_2, ..., w_{n-1})^T$$

and

$$r = (r_1, r_2, ..., r_{n-1})^T \quad \text{with } r_i = (h_i + h_{i+1})[x_{i-1}, x_i, x_{i+1}]_y \qquad (4.32)$$

Note also that the matrix $A = \{a_{ij}\}$ has the following properties:

- A is symmetric, since $a_{ij} = a_{ji}$

- A is tridiagonal, since $a_{ij} = 0$ for all i, j with $|i - j| > 1$.

- A is strictly diagonally dominant, since $|a_{ii}| > \sum_{j \neq i} |a_{ij}|, \forall i$.

Under these conditions, the system (4.30) has a unique solution that can be obtained through a straightforward Gauss reduction process that does not necessitate any pivoting strategy. We can now write a **pseudocode for the natural cubic spline.**

Algorithm 4.7 Cubic Spline

```
% Input the data  D_n
% Output: cubic spline  s(x)    interpolating on  D_n

% Obtain first w by solving  Aw=r by performing the following steps:
```

1. Generate $r = [r_1, ..., r_{n-1}]^T$ with $r_i = (h_i + h_{i+1})[x_{i-1}, x_i, x_{i+1}], i = 1, ..., n-1$.

2. Generate the matrix A.

3. Perform Gauss reduction on $[A|r]$.

4. Perform back substitution on reduced system to get w with $w_0 = w_n = 0$.

5. Compute $z_0 = [x_0, x_1] - (2w_0 + w_1)h_1/6$

6. Compute $z_{i+1} = z_i + h_{i+1}(w_{i+1} + w_i)/2, i = 0, 1, ..., n-1$.

7. Generate $s(x)$ through generating $s_i(x)$:
   $s_i(x) = y_i + z_i(x - x_i) + w_i(x - x_i)^2/2 + ((w_{i+1} - w_i/6h_{i+1})(x - x_i)^3, i = 0, 1, ..., n-1$.

As an illustration, let us reconsider the Runge function $f(x) = \frac{1}{1+x^2}$ introduced in Example 4.5, and its natural cubic spline interpolant. The curves representing these 2 functions over the interval $[-5, +5]$ match completely at all points. The results are summarized in Figure 4.5.

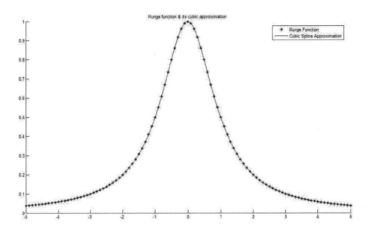

FIGURE 4.5: Natural cubic spline approximation for the Runge function $f(x) = \frac{1}{1+x^2}$

Example 4.8 The Runge Function and the Natural Cubic Spline Interpolant

Example 4.9 *Determine the natural cubic spline interpolating the following set of data:*

$$D_3 = \{(-1, 1); (1, 2); (2, -1); (2.5, 0)\}$$

- Since $w_0 = w_3 = 0$
  and $h_1 = x_1 - x_0 = 2$; $h_2 = x_2 - x_1 = 1$; $h_3 = x_3 - x_2 = 0.5$,
  the system (4.30) is:

$$\begin{pmatrix} 1 & 1/6 \\ 1/6 & 1/2 \end{pmatrix} \begin{pmatrix} w_1 \\ w_2 \end{pmatrix} = \begin{pmatrix} -7/2 \\ 5 \end{pmatrix}$$

- Applying the naive Gauss reduction on that system followed by back substitution:

$$w = [w_0 = 0, w_1 = -93/17, w_2 = 201/17, w_3 = 0]'$$

- Using (4.28) with $i = 0$ leads to

$$z_0 = [x_0, x_1] - (2w_0 + w_1)h_1/6 = -45/34$$

- Once the value of z_0 and (4.26), the vector of slopes is fully determined with:

$$z = [z_0 = -45/34, z_1 = --231/34/17, z_2 = -123/34, w_3 = -45/68]'$$

- Using (4.27) for successively $i = 0, 1, 2$, the equations of the cubic spline are then as follows:

$$S(x) = \begin{cases} S_0(x) = 1 - \frac{9}{7}(x+1) - \frac{31}{68}(x+1)^3 \, ; \, -1 \leq x \leq 1 \\ S_1(x) = 2 - \frac{231}{34}(x-1) - \frac{93}{17}(x-1)^2 + \frac{49}{17}(x-1)^3 \, ; \, 1 \leq x \leq 2 \\ S_2(x) = -1 - -\frac{123}{34}(x-2) + \frac{201}{17}(x-2)^2 - \frac{67}{17}(x-2)^3 \, ; \, 2 \leq x \leq 2.5 \end{cases}$$

4.6.4   Solving a Tridiagonal System

Note that, in case D_n, the x-data are equidistant, i.e.,

$$h_{i+1} = x_{i+1} - x_i = h, \, \forall i = 0, ..., n-1$$

the matrix A in (4.31) becomes:

$$A = h \begin{pmatrix} 2/3 & 1/6 & 0 & 0... & 0 \\ 1/6 & 2/3 & 1/6 & 0.. & 0 \\ 0 & 1/6 & 2/3 & 1/6.. & 0 \\ ... & ... & ... & ... & ... \\ ... & ... & ... & ... & ... \\ 0 & ... & 0 & 1/6 & 2/3 \end{pmatrix}$$

Since also $[x_{i-1}, x_i, x_{i+1}] = \frac{y_{i+1} - 2y_i + y_{i-1}}{2h^2}$, the right hand side r in (4.32) simplifies, and the system (4.30) becomes:

$$h \begin{pmatrix} 2/3 & 1/6 & 0 & 0... & 0 \\ 1/6 & 2/3 & 1/6 & 0.. & 0 \\ 0 & 1/6 & 2/3 & 1/6 & 0.. \\ ... & ... & ... & ... & ... \\ ... & ... & ... & ... & ... \\ 0 & ... & 0 & 1/6 & 2/3 \end{pmatrix} \begin{pmatrix} w_1 \\ w_2 \\ ... \\ ... \\ ... \\ w_{n-1} \end{pmatrix} = \frac{1}{h} \begin{pmatrix} y_2 - 2y_1 + y_0 \\ y_3 - 2y_2 + y_1 \\ y_4 - 2y_3 + y_2 \\ ...... \\ y_n - 2y_{n-1} + y_{n-2} \end{pmatrix}.$$

The elements of the matrix A can be made independent of h, through dividing each of the equations by h, thus yielding the following tridiagonal system:

$$\begin{pmatrix} 2/3 & 1/6 & 0 & 0... & 0 \\ 1/6 & 2/3 & 1/6 & 0.. & 0 \\ 0 & 1/6 & 2/3 & 1/6 & 0.. \\ ... & ... & ... & ... & ... \\ ... & ... & ... & ... & ... \\ 0 & ... & 0 & 1/6 & 2/3 \end{pmatrix} \begin{pmatrix} w_1 \\ w_2 \\ ... \\ ... \\ ... \\ w_{n-1} \end{pmatrix} = \frac{1}{h^2} \begin{pmatrix} y_2 - 2y_1 + y_0 \\ y_3 - 2y_2 + y_1 \\ y_4 - 2y_3 + y_2 \\ ... \\ ... \\ y_n - 2y_{n-1} + y_{n-2} \end{pmatrix}.$$

In what follows, we consider the general triadiagonal system of equations:
$Aw = r$:

$$\begin{pmatrix} a_1 & b_1 & 0 & 0... & 0 \\ c_1 & a_2 & b_2 & 0.. & 0 \\ 0 & c_2 & a_3 & b_3 & 0.. \\ ... & ... & ... & ... & ... \\ ... & ... & c_{N-2} & a_{N-1} & b_{N-1} \\ 0 & ... & 0 & c_{N-1} & a_N \end{pmatrix} \begin{pmatrix} w_1 \\ w_2 \\ ... \\ ... \\ ... \\ w_N \end{pmatrix} = \begin{pmatrix} r_1 \\ r_2 \\ ... \\ ... \\ ... \\ r_N \end{pmatrix},$$

where the "diagonal" entries of the matrix A are generated by:
- The "main diagonal" vector, $a = [a_i : 1 \leq i \leq N]$
- The "upper diagonal" vector, $b = [b_i : 1 \leq i \leq N - 1]$
- The "lower diagonal" vector, $c = [c_i : 1 \leq i \leq N - 1]$
satisfying the following properties:

$$\begin{cases} |a_i| > |b_i| + |c_{i-1}| : 2 \leq i \leq N - 1 \\ |a_1| > |b_1|, \ |a_N| > |c_{N-1}| \end{cases}$$

The following algorithm solves this given system:

Algorithm 4.8 Diagonally Dominant Triangular Systems
```
function w=SolveTridiag(a,b,c,r)
% N is the dimension of a and r; N-1 is the dimension of b and c
% Start with the Gauss reduction process then use back-substitution
for k=1:N-1
    m=c(k)/a(k);
    a(k+1)=a(k+1)-m*b(k);
    r(k+1)=r(k+1)-m*r(k);
end
for k=N:-1:1
    w(k)=r(k)/a(k);
    if k>1
        r(k-1)=r(k-1)-w(k)*b(k-1);
    end
end
```

This algorithm takes $2N - 1$ divisions, $3(N - 1)$ multiplications and as many algebraic additions, thus a total of $8N - 7$ flops.

4.6.5 Errors in Spline Interpolation

From ([27], pages 14 and 61), we can state the following convergence result:

Theorem 4.6 *Let f be a function in $C^{k+1}[a, b]$, and S_k the spline function that interpolates f based on the set of data D_n, where $k = 1, 2, 3$. Then,*

$$\max_{[a,b]} |f(x) - S_k(x)| \leq C_k h^{k+1} \max_{[a,b]} |f^{(k+1)}(x)|$$

where $h = \max |x_i - x_{i-1}|$, for $1 \leq i \leq n$.

For example:

- If $k = 1$, then $\max_{[a,b]} |f(x) - S_1(x)| = O(h^2)$

- If $k = 2$, then $\max_{[a,b]} |f(x) - S_2(x)| = O(h^3)$

- If $k = 3$, then $\max_{[a,b]} |f(x) - S_3(x)| = O(h^4)$

Note that, in spline interpolation, increasing the number of nodes for a fixed value of k will definitely lead to convergence. One can prove that:

$$\forall x \in [a, b], \;] \lim_{n \to \infty} S_k(x) = f(x).$$

This property is noticeably absent for global Lagrange interpolation (recall Runge example).

4.7   Concluding Remarks

1. Based on a set of data D_n, considering higher degree Lagrange interpolating polynomials does not guarantee reaching more accurate approximations of the unknown function f; this problem can be overcome by spline functions, particularly cubic splines. However, neither are suitable to extrapolate information from the available set of data D_n. To generate new values at points lying outside the interval $[x_0, x_n]$, one could use, for example, regression analysis based on least squares approximations.

2. Polynomial interpolation can also be used to approximate multi-dimensional functions. In particular, spline function interpolation is well suited when the region is partitioned into polygons in 2D (triangles or quadrilaterals) and polyhedra in 3D (tetrahedra or prisms). See([26]).

4.8 Exercises

1. Use the Lagrange form of the Lagrangian interpolating polynomial to obtain a polynomial of least degree that satisfies the following set of data: $D_3 = \{(0,7), (2,10), (3,25), (4,50)\}$.

2. Consider the following four interpolation nodes: $-1, 1, 2, 4$. Find the l_i functions required in the Lagrange interpolation process and draw their graphs. Use the Lagrange interpolation form to obtain a polynomial of least degree that satisfies the following set of data: $D_3 = \{(-1,1), (1,0), (2,2), (4,-3)\}$.

3. Write the Lagrange form of the interpolating polynomial of degree ≤ 3 that interpolates $f(x)$ at x_0, x_1, x_2 and x_3, where the nodes are sorted from the smallest to the greatest.

4. Write Newton's interpolating polynomial on the following set of data:

$$\{(0,7), (2,10), (3,12), (4,15)\}$$

5. Given the data

$$D_4 = \{(1,-1), (2,-1/3), (2.5,3), (3,4), (4,5)\}$$

   (a) Construct its divided difference table.

   (b) Use the "best" quadratic then cubic Newton's interpolating polynomial, to find an approximation to $f(2.7)$.

6. Using a difference table, derive the polynomial of least degree that assumes the values $2, 14, 10$ and 2 respectively for $x = -2, -1, 0, 1$ and 2. Use that result, to find a polynomial that takes the values shown and has at $x = 2$ the value 4.

7. The polynomial $p(x) = 2x^4 - x^3 - x$ satisfies the following set of data:

i	x_i	y_i
0	-1	4
1	0	0
2	1	0
3	1	22
4	2	138
5	3	444

Find a polynomial q that takes these values:

i	x_i	y_i
0	−1	4
1	0	0
2	1	0
3	2	22
4	3	138
5	4	1

8. Construct a divided difference (or difference) table based on the two given sets of data in the preceding exercise, then use Newton's polynomials of all orders to approximate $f(2.5)$, in each case.

9. Determine the polynomial of degree 2 or less whose graph passes through the points $(0, 1), (1, 2)$, and $(2, 4.2)$. Use two different methods. Verify that they lead to the same polynomial.

10. Create the table of all Neville's polynomials in P_4 satisfying the following set of data:

i	x_i	y_i
0	1.0	−1.5
1	2.0	−0.5
2	2.5	0.0
3	3.0	0.5
4	4.0	1.0

11. (a) Consider the following set of data:

$$D_5 = \{(-2, 1); (-1, 4); (0, 11); (1, 16); (2, 13); (3, -4)\}$$

Show that the interpolating polynomial based on D_5 is cubic.
(b) The set D_5 is altered as follows:

$$D_5' = \{(-2, 1); (-1, 4); (0, 11); (1, 16); (2, 10); (3, -4)\},$$

so that $y_4 = 10$. Based on D_5' and using the polynomial found in part (a), find $q_{01234}(x)$, without computing new divided differences.

12. The polynomial $p(x) = x^4 + 3x^3 - 2x^2 + x + 1$ interpolates the set of data

i	0	1	2	3	4
x_i	−1	−2	0	1	2
y_i	−4	−17	1	4	35

Without computing any difference or divided difference, use Newton's form to determine the polynomial $q(x)$ interpolating the following set of data:

i	0	1	2	3	4
x_i	-1	-2	0	1	2
y_i	-4	0	1	4	35

13. Consider the set of data: $D_3 = \{(-2, -1); (-1, 1); (0, 4); (1.5, 0)\}$

(a) Based on D_3, fill in the following divided difference table, then write Newton' form of the interpolating polynomial $p_{123}(x)$, reproducing part of the data:

i	x_i	y_i	$[\cdot, \cdot]$	$[\cdot, \cdot, \cdot]$	$[\cdot, \cdot, \cdot, \cdot]$
0	-2	1			
1	-1	1			
2	0	4			
3	1.5	0			

(b) Based on the equation of the polynomial $p_{123}(x)$ obtained in (a) and using the coefficients in the divided differences table, determine the equation of the interpolating polynomial $p_{0123}(x)$. Explain and justify all your steps.

(c) The initial set of data D_3 is modified by inserting a new point between its elements, with coordinates $(x_A, y_A) = (-0.5, 2)$. Consider now the set:

$$D_4 = \{(-2, -1); ; (-1, 1); (-0.5, 2); (0, 4); (1.5, 0)\},$$

Using the polynomial found in (b), determine the polynomial $q(x)$ interpolating D_4, without computing new divided differences. Explain and justify all your steps.

14. Are these functions linear splines? Explain why or why not.

(a)
$$S(x) = \begin{cases} x\,; & -1 \leq x \leq 0 \\ 1 - x\,; & 0 \leq x < 1 \\ 2x - 2\,; & 1 \leq x \leq 2 \end{cases}$$

(b)

$$S(x) = \begin{cases} x \, ; \; -1 \le x \le 0.5 \\ 0.5x + 2(x - 0.5) \, ; \; 0.5 \le x \le 2 \\ x + 1.5 \, ; \; 2 \le x \le 4 \end{cases}$$

15. Determine the linear spline function $s(x)$ interpolating the set of data D_3 and plot its graph. Interpolate $f(2.4)$.

$$D_3 = \{(0,1); (1.5, 2); (2, 6); (2.5, 3)\}$$

16. Could the function $S(x) = |x|$ be a linear spline on the interval $[-2, 1]$? Justify your answer.

17. Find the natural quadratic spline interpolant for the following data

i	x	y
0	−1	3
1	0	0
2	1	1
3	2	2

18. Find the natural quadratic spline interpolant for the following data

i	x	y
0	1	2
1	2	1
2	5	0
3	3	−1
4	4	4

.

19. Find the quadratic spline interpolant for the following data:

i	x	y
0	−1	0
1	0	1
2	1/2	0
3	1	1
4	2	0

20. Are these functions quadratic splines? Explain why or why not.

(a)

$$Q(x) = \begin{cases} -x^2 \, ; \; 0 \le x \le 1 \\ x \, ; \; 0 \le x \le 100 \end{cases}$$

(b)

$$Q(x) = \begin{cases} x \; ; \; -50 \leq x \leq 1 \\ x^2 \; ; \; 1 \leq x \leq 2 \\ 4 \; ; \; 2 \leq x \leq 50 \end{cases}$$

21. Do there exist a,b, c and d so that the function

$$S(x) = \begin{cases} ax + e \; ; \; -5 \leq x \leq 1 \\ bx^2 + cx \; ; \; 1 \leq x \leq 2 \\ dx^2 \; ; \; 2 \leq x \leq 3 \end{cases}$$

is a quadratic spline function?

22. Determine the natural cubic spline that interpolates the function $f(x) = 2x^7$ over the interval $[0, 2]$ using nodes $0, 1$ and 2.

23. Find the natural cubic spline interpolant for this table:

i	x	y
0	1	0
1	2.5	1
2	3	0
3	4.5	1
4	5	0

24. Find the natural cubic spline interpolant for this table:

i	x	y
0	1	0
1	2	1
2	3	0
3	4	1
4	5	0

25. Consider the following set of data generated using the function:

$$f(x) = x^2 \sin x - 2x^2 + x + 1$$

i	x	y
0	-0.2	0.8790
1	-0.1	0.7121
2	0.1	1.0810
3	0.2	0.1279

(a) Construct the natural cubic spline for the data above.

(b) Use the cubic spline constructed above to approximate $f(0.25)$ and $f'(0.25)$, and calculate the absolute error.

26. Give an example of a cubic spline with nodes $0, 1, 2$, and 3 that is linear in $[0, 1]$, but of degree 3 in at least one of the other two intervals.

27. Give an example of a cubic spline with nodes $0, 1, 2$, and 3 that is quadratic in $[0, 1]$ and in $[1, 2]$, and is cubic in $[2, 3]$.

28. Are these functions cubic splines? Explain why or why not.

    (a)
    $$S(x) = \begin{cases} x + 1; & -2 \le x \le -1 \\ 2x^3 - 5x + 1; & -1 \le x \le 1 \\ 9x - 1; & 1 \le x \le 2 \end{cases}$$

    (b)
    $$S(x) = \begin{cases} x^3 + 3x - 2; & -1 \le x \le 0 \\ x^3 - 2x - 1; & 0 \le x \le 1 \end{cases}$$

29. Construct a natural cubic spline to approximate $f(x) = e^{-x}$ based on the nodes $x = 0, 0.25, 0.75$ and 1. Integrate the spline over the interval $[0, 1]$ and compare the results to $\int_0^1 e^{-x}\, dx$. Use the derivatives of the spline to approximate $f'(0.5)$ and $f''(0.5)$. Compare the approximations to the actual values.

30. Use the data points $(0, 1), (1, e), (2, e^4), (3, e^9)$ to form a natural cubic spline that approximates $f(x) = e^{x^2}$. Use then the cubic spline to approximate $I = \int_0^3 e^x\, dx$.

31. How many additional conditions are needed to specify uniquely a spline of degree 4 over n knots ? Justify your answer.

32. Let S be a cubic spline that has knots $t_0 < t_1 < t_2 < t_3$. Suppose that on the 2 intervals $[t_0, t_1]$ and $[t_2, t_3]$, S reduces to linear polynomials. What can be said of S on $[t_1, t_2]$?

33. Provide an upper bound of the Lagrange interpolation for the Runge function defined over the interval $[-5, +5]$ with 11 equally spaced nodes:

    (a) Using Lagrange polynomials.

    (b) Using linear, quadratic then cubic spline interpollants.

34. Determine the equations of the cubic spline based on the set of data D_n, if the 2 additional constraints are set to:

    $$s''(x_0) = 2[x_0, x_1, x_2], \text{ and } s''(x_n) = 2[x_{n-2}, x_{n-1}, x_n]$$

    Use these equations to determine the cubic spline based on the set of data given in Exercise 23.

35. (a) Write the equations of the natural cubic spline $S(x)$ that interpolates the set of data $D_3 = \{(2,1), (3,0), (4,1), (5,-1)\}$, and fill in the following table:

i	x_i	y_i	z_i	w_i
0	2.00	1	.	0
1	3.00	0	.	.
2	4.00	1	.	.
3	5.00	-1	.	0

   (b) Assume now that the cubic spline is "not natural" and that the 2 additional supplied conditions are:

$$s''(x_0) = 2 \text{ and } s''(x_3) = 0$$

   Determine in that case, the equations of the cubic spline and fill in the following table:

i	x_i	y_i	z_i	w_i
0	2.00	1	.	2
1	3.00	0	.	.
2	4.00	1	.	.
3	5.00	-1	.	0

4.9   Computer Projects

Exercise 1: Polynomial Interpolation

Let $x = (x_1, x_2, ..., x_n)$ and $y = (y_1, y_2, ..., y_n)$ be **2 vectors of equal length** n, representing a set of n points in the plane:

$$D_n = \{(x_i, y_i) | x_1 < x_2 < ... < x_n \; ; i = 1, 2, ..., n\}$$

where $y_i = f(x_i)$ for some real valued function f.

To solve Exercise 1, use the MATLAB function p =NevillePolynomial(x, y, r) given in the lecture notes without checking the validity of inputs. This function takes as input the 2 vectors x and y and a real number r, with $x_1 < r < x_n$, and computes

$$p = p_{1,2,...,n}(r)$$

where $p_{1,2,...,n}(.)$ is Neville's form of the interpolating polynomial based on the set of data D_n.

1. Write a MATLAB
   function v =VectorNevillePolynomial(x, y, w)
   that takes as input the 2 vectors x and y, and a vector w of any length, and computes the values of Neville's polynomial at each component of w. The output of this function is a vector v whose components are:

   $$v(i) = p_{1,2,...,n}(w(i)), \; \forall \, i = 1, ..., length(w)$$

   (Assume that $x_1 < w(i) < x_n \; \forall i$).

2. Consider the Runge function $f(x) = \frac{1}{1+x^2}$ on the interval $[-5, +5]$.
   Write a MATLAB
   function [x, fx, s, fs] =GenerateVectors(n, f)
   that takes as input an integer n and the Runge function f. Your function:

   - First: generates a vector x of length n, whose components are n equally spaced points in the interval $[-5, +5]$ including the end-points, evaluates f at these points, and saves these values in a vector fx.
     Hint: The MATLAB built-in function **linspace(a,b,n)** generates a row vector of n equally spaced points between a and b, including the end-points.

   - Secondly: generates a vector s of length $(n-1)$ whose i^{th} component is the midpoint of the interval $[x_i, x_{i+1}]$, that is:

     $$s = [s_1 = \frac{x_1 + x_2}{2}, ..., s_i = \frac{x_i + x_{i+1}}{2}, ..., s_{n-1} = \frac{x_{n-1} + x_n}{2}]$$

     Your function then evaluates f at all components of s and saves these values in a vector fs.

3. Write a MATLAB
   function PlotPolynomial(n, f)
   that takes as input an integer n and the Runge function f and plots in
   the same figure window, the **graphs of** f **and** $p_{1,2,...,n}$ over the set of
   ordered points in

$$X = x \ U \ s = \{x_i, s_i, x_{i+1} \mid i = 1, ..., n-1\}$$

   Note that $p_{1,2,...,n}$ is Neville's form of the interpolating polynomial based
   on the set of data represented by x and fx.

4. Write a MATLAB
   function EP =ErrorPolynomial(n, f)
   that takes as input an integer n and the Runge function f. Your function
   outputs a matrix EP of size $(n-1) \times 4$, whose 4 columns are successively
   the vectors:

$$f(s) \quad p_{1,2,...,n}(s) \quad err = |p_{1,2,...,n}(s) - f(s)| \quad relerr = \frac{|p_{1,2,...,n}(s) - f(s)|}{|f(s)|}$$

5. Test each of the functions of this exercise on 2 different test cases $n > 10$,
   (n is an odd integer). Save your results and graphs in a Word document.

Exercise 2: Spline Interpolation
All questions are as in Exercise 2, but applied to the quadratic spline instead
of the interpolating polynomial.
Let $x = [x_1, x_2, ..., x_n]$ and $y = [y_1, y_2, ..., y_n]$ be **2 vectors of equal length**
n, representing a set of n points in the plane:

$$D_n = \{(x_i, y_i) | x_1 < x_2 < ... < x_n \ ; i = 1, 2, ..., n\}$$

where $y_i = f(x_i)$ for some real valued function f.
To solve Exercise 2, use the MATLAB function q =QuadraticSpline(x, y,
r) given in the lecture notes. (Do not check validity of inputs).
This function takes as input the 2 vectors x and y and a real number r, with
$x_1 < r < x_n$, and computes

$$q = Q(r)$$

where $Q(.)$ **is the quadratic spline interpolating the set of data** D_n.

1. Write a MATLAB
   function v =VectorQuadraticSpline(x, y, w)
   that takes as input the 2 vectors x and y, and a vector w of any length,
   and computes the values of the quadratic spline function at each component of w.
   (Assume that $x_1 < w(i) < x_n \ \forall i = 1, ..., length(w)$).

2. Consider the Runge function $f(x) = \frac{1}{1+x^2}$ on the interval $[-5, +5]$. Write a MATLAB
   function PlotSpline(n, f)
   that takes as input an integer n and the Runge function f and plots in the same figure window, the **graphs of f and Q** over the set of ordered points in:
   $$X = x \ U \ s = \{x_i, s_i, x_{i+1} \mid i = 1, ..., n-1\}$$
   Note that Q is the quadratic spline interpolating the set of data represented by x and fx.
   Hint: Call for the function **GenerateVectors(n, f)** programmed in Exercise 2.

3. Write a MATLAB
   function ES =ErrorSpline(n, f)
   that takes as input the integer n and the Runge function f. Your function outputs a matrix ES of size $(n+1) \times 4$ whose 4 columns are successively the vectors:

   $$f(s) \qquad Q(s) \qquad err = |Q(s) - f(s)| \qquad relerr = \frac{|Q(s) - f(s)|}{|f(s)|}$$

4. Test each of the functions of this exercise on 2 different test cases $n > 20$, (n is an odd integer.) Save your results and graphs in a Word document.

Exercise 3: Quadratic Spline Interpolation
Let $D_n = \{(x_i, y_i) \mid i = 1, 2, ..., n ; x_1 < x_2 < \ ..... \ < x_n; y_i = f(x_i), \text{f : unknown}\}$ be a given set of n points in the plane. The objective of this exercise is to determine the quadratic spline onterpolant $S(x)$, based on D_n. For this purpose:

1. Write a MATLAB
   function z = QuadrSplineDerivatives(x,y)
   that takes as input a set of 2 vectors $x = [x_1, x_2, ..., x_n]$ and $y = [y_1, y_2, ..., y_n]$ as given by D_n, and returns a vector z which components are the derivatives of the quadratic spline at all nodes of the interpolation. Select $z(1)$ arbitrarily.

2. Write a MATLAB
   function C = QuadrSplineCoefficients(x,y)
   that takes as input a set of 2 vectors x and y, finds the derivatives of the corresponding quadratic spline at all the nodes of the interpolation, and returns a matrix C of size $3 \times (n-1)$ representing the coefficients $(y_i, z_i, \frac{z_{i+1}-z_i}{x_{i+1}-x_i})$ of the quadratic spline over each subinterval $[x_i, x_{i+1}]$.

3. Write a MATLAB
   function E = EvaluateQuadrSpline(x,y,u)
   that computes the value of $S(u)$ by locating first u in the appropriate

subinterval $[x_i, x_{i+1}]$. Your function should also display an error message if $u \notin [x_1, x_n]$. (For example "The value of S(2.5) cannot be evaluated")

4. Write a MATLAB
   `function V = EvaluateQuadrSpline1(x,y,w)`
   that computes the value of the quadratic spline at each component of a given vector w of any length.

5. Write a MATLAB
   `function PlotQuadrSpline(x,y)`
   that takes as input a set of 2 vectors x and y and plots the graph of $S(x)$ over each subinterval $[x_i, x_{i+1}]$.

6. Test each one of the functions above for 2 different test cases, and save the results in a Word document.

Chapter 5

Numerical Differentiation and Integration

5.1   Introduction

As in the previous chapter, let D_n be a set of $n+1$ given points in the (x, y) plane:

$$D_n = \{(x_i, y_i) | \, 0 \le i \le n; a = x_0 < x_1 < ... < x_n = b; \, y_i = f(x_i)\}, \quad (5.1)$$

for some function $f(x)$. Based on D_n, our basic objective is to seek accurate "approximations" for:

1. $f'(x_i)$ and $f''(x_i) : i = 0, 1, ..., n$ (Numerical Differentiation),

2. $I = \int_a^b f(x)\,dx$ (Numerical Integration).

In what follows and unless stated otherwise, we shall assume that the x-data in D_n are {**equally spaced**, with:

$$h = x_{i+1} - x_i.$$

The topic of numerical differentiation and integration is usually based on the theory of Lagrange interpolation (see Chapter 4). However, it uses also some standard calculus tools such as Taylor's formula, the Intermediate Value Theorem and the Mean Value formulae (first and second). We start by a quick review of these basic results. (For references, see [4], [9], [21].)

5.2   Mathematical Prerequisites

1. **Taylor's formula**

   Let $h_0 > 0$ and $m \in \mathbb{R}$. Assume the function $f(x) \in C^{k+1}[(m - h_0, m + h_0)]$ that is, its derivatives:

$$\{f^{(j)}(x) : j = 1, ..., k, k+1\}$$

   are continuous in the interval $(m - h_0, m + h_0)$. Then for all $h < h_0 \in \mathbb{R}$, there exists $t \in (0, 1)$, such that:

$$f(m + h) = f(m) + f'(m)h + f^{(2)}(m)\frac{h^2}{2} + ... \qquad (5.2)$$

$$... + f^{(k)}(m)\frac{h^k}{k!} + f^{(k+1)}(c)\frac{h^{k+1}}{(k+1)!},$$

   with $c = m + th$. Formula (5.2) will be refered to as "Taylor's development about m" up to the kth-order, the "remainder term" being $R_k = f^{(k+1)}(c)\frac{h^{k+1}}{(k+1)!}$. Using the big-$O(.)$ notation, we abbreviate the formula as follows:

$$f(m + h) = f(m) + f'(m)h + f^{(2)}(m)\frac{h^2}{2} + ... + f^{(k)}(m)\frac{h^k}{k!} + O(h^{k+1}). \qquad (5.3)$$

   For the case where f is **analytical**, that implies continuity of derivatives up to any order k, the finite Taylor series can be transformed into an infinite convergent series, for $|h| < h_0$:

$$f(m + h) = f(m) + f'(m)h + f^{(2)}(m)\frac{h^2}{2} + ... + f^{(k)}(m)\frac{h^k}{k!} + ... \quad (5.4)$$

   Hence, we will be using subsequently each of (5.2), (5.3) or (5.4).

2. **Use of the Intermediate Value Theorem**

   Let g be a continuous function defined on $\mathbb{R}$. Then for every finite subset $\{m_1, m_2, ..., m_k\}$ of $\mathcal{D}_g$, the domain of g, then there exists a number $c \in \mathcal{D}_g$, such that:

$$\sum_{i=1}^{k} g(m_i) = kg(c). \qquad (5.5)$$

This identity is a straightforward application of the well-known "Intermediate Value Theorem, "based on the continuity of g and on the fact that:

$$\min_{x \in \mathcal{D}_g} g(x) \leq \frac{1}{k} \sum_{i=1}^{k} g(m_i) \leq \max_{x \in \mathcal{D}_g} g(x).$$

3. **Mean Value Theorems**

(a) **First Mean Value Theorem**
This theorem results from the application of Taylor's formula where the error term is expressed in terms of a first derivative, specifically:

$$f(m+h) - f(m) = hf'(c), \; c \in (m, m+h),$$

which is equivalent to:

$$\int_{m}^{m+h} f'(x)dx = f'(c)h. \tag{5.6}$$

(b) **Second Mean Value Theorem**
This one generalizes the previous one, (5.6) becoming:

$$\int_{m}^{m+h} w(x)g(x)dx = g(c) \int_{m}^{m+h} w(x)dx, \tag{5.7}$$

where $g(x)$ and $w(x)$ are continuous functions with $w(x) \geq 0$ (or $w(x) \leq 0$).

5.3 Numerical Differentiation

The basic principle in approximation of derivatives is the systematic use of divided differences as suggested by the following result.

Theorem 5.1 *Assume that the function f is k-times continuously differentiable. Then for every subset of distinct points $\{x_i, x_{i+1}, ..., x_{i+k}\}$ in the domain of f, there exists $c_k \in (x_i, x_{i+k})$, such that:*

$$[x_i, x_{i+1}, ..., x_{i+k}] = \frac{f^{(k)}(c_k)}{k!}.$$

Proof. To obtain such result, one considers the function $g(x)$ defined by:

$$g(x) = f(x) - p_{i\,i+1\,...\,i+k}(x).$$

where $p_{i\,i+1\,...\,i+k}(x)$ is the Lagrange interpolating polynomial based on the nodes $\{x_i, x_{i+1}, ..., x_{i+k}\}$. As $g(x) = 0$ at all these nodes, then according to Rolles' theorem and the regularity assumptions on $f(x)$, one concludes the existence of "intermediate points": $\{x_{i+j}^1 \,|\, j = 0, 1, ..., k-1\}$, for which $g'(x) = 0$. Repeating the argument k times, one reaches one "last intermediate point" $c_k = x_i^k$, such that $g^{(k)}(c_k) = 0$. Since according to Newton's formula:

$$p_{i\,i+1\,...\,i+k}^{(k)}(x) = k![x_i, x_{i+1}, ..., x_{i+k}],$$

then:

$$g^{(k)}(x) = f^{(k)}(x) - k![x_i, x_{i+1}, ..., x_{i+k}].$$

Setting in this last equation $x = c_k$, yields the result of the theorem. ∎

Based on the set of points (5.1), Divided Differences appear to provide efficient "discrete" tools to approximate derivatives. Specifically, for $0 < l \le k$, we approximate $f^{(l)}(x_j)$, for $j = i, i + 1, ..., i + k$ by $l! \times$ (some appropriate Divided Difference of order l). Specifically:

$$f^{(l)}(x_j) \approx l![x_{i_0}, x_{i_1}, ..., x_{i_l}],$$

for distinct indices $i_m \in \{i, i+1, ..., i+k\}$. In what follows, we will only handle the cases of first and second derivatives, i.e., $l = 1, 2$.

5.3.1   Approximation of First Derivatives: Error Analysis

Theorem 5.1 suggests the following approximation formulae for first order derivatives:

$$f'(x_i) \approx \begin{cases} [x_i, x_{i+1}] = \frac{y_{i+1}-y_i}{h} = \frac{\Delta_h y_i}{h} & (5.8.1) \\ [x_{i-1}, x_i] = \frac{y_i-y_{i-1}}{h} = \frac{\nabla_h y_i}{h} & (5.8.2) \\ [x_{i-1}, x_{i+1}] = \frac{y_{i+1}-y_{i-1}}{2h} = \frac{\delta_h y_i}{2h} & (5.8.3) \end{cases} \qquad (5.8)$$

These approximations to the first derivative are successively:

- the Forward Divided Difference approximation (5.8.1)

- the Backward Divided Difference approximation (5.8.2)

- the Central Divided Difference approximation (5.8.3)

Obviously, the Forward approximation formula (5.8.1) for the derivative is particularly suitable when computing $f'(x_0)$, while (5.8.2) would be used when approximating $f'(x_n)$. The Central Divided Difference (5.8.3) is suitable for approximating $f'(x_i)$ for all $i = 1, ..., n - 1$.

Error Analysis and Order of the Methods
Let h be a positive number, such that $0 < h \le 1$. We analyze the error estimate in each of the above three approximations.

- Forward Difference approximation:
  Using Taylor's formula up to order 1, we can write:

  $$f(x + h) = f(x) + hf'(x) + \frac{1}{2}h^2 f''(c) \tag{5.9}$$

  where c is in the interval $(x, x + h)$, and which leads to:

  $$f'(x) = \frac{1}{h}[f(x + h) - f(x)] - \frac{1}{2}hf^{(2)}(c)$$

  Hence the approximation:

  $$f'(x) \approx \frac{1}{h}[f(x + h) - f(x)] = \frac{\Delta_h f(x)}{h}$$

  is the first order **Forward Difference** approximation (5.8.1) to $f'(x)$, with a truncation error term $E = -\frac{1}{2}f''(c)h$ of order h. We write then: $E = O(h)$.

- Backward Difference approximation:
  Likewise, replacing h by $(-h)$, equation (5.9) implies then:

  $$f(x - h) = f(x) - hf'(x) + \frac{1}{2}h^2 f^{(2)}(c') \tag{5.10}$$

  where c is in the interval $(x - h, x)$, leading to:

  $$f'(x) = \frac{1}{h}[f(x) - f(x - h)] + \frac{1}{2}hf^{(2)}(c)$$

  Hence the approximation:

  $$f'(x) \approx \frac{1}{h}[f(x) - f(x - h)] = \frac{\nabla_h f(x)}{h}$$

  is the first order **Backward Difference** approximation (5.8.2) to $f'(x)$, and its truncation error term $E = +\frac{1}{2}hf''(c')$ is of order h.

- Central Difference approximation:
  However, it is advantageous to have the convergence of numerical processes occur with higher orders. In the present situation, we want an approximation to $f'(x)$ in which the error behaves like $O(h^2)$. One such result is easily obtained based on the Central Divided Difference approximation with the aid of Taylor's series where f is assumed to have continuous order derivatives up to order 3. Thus:

  $$f(x+h) = f(x)+hf'(x)+\frac{1}{2!}h^2 f^{(2)}(x)+\frac{1}{3!}h^3 f^{(3)}(c_1) \; ; \; x < c_1 < x+h \tag{5.11}$$

and similarly:

$$f(x-h) = f(x) - hf'(x) + \frac{1}{2!}h^2 f^{(2)}(x) - \frac{1}{3!}h^3 f^{(3)}(c_2) \ ; \ x-h < c_2 < x$$
$$(5.12)$$

By subtraction, and using the Intermediate Value Theorem, we obtain:

$$f(x+h) - f(x-h) = 2hf'(x) + \frac{2}{3!}h^3(f^{(3)}(c)) \ ; \ x-h < c < x+h \ (5.13)$$

This leads to a new approximation for $f'(x)$:

$$f'(x) = \frac{1}{2h}[f(x+h) - f(x-h)] - \frac{1}{3!}h^2 f^{(3)}(c) \qquad (5.14)$$

where the approximation

$$f'(x) \approx \frac{f(x+h) - f(x-h)}{2h} = \frac{\delta_h f(x)}{2h}$$

is the first order **Central Difference** approximation to $f'(x)$, with its truncation error $E = O(h^2)$.

Based on the formulae above, we can therefore write the first order approximations to the first derivative with their respective order of convergence as follows:

Proposition 5.1 *Let* $0 < h \le 1$. *Then*

$$f'(x) = \begin{cases} \frac{\Delta_h f(x)}{h} - f^{(2)}(c)\frac{h}{2} = \frac{\Delta_h f(x)}{h} + O(h), \ f(x) \in C^2, & (5.15.1) \\ \frac{\nabla_h f(x)}{h} + f^{(2)}(c')\frac{h}{2} = \frac{\nabla_h f(x)}{h} + O(h), \ f(x) \in C^2, & (5.15.2) \\ \frac{\delta_h f(x)}{h} - f^{(3)}(c'')\frac{h^2}{6} = \frac{\delta_h f(x)}{2h} + O(h^2), \ f(x) \in C^3, & (5.15.3) \end{cases}$$
$$(5.15)$$

where $c \in (x, x+h)$, $c' \in (x-h, x)$ and $c'' \in (x-h, x+h)$. Obviously, for the first 2 approximations it is sufficient that f be a C^2 function, while for the third one f is required to be C^3 function over its domain.

For $x = x_i$, the above formulae can be rewritten in terms of first order finite differences:

$$f'(x_i) = \begin{cases} \frac{\Delta_h f(x_i)}{h} + O(h) = [x_i, x_i + h] + O(h) \\ \frac{\nabla_h f(x_i)}{h} + O(h) = [x_i - h, x_i] + O(h) \\ \frac{\delta_h f(x_i)}{2h} + O(h^2) = [x_i - h, x_i + h] + O(h^2) \end{cases}$$

To illustrate, consider the following table of data associated with the 0 -order Bessel's function of the first kind $f(x) = J_0(x)$ and 9 equidistant points (8 intervals) where $h = 0.25$:

Example 5.1 *Based on Table 5.1, find approximations to* $J_0'(0) = 0$ *using the Forward Difference approximation formula.*

i	x_i	y_i
0	0.00	1.0000000
1	0.25	0.98443593
2	0.50	0.93846981
3	0.75	0.86424228
4	1.00	0.76519769
5	1.25	0.64590609
6	1.50	0.51182767
7	1.75	0.36903253
8	2.00	0.22389078

TABLE 5.1: Data for Bessel function $J_0(x)$, $x = 0.0\,0.25, ..., 2.00$

h	$\frac{1}{h}\Delta_h f(0)$
0.25	-0.06225628
0.50	-0.12306039
0.75	-0.18101020
1.00	-0.23480231

TABLE 5.2: Approximations for $J_0'(0) = 0$, for $h = 0.25, 0.50, 0.75, 1.00$

Applying formula (5.15.1), we obtain results of such approximations in Table 5.2.

Example 5.2 *Based on Table 5.1, find approximations to $J_0'(0.25) = -0.12402598$ using the Forward, the Backward and the Central Difference approximation formulae.*

Table 5.3 summarizes the results of such approximations.

Example 5.3 *Find approximations to $J_0'(1) = -0.44005059$ using the central difference approximation formula.*

Table 5.4 provides the results obtained by applying formula (5.15.3).

h	$\frac{1}{h}\Delta_h f(0.25)$	$\frac{1}{2h}\delta_h f(0.25)$	$\frac{1}{h}\nabla_h f(0.25)$
0.25	-0.18386449	-0.12306039	-0.06225628

TABLE 5.3: Approximations to $J_0'(0.25) = -0.12402598$ using central, backward and forward differences

h	$\frac{1}{2h}\delta_h f(1)$
0.25	-0.43667238
0.50	-0.42664214
1.00	-0.38805461

TABLE 5.4: Approximations for $J_0'(1) = -0.44005059$, using central difference formula

5.3.2 Approximation of Second Derivatives: Error Analysis

A direct application of Theorem 5.1, with $k = 2$ suggests the following approximation formulae for second order derivatives:

$$f''(x_i) \approx \begin{cases} 2[x_i, x_{i+1}, x_{i+2}] = \frac{y_{i+2} - 2y_{i+1} + y_i}{h^2} = \frac{\Delta_h^2 y_i}{h^2}; & \text{Forward difference} \\ 2[x_{i-2}, x_{i-1}, x_i] = \frac{y_i - 2y_{i-1} + y_{i-2}}{h^2} = \frac{\nabla_h^2 y_i}{h^2}; & \text{Backward difference} \\ 2[x_{i-1}, x_i, x_{i+1}] = \frac{y_{i+1} - 2y_i + y_{i-1}}{h^2} = \frac{\delta_h^2 y_i}{h^2}; & \text{Central difference} \end{cases}$$

Error Analysis and Order of the Methods

- Forward Difference approximation
  Consider the 2 Taylor's series expansions of f up to second order given by:

  (i) $f(x + h) = f(x) + \frac{h}{1!}f'(x) + \frac{h^2}{2!}f''(x) + \frac{h^3}{3!}f^{(3)}(c_1)$; $c_1 \in (x, x + h)$

  (ii) $f(x + 2h) = f(x) + \frac{(2h)}{1!}f'(x) + \frac{(2h)^2}{2!}f''(x) + \frac{(2h)^3}{3!}f^{(3)}(c_2)$; $c_2 \in (x, x + 2h)$

  where f is assumed to be a C^3-function.
  The algebraic operation: $f(x + 2h) - 2f(x + h)$ leads to the **Forward Difference** approximation to $f''(x)$, which satisfies the following:

  $$f''(x) = \frac{f(x + 2h) - 2f(x + h) + f(x)}{h^2} + O(h) = \frac{\Delta_h^2 f(x)}{h^2} + O(h) \tag{5.16}$$

- Backward Difference approximation
  Furthermore, replacing h by $-h$ in equations (i) and (ii) above, one also has:

  (iii) $f(x - h) = f(x) - \frac{h}{1!}f'(x) + \frac{h^2}{2!}f''(x) - \frac{h^3}{3!}f^{(3)}(c_3)$; $c_3 \in (x - h, x)$

  (iv) $f(x - 2h) = f(x) - \frac{(2h)}{1!}f'(x) + \frac{(2h)^2}{2!}f''(x) - \frac{(2h)^3}{3!}f^{(3)}(c_4)$; $c_4 \in (x - 2h, x+)$

  The algebraic operation: $f(x - 2h) - 2f(x - h)$ leads to the **Backward**

Divided Difference approximation to $f''(x)$. This one satisfies the following:

$$f''(x) = \frac{f(x-2h) - 2f(x-h) + f(x)}{h^2} + O(h) = \frac{\nabla_h^2 f(x)}{h^2} + O(h)$$

$$\text{(5.17)}$$

- Central Difference approximation
  In this case we start by writing Taylor's series expansions up to the third order successively for $f(x+h)$ and $f(x-h)$. This leads to:

$$f(x+h) + f(x-h) = 2f(x) + f''(x)h^2 + \frac{h^4}{4!}(f^{(4)}(c_1) + f^{(4)}(c_2))$$

Dividing by h^2 and using the Intermediate Value Theorem, one concludes that:

$$f''(x) = \frac{f(x+h) - 2f(x) + f(x-h)}{h^2} - \frac{h^2}{12}f^{(4)}(c) = \frac{\delta_h^2}{h^2}(f(x)) + O(h^2)$$

$$\text{(5.18)}$$

which is **Central Difference** approximation to $f''(x)$.
Based on the results above, the following proposition is satisfied:

Proposition 5.2 *Let $0 < h \leq 1$. Then*

$$f''(x) = \begin{cases} \frac{\Delta_h^2 f(x)}{h^2} + O(h), & f(x) \in C^3, & \text{(5.19.1)} \\ \frac{\nabla_h^2 f(x)}{h^2} + O(h), & f(x) \in C^3, & \text{(5.19.2)} \\ \frac{\delta_h^2 f(x)}{h^2} + O(h^2), & f(x) \in C^4, & \text{(5.19.3)} \end{cases} \quad \text{(5.19)}$$

where f is assumed to be a C^3 function for the first 2 approximations, and a C^4 function for the third approximation.
In terms of divided differences, the second order derivatives at $x = x_i$ satisfy the following estimates:

$$f''(x_i) = \begin{cases} \frac{\Delta_h^2 y_i}{h^2} + O(h) = 2[x_i, x_{i+1}, x_{i+2}] + O(h); & \text{Forward Difference} \\ \frac{\nabla_h^2 y_i}{h^2} + O(h) = 2[x_{i-2}, x_{i-1}, x_i] + O(h); & \text{Backward Difference} \\ \frac{\delta_h^2 y_i}{h^2} + O(h^2) = 2[x_{i-1}, x_i, x_{i+1}] + O(h^2); & \text{Central Difference} \end{cases}$$

Remark 5.1 *Note that, based on Theorem 5.1, the following approximation formulae for the third derivative of f can also be obtained:*

$$f'''(x_i) \approx \begin{cases} \frac{\Delta_h^3 y_i}{h^3} = 6[x_i, x_{i+1}, x_{i+2}, x_{i+3}]; & \text{Forward Difference} \\ \frac{\nabla_h^3 y_i}{h^3} = 6[x_{i-3}, x_{i-2}, x_{i-1}, x_i]; & \text{Backward Difference} \\ \frac{\delta_h^3 y_i}{h^3} = 6[x_{i-2}, x_{i-1}, x_{i+1}, x_{i+2}]; & \text{Central Difference} \end{cases}$$

To improve accuracy on the basis of the formulae obtained for first and second derivatives, we turn to the subtle tool of Richardson extrapolation.

5.4   Richardson Extrapolation

In order to obtain higher order approximations to a target quantity Q, it is possible to use a powerful technique known as **Richardson Extrapolation**. Such technique is a powerful tool in numerical computing. Its purpose is to accelerate convergence to Q of sequences $\{Q(h)\}$ when $h \to 0$, without a need to consider too small values of h (or equivalently too large values of n as introduced in (5.1)). Specifically, it assumes an a-priori knowledge of the behavior of the error in the case where one is approximating the quantity Q by $Q(h)$, whereas $\lim_{h \to 0} Q(h) = Q$, and:

$$Q = Q(h) + c_1 h^\alpha + O(h^\beta), \ \beta > \alpha, \ \text{(a-priori estimate)} \qquad (5.20)$$

where c_1 is independent from h. An improved Richardson formula can then be derived based on the two approximations $Q(h)$ and $Q(h/2)$. For that purpose, we rewrite (5.20) with h replaced by $h/2$. This leads to:

$$Q = Q(h/2) + c_1 (\frac{h}{2})^\alpha + O(h^\beta) \qquad (5.21)$$

By considering the algebraic combination:

$$2^\alpha \times \text{Equation (5.21)} - \text{Equation (5.20)},$$

one obtains:

$$(2^\alpha - 1)Q = 2^\alpha Q(h/2) - Q(h) + O(h^\beta).$$

Such equation is equivalent to:

$$Q = [\frac{2^\alpha Q(h/2) - Q(h)}{2^\alpha - 1}] + O(h^\beta).$$

hence leading to $Q^1(h/2)$, a first-order Richardson approximation to Q, verifying:

$$\begin{cases} Q^1(h/2) = \frac{2^\alpha Q(h/2) - Q(h)}{2^\alpha - 1}, & (5.22.1) \\ Q = Q^1(h/2) + O(h^\beta), & (5.22.2) \end{cases} \qquad (5.22)$$

Therefore, by using simple algebra and eliminating (or "killing") the most dominant term in the error expression of $Q - Q(h)$, one reaches a more accurate approximation $Q^1(h/2)$ as defined in (5.22.1) and satisfying (5.22.2). Equivalently, (5.22) is also written as follows:

$$\begin{cases} Q^1(h) = \frac{2^\alpha Q(h) - Q(2h)}{2^\alpha - 1}, & (5.23.1) \\ Q = Q^1(h) + O(h^\beta), & (5.23.2) \end{cases} \qquad (5.23)$$

where it is understood that h represents the **"last value"** reached by that

h	$Q(h)$	$Q^1(h)$	$Q^2(h)$
h_0	$Q(h_0)$	.	
$h_0/2$	$Q(h_0/2)$	$Q^1(h/2)$	.
$h_0/4$	$Q(h_0/4)$	$Q^1(h/4)$	$Q^2(h/4)$

TABLE 5.5: Description of a Richardson's process for $Q = Q(h) + c_1 h^\alpha + O(h^\beta) + ...$

parameter.

Obviously, in case an a-priori knowledge is given also on $Q - Q^1(h)$, such as:

$$Q - Q^1(h) = c_1' h^\beta + O(h^\gamma), \text{ with } \gamma > \beta,$$

then a second Richardson extrapolation can be carried out. Specifically if we let:

$$Q^2(h/2) = \frac{2^\beta Q^1(h/2) - Q^1(h)}{2^\beta - 1},$$

then we show that:

$$Q - Q^2(h) = O(h^\gamma).$$

Such formula is supposed to provide better approximations to Q than would $Q(.)$ and $Q^1(.)$.

Remark 5.2 *When dealing with Richardson extrapolation, one starts by computing a first set of values of $Q(h)$, for $h = h_0, h_0/2, h_0/4, ...$*

Although theoretically the values

$$\{Q^k(h) \mid k \geq 0 \in N, \text{ and } h \to 0\}$$

get closer to Q as k increases or h decreases to 0, one observes that in practice, **due to the propagation of rounding errors**, these computed values tend to become less reliable. Henceforth a threshold $h_m = \frac{h_0}{2^m}$ for h can be reached, whereas all calculated values for $h < h_m$ are to be rejected, keeping only:

$$\{Q^k(h) \mid h = h_0, h_0/2, ..., h_0/2^m\}.$$

Given that fact, Richardson extrapolations would result using this last set of valid data. This is indicated in Table 5.5, for the case where $m = 2$.

Remark 5.3 *Note also that one can carry a Richardson extrapolation without necessarily dividing h by 2, but more generally by a factor $q > 1$.*

In such case (5.21) becomes:

$$Q = Q(h/q) + c_1 \left(\frac{h}{q}\right)^\alpha + O(h^\beta). \tag{5.24}$$

Thus:
$$q^\alpha \times \text{Equation (5.24)} - \text{Equation (5.20)}$$

yields a first order extrapolation formula:

$$Q^1(h/q) = \frac{q^\alpha Q(h/q) - Q(h)}{q^\alpha - 1} \text{ with } Q = Q^1(h/q) + O(h^\beta). \qquad (5.25)$$

5.5 Richardson Extrapolation in Numerical Differentiation

We start by illustrating this process on the approximation formulae obtained for the first and second derivatives in Section 5.3.

5.5.1 Richardson Extrapolation for First Derivatives

Forward and Backward Differences

Recall that for a function $f \in C^\infty$, the infinite Taylor's series expansion formula of $f(x+h)$ is as follows:

$$f(x+h) = f(x) + hf'(x) + \frac{h^2}{2!}f^{(2)}(x) + \frac{h^3}{3!}f^{(3)}(x) + \dots$$

leading to:

$$f'(x) = \frac{\Delta_h f(x)}{h} + a_1 h + a_2 h^2 + a_3 h^3 + \dots \qquad (5.26)$$

where the $\{a_i\}$'s are constants that are independent of h and depend on the derivatives of f at x.

Considering that $Q = f'(x)$ is the quantity to be approximated, let now:

$$Q(h) = \phi_h(f(x)) = \frac{\Delta_h(f(x))}{h} \qquad (5.27)$$

Considering successively h then $h/2$ in (5.26), one has:

(5.26.a) $f'(x) = \phi_h(f(x)) + a_1 h + a_2 h^2 + a_3 h^3 + \dots$

(5.26.b) $f'(x) = \phi_{h/2}(f(x)) + a_1 h/2 + a_2 (h/2)^2 + a_3 (h/2)^3 + \dots$

The algebraic operation $2 \times (5.26.b) - (5.26.a)$ yields then:

$$f'(x) = [2\phi_{h/2}(f(x)) - \phi_h(f(x))] + (a_2/2)h^2 + O(h^3).$$

Introducing the first-order Forward Richardson extrapolation operator, let:

$$\phi^1_{h/2}(f(x)) = 2\phi_{h/2}(f(x)) - \phi_h(f(x)) \qquad (5.28)$$

One obtains as clarified in (5.23):

$$f'(x) = \begin{cases} \phi_h^1(f(x)) + a_2'h^2 + a_3'h^3 + \dots \\ \phi_h^1(f(x)) + O(h^2) \end{cases} \tag{5.29}$$

with the constants a_2', a_3',..., independent of h.

The process can be further continued, i.e., one can consider **second-order Richardson extrapolations**. From equation (5.29), one has simultaneously:

(5.29.a) $f'(x) = \phi_h^1(f(x)) + a_2'h^2 + a_3'h^3 + \dots$

(5.29.b) $f'(x) = \phi_{h/2}^1(f(x)) + a_2'(h/2)^2 + a_3'(h/2)^3 + \dots$

The algebraic operation $4 \times (5.29.b) - (5.29.a)$ eliminates the most dominant term in the error series and yields:

$$f'(x) = [\frac{4\phi_{h/2}^1(f(x)) - \phi_h^1(f(x))}{3}] - \frac{1}{2}a_3'h^3 + O(h^4)$$

Introducing the second-order Richardson extrapolation operator, let

$$\phi_{h/2}^2(f(x)) = \frac{4\phi_{h/2}^1(f(x)) - \phi_h^1(f(x))}{3} \tag{5.30}$$

One obtains:

$$f'(x) = \begin{cases} \phi_h^2(f(x)) - \frac{1}{2}a_3'h^3 + \dots \\ \phi_h^2(f(x)) + O(h^3) \end{cases} \tag{5.31}$$

This is yet another improvement with a precision of $O(h^3)$, i.e., $\phi_{h/2}^2(f(x))$ provides a third order approximation to $f'(x)$.

The successive Richardson extrapolation formulae and error estimates are then as follows:

$$f'(x) = \begin{cases} \phi_h(f(x)) + O(h) \\ \phi_h^1(f(x)) + O(h^2) \\ \phi_h^2(f(x)) + O(h^3) \\ \phi_h^3(f(x)) + O(h^4) \\ \dots\dots \\ \phi_h^k(f(x)) + O(h^{k+1}) \end{cases} \tag{5.32}$$

where $\phi_h(.) = \frac{\Delta_h(.)}{h}$ and:

$$phi_h^1 = \frac{2^1\phi_h(.) - \phi_{2h}(.)}{2^1 - 1}; \quad \phi_h^2(.) = \frac{2^2\phi_h^1(.) - \phi_{2h}^1(.)}{2^2 - 1}; \quad \phi_h^3(.) = \frac{2^3\phi_h^2(.) - \phi_{2h}^2(.)}{2^3 - 1}.$$

The kth-**order Forward Richardson operator** being defined as follows:

$$\phi_h^k(.) = \frac{2^k\phi_h^{k-1}(.) - \phi_{2h}^{k-1}(.)}{2^k - 1}$$

with the error term of $O(h^{k+1})$.

h	$\phi_h = \frac{1}{h}\Delta_h(f(0))$	$\phi_h^1(f(0))$	$\phi_h^2(f(0))$
1.00	-0.23480231	·	·
0.50	-0.12306039	0.06694897	·
0.25	-0.06225628	-0.00145217	-0.02425255

TABLE 5.6: Refined approximations to $J_0'(0)$ using Richardson's extrapolation

Example 5.4 *On the basis of Table 5.1, find improvements to Forward Difference approximations to $f'(x)$, using Richardson extrapolation operators of the 1st and second order.*

We apply (5.28) and (5.30) yielding the results in Table 5.6. The following MATLAB algorithm is based on the Forward difference scheme. It approximates the 1st order derivative $f'(a)$ by k successive applications of Richardson process.

Algorithm 5.1 Implementation of Richardson Extrapolation for Forward Difference Formula to First Derivative

```
function D = Richardson(f,k,h,a)
% Inputs:
% k: order of extrapolation algorithm, k>=2
% f: function for which one approximates f'(a)
% h: 0 < h <=1, mesh value
% Output: D approximation of f'(a)
A = zeros(k+1,k+2) ;
% Matrix A stores
%    1st column: values of h;
%    2nd column: first forward differences
%    columns 3 to k+2: Richardson's extrapolations
for i=1:k+1
      A(i,1)=h/ 2^(i-1) ;
      A(i,2)=(f(a + A(i,1))-f(a))/A(i,1)   ;
end
for j=3: k+2
        it=j-2 ;
        for i=j-1: k+1
              A(i,j)=(2^it*A(i,j-1)-A(i-1,j-1))/(2^it-1) ;
        end
end
D = A(k+1,k+2) ;
```

Note that we can also derive Richardson extrapolation formulae based on the Backward difference approximation to $f'(x)$ as in (5.8.2), i.e., starting with

$$f'(x) = \frac{\nabla_h(f(x))}{h} + b_1 h + b_2 h^2 + \dots$$

where the $\{b_i\}$ are constants independent of h. We let then:

$$Q(h) = \chi_h(.) = \frac{\nabla_h(.)}{h} \tag{5.33}$$

It is easy to verify that the successive **Backward Difference Richardson operators** satisfy the following estimates:

$$f'(x) = \begin{cases} \chi_h(f(x)) + O(h) \\ \chi_h^1(f(x)) + O(h^2) \\ \chi_h^2(f(x)) + O(h^3) \\ \dots\dots \\ \chi_h^k(f(x)) + O(h^{k+1}) \end{cases} \tag{5.34}$$

where: $\chi_h(.) = \frac{\nabla_h(.)}{h}$ and $\chi_h^k(.) = \frac{2^k \chi_h^{k-1}(.) - \chi_{2h}^{k-1}(.)}{2^k - 1}$ with the error term of $O(h^{k+1})$.

Central Difference

As derived in Section 5.3.1, if the function $f \in C^\infty$, the Central Difference approximation to $f'(x)$ satisfies the following equation:

$$f'(x) = \frac{\delta_h(f(x))}{2h} + d_1 h^2 + d_2 h^4 + \dots$$

With such information, it is possible to rely again on Richardson extrapolation to bring more accuracy out of the method in the approximation formulae of $f'(x)$. Specifically, letting now:

$$Q(h) = \psi_h(.) = \frac{\delta_h(.)}{2h}$$

obviously, then:

$$f'(x) = \psi_h(f(x)) + d_1 h^2 + d_2 h^4 + \dots = \psi_h(f(x)) + O(h^2) \tag{5.35}$$

Taking successively h then $h/2$ in the equation above, one has:

(5.35.a) $f'(x) = \psi_h(f(x)) + d_1 h^2 + d_2 h^4 + \dots$

(5.35.b) $f'(x) = \psi_{h/2}(f(x)) + d_1 (h/2)^2 + d_2 (h/2)^4 + \dots$

The algebraic operation $4 \times (5.35.b) - (5.35.a)$ yields:

$$f'(x) = [\frac{4\psi_{h/2}(f(x)) - \psi_{2h}(f(x))}{3}] + O(h^4)$$

Let the **first-order Richardson extrapolation** operator be defined by

$$\psi^1_{h/2}(.) = \frac{2^2\psi_{h/2}(.) - \psi_h(.)}{2^2 - 1}$$

One can write then:

$$f'(x) = \psi^1_h(f(x)) + O(h^4) \tag{5.36}$$

Reapplying the same process on this result leads therefore to the following identities:

(5.36.a) $f'(x) = \psi^1_h(f(x)) + d'_2 h^4 + d'_3 h^6 + ...$

(5.36.b) $f'(x) = \psi^1_{h/2}(f(x)) + d'_2 f^{(5)}(x)(h/2)^4 + d'_3(h/2)^6 + ...$

The algebraic operation $16 \times (5.36.\text{b}) - (5.36.\text{a})$ yields:

$$f'(x) = \frac{2^4\psi^1_{h/2}(f(x)) - \psi^1_h(f(x))}{2^4 - 1} + O(h^6)$$

or equivalently:

$$f'(x) = \psi^2_h(f(x)) + O(h^6).$$

Therefore, the first **Central Difference Richardson extrapolation formulae** obtained are as follows:

$$f'(x) = \begin{cases} \psi_h(f(x)) + O(h^2) \\ \psi^1_h(f(x)) + O(h^4) \\ \psi^2_h(f(x)) + O(h^6) \\ \quad ...... \\ \psi^k_h(f(x)) + O(h^{2k+2}) \end{cases} \tag{5.37}$$

where

$$\psi_h(.) = \frac{\delta_h(.)}{2h}, \quad \psi^1_h(.) = \frac{2^2\psi_h(.) - \psi_{2h}(.)}{2^2 - 1}, \quad \psi^2_h(.) = \frac{2^4\psi^1_h(.) - \psi^1_{2h}(.)}{2^4 - 1}$$

with the kth-order operator defined as follows:

$$\psi^k_h(.) = \frac{2^{2k}\psi^{k-1}_h(.) - \psi^{k-1}_{2h}(.)}{2^{2k} - 1}$$

where the error term is $O(h^{2k+2})$

5.5.2   Richardson Extrapolation for Second Derivatives

Consider now some function $f \in C^\infty$. In order to improve the accuracy of the approximations to the second derivative $f''(x)$, we also rely on Richardson extrapolation process that could be applied successively to the Forward,

Backward and Central Difference formulae.

In this section, we analyze briefly the **Richardson extrapolation central difference approximations** to $f''(x)$, as the steps are similar to those of the first derivative detailed in 5.5.1 above.

Starting by adding the infinite Taylor's series expansions for $f(x+h)$ and $f(x-h)$ and based on (5.18), let now:

$$Q = f''(x) \approx \frac{f(x+h) - 2f(x) + f(x-h)}{h^2} = \frac{\delta_h^2 f(x)}{h^2} \qquad (5.38)$$

with:

$$Q(h) = \psi_h(.) = \frac{\delta_h^2(.)}{h^2}$$

It is easy to verify that:

$$f''(x) = \psi_h(f(x)) + d_1 h^2 + d_2 h^4 + \dots \qquad (5.39)$$

leading then successively to the following estimates equivalent to (5.37):

$$f''(x) = \begin{cases} \psi_h(f(x)) + O(h^2) \\ \psi_h^1(f(x)) + O(h^4) \\ \psi_h^2(f(x)) + O(h^6) \\ \quad \dots \dots \\ \psi_h^k(f(x)) + O(h^{2k+2}) \end{cases} \qquad (5.40)$$

where:

$$\psi_h(.) = \frac{\delta_h^2(.)}{2h}; \quad \psi_h^1(.) = \frac{2^2 \psi_h(.) - \psi_{2h}(.)}{2^2 - 1}; \quad \psi_h^2(.) = \frac{2^4 \psi_h^1(.) - \psi_{2h}^1(.)}{2^4 - 1}$$

with the kth-order operator defined as follows:

$$\psi_h^k(.) = \frac{2^{2k} \psi_h^{k-1}(.) - \psi_{2h}^{k-1}(.)}{2^{2k} - 1}$$

where the error term is $O(h^{2k+2})$

5.6   Numerical Integration

Based on the data (5.1)

$$D_n = \{(x_i, y_i) \mid 0 \le i \le n; a = x_0 < x_1 < \dots < x_n = b; y_i = f(x_i)\}$$

we consider the approximation of

$$I = I(a, b; f) = \int_a^b f(x)dx.$$

Unlike numerical differentiation, the $\{x_i\}$ need not be equidistant. However, unless stated otherwise, we shall assume to start with:

1. Equidistance of nodes, i.e., $h = x_{i+1} - x_i \, \forall i$, with $nh = b - a$.

2. Continuity of the function f over the interval of integration, i.e., $f \in C(a, b)$.

To derive all numerical integration formulae in this chapter, we proceed systematically by decomposing first the integral I into the sum of simple integrals. Specifically, one has $\forall n$ (even or odd):

$$I = \int_{x_0}^{x_1} f(x)dx + \int_{x_1}^{x_2} f(x)dx + ... + \int_{x_{n-1}}^{x_n} f(x)dx = \sum_{k=0}^{n-1} \int_{x_k}^{x_{k+1}} f(x)dx$$

and in particular when n is even, i.e., $n = 2m$:

$$I = \int_{x_0}^{x_2} f(x)dx + \int_{x_2}^{x_4} f(x)dx + ... + \int_{x_{2m-2}}^{x_{2m}} f(x)dx = \sum_{k=0}^{m-1} \int_{x_{2k}}^{x_{2k+2}} f(x)dx$$

Thus, we will be dealing with 2 types of formulae:

1. **Simple Numerical integration** formulae

$$I_k = \int_{x_k}^{x_{k+1}} f(x)dx \, \forall \, n \text{ , or } I_k' = \int_{x_{2k}}^{x_{2k+2}} f(x)dx \, \forall \, n = 2m$$

   Subsequently, we derive:

2. **Composite Numerical integration** formulae

$$I = \int_a^b f(x)dx = \sum_{k=0}^{n-1} I_k \, \forall \, n \text{ , or } I = \int_a^b f(x)dx = \sum_{k=0}^{m-1} I_k' \, , \forall \, n = 2m$$

Since a definite integral is usually defined as a limit of a Riemann sum, and more explicitly a sum of signed areas of rectangles, it is therefore natural to assume that any summation of the form:

$$C_n = \sum_{k=0}^{n-1} hf(c_k), \, x_k \le c_k \le x_{k+1},$$

could approximate I. The simplest choice for the sequence $\{c_k\}$ is one of the following, leading to the **rectangular rules**.

5.6.1 The Rectangular Rules

The **rectangular rules** can be used for all positive integer values of n.

The Formulae

1. The left rectangular rule: for $c_k = x_k$, let $A_k = hf(x_k)$, then:

$$A(h) = \sum_{k=0}^{n-1} A_k = \sum_{k=0}^{n-1} hf(x_k) = h \sum_{k=0}^{n-1} y_k, \qquad (5.41)$$

2. The right rectangular rule: for $c_k = x_{k+1}$, let $B_k = hf(x_{k+1})$, then:

$$B(h) = \sum_{i=0}^{n-1} B_k = \sum_{i=0}^{n-1} hf(x_{k+1}) = h \sum_{k=0}^{n-1} y_{k+1}, \qquad (5.42)$$

Error Analysis
It can be easily shown that such formulae provide $O(h)$ approximations, in the sense that:

$$f \in C^1(a, b) : |I - A_n| = O(h) \text{ and } |I - B_n| = O(h).$$

More specifically, through integration by parts formulae, one easily shows that:

$$I_k = A_k + \int_{x_k}^{x_{k+1}} (x_{k+1} - t)f'(t)dt \qquad (5.43)$$

and similarly that:

$$I_k = B_k + \int_{x_k}^{x_{k+1}} (x_k - t)f'(t)dt \qquad (5.44)$$

leading therefore to the following error estimates:

Proposition 5.3 *For $f \in C^1$, the simple and composite rectangular rules satisfy:*

- $I_k - A_k = \frac{h^2}{2}f'(c_k)$, $c_k \in (x_k, x_{k+1})$ *and* $I - A(h) = \frac{(b-a)h}{2}f'(c)$, $c \in (a, b)$.

- $I_k - B_k = -\frac{h^2}{2}f'(d_k)$, $d_k \in (x_k, x_{k+1})$ *and* $I - B(h) = \frac{(b-a)h}{2}f'(d)$, $d \in (a, b)$.

Proof. The results follow from first using the second mean-value theorem on both $\int_{x_k}^{x_{k+1}} (x_{k+1} - t)f'(t)dt$ and $\int_{x_k}^{x_{k+1}} (x_k - t)f'(t)dt$ and applying subsequently the intermediate value theorem when evaluating:

$$I - A(h) = \sum_{k=1}^{n} (I_k - A_k) \text{ and } I - B(h) = \sum_{k=1}^{n} (I_k - B_k).$$

∎

5.6.2   The Trapezoidal Rule

The **trapezoid rule** can be used for all positive integer values of n.

The Formulae
A simple geometric argument consists in approximating the surface between the x-axis, the curve $y = f(x)$ and the vertical lines $x = x_k$ and $x = x_{k+1}$ by the area of the rectangular trapezoid which vertices are $(x_k, 0)$, $(x_{k+1}, 0)$, $(x_k, f(x_k))$ and $(x_{k+1}, f(x_{k+1}))$. This leads first to the **simple trapezoidal rule**, given by:

$$g = \int_{x_k}^{x_{k+1}} f(x)dx \approx T_k = \frac{h}{2}(f(x_k) + f(x_{k+1})), \qquad (5.45)$$

and subsequently to the **composite trapezoid rule** given by:

$$I \equiv I(a, b) = \int_a^b f(x)dx \approx T(h) = \frac{h}{2} \Sigma_{k=0}^{n-1} T_k, \qquad (5.46)$$

More precisely:

$$T(h) = \Sigma_{k=0}^{n-1}(f(x_k) + f(x_{k+1})) = \frac{h}{2}\left(y_0 + 2(y_1 + ... + y_{n-1}) + y_n\right).$$

Error Analysis
Note that:

$$T_k = \int_{x_k}^{x_{k+1}} p_{k,k+1}(x)dx \qquad (5.47)$$

where $p_{k,k+1}(x) = y_k + [x_k, x_{k+1}](x - x_k)$, is the linear interpolating polynomial to $f(x)$ at x_k and x_{k+1}. Furthermore, it is well known (Section 4.5) that:

$$f(x) = p_{k,k+1}(x) + \frac{1}{2}(x - x_k)(x - x_{k+1})f''(c(x)),$$

with $c(x) \in (x_k, x_{k+1})$ depending continuously on x. By integration of this identity over the interval (x_k, x_{k+1}), one has:

$$I_k = T_k + \frac{1}{2}\int_{x_k}^{x_{k+1}}(x - x_k)(x - x_{k+1})f''(c(x)).dx \qquad (5.48)$$

using then the second Mean Value Theorem, one gets:

$$I_k = T_k + \frac{f''(c_k)}{2}\int_{x_k}^{x_{k+1}}(x - x_k)(x - x_{k+1})\, dx$$

leading to:

$$I_k = T_k - \frac{h^3}{12}f''(c_k), \qquad (5.49)$$

where $c_k \in (x_k, x_{k+1})$. Turning up now to the composite trapezoid rule: by summing up (5.49) over k and use of the intermediate value theorem, one gets then an expression for the error term as follows:

$$I = I(a, b) = \sum_{k=0}^{n-1} I_k = T(h) - \frac{(b-a)}{12} f''(c) h^2 \tag{5.50}$$

where $c \in (a, b)$.

Proposition 5.4 *Let the data* $D_n = \{(x_k, f(x_k)) | k = 0, 1, ..., n\}$, *be a set representing a function* f *in* $C^2([a, b])$, *then:*

$$I = \int_a^b f(x) dx = T(h) + O(h^2),$$

with:

$$T(h) = \frac{h}{2} \sum_{k=0}^{n-1} (f(x_k) + f(x_{k+1})) = h[\frac{(y_0 + y_n)}{2} + \sum_{k=1}^{n-1} y_k].$$

Remark 5.4 *Note that the error analysis on* $I_k - T_k$ *and* $I - T(h)$ *can be also done through thethe the two rectangular rules introduced in (5.41) and (5.42).*

More specifically since:

$$T_k = \frac{A_k + B_k}{2}.$$

by averaging (5.43) and (5.44) one reaches:

$$I_k = \frac{A_k + B_k}{2} + \int_{x_k}^{x_{k+1}} (m_k - t) f'(t) dt,$$

with $m_k = \frac{x_k + x_{k+1}}{2}$. Equivalently, one has:

$$I_k = T_k + \int_{x_k}^{x_{k+1}} (m_k - t) f'(t) dt.$$

Assuming $f \in C^2$, as in the above error analysis for the trapezoidal rule, we show, through integration by parts, that the last identity yields:

$$I_k = T_k + \frac{1}{2} \int_{x_k}^{x_{k+1}} ((m_k - t)^2 - \frac{h^2}{4}) f''(t) dt,$$

which can be rewritten in the same form as (5.48):

$$I_k = T_k + \frac{1}{2} \int_{x_k}^{x_{k+1}} (x_k - t)(x_{k+1} - t) f''(t) dt. \tag{5.51}$$

Use of the second mean value theorem yields the same result as (5.49). ∎

The MATLAB code of the composite trapezoid rule is as follows:

i	x_i	$f(x_i)$
0	0.00	1.0000000
1	0.25	0.98443593
2	0.50	0.93846981
3	0.75	0.86424228
4	1.00	0.76519769
5	1.25	0.64590609
6	1.50	0.51182767
7	1.75	0.36903253
8	2.00	0.22389078

TABLE 5.7: A copy of data for the function $J_0(x)$, $x = 0.00, 0.25, ..., 2.00$

Algorithm 5.2 Composite Trapezoid Rule
```
function I = CompositeTrapezoid(x,y)
% Input x = [a=x(1),...,x(n+1)=b] and y =[y(1),...,y(n+1)]
% where y represents the (n+1) values of a function f(x) at (n+1)
distinct points
% The x data is assumed equidistant
n = length(y) - 1 ; a=x(1) ; b=x(n+1) ; h=(b-a)/n;
I= h*(y(1) + y(n+1))/2 ;
Y = y(2:n);
I = I + 2*h*sum(Y);
```

5.6.3   The Midpoint Rectangular Rule

Such rule applies only in the case when the number of subintervals is even, that is when $\underline{n = 2m}$.

The Formulae
A simple geometric argument consists in considering the simple integral $I'_k = \int_{x_{2k}}^{x_{2k+2}} f(x)\,dx$, as being the area of the region between the x- axis, the curve $y = f(x)$ and the vertical lines $x = x_{2k}$ and $x = x_{2k+2}$. Such area is then approximated by the surface of the rectangle which vertical sides are $x = x_{2k}$ and $x = x_{2k+2}$, and horizontal sides $y = 0$ and $y = f(x_{2k+1})$. In such case, the function values at the midpoints are known. For example, we consider the case of the data in Table 5.1 which we reproduce for simplicity of reading in Table 5.7. The set of midpoints is $\{x_1, x_3, x_5, x_7\}$. This leads first to the **simple midpoint rectangular rule**, given by:

$$I'_k = \int_{x_{2k}}^{x_{2k+2}} f(x)dx \approx M_k = 2hf(x_{2k+1}), k = 0, 1, ..., m - 1. \quad (5.52)$$

and subsequently to the **composite midpoint rule** given by:

$$I \equiv I(a, b) = \int_a^b f(x)dx \approx M(h) = \Sigma_{k=0}^{m-1} 2hf(x_{2k+1}) \quad (5.53)$$

Error Analysis

The error analysis of this method is based on either one of Taylor's formulae where the expansion is made about the point $x = x_{2k+1}$, yielding when the function f is at least in $C^2[a,b]$:

$$f(x) = f(x_{2k+1}) + f'(x_{2k+1})(x - x_{2k+1}) + f''(c_k(x))\frac{(x - x_{2k+1})^2}{2} \quad (5.54)$$

where $c_k(x) = x_{2k+1} + t(x - x_{2k+1})$, $0 < t < 1$. Integration of equation (5.54) from x_{2k} to x_{2k+2} and the use of the second mean value theorem leads to:

$$\int_{x_{2k}}^{x_{2k+2}} f(x)dx = 2hf(x_{2k+1}) + f''(c_k)\int_{x_{2k}}^{x_{2k+2}} \frac{(x - x_{2k+1})^2}{2}dx \quad (5.55)$$

where c_k is a point in (x_{2k}, x_{2k+2}). Hence:

$$I'_k = M_k + f''(c_k)\frac{h^3}{3} \quad (5.56)$$

Summing up (5.56) over k yields:

$$I(a,b) = \sum_{k=0}^{m-1} I'_k = \sum_{k=0}^{m-1} M_k + \frac{h^3}{3}\sum_{k=0}^{m-1} f''(c_k) = M(h) + \frac{h^3}{3}\sum_{k=0}^{m-1} f''(c_k).$$

Using the intermediate value theorem, one has:

$$\sum_{k=0}^{m-1} f''(c_k) = mf''(d) = \frac{b-a}{2h}f''(d), \ d \in (a,b)$$

and therefore, noting that the length of the interval of integration is

$$nh = (2m)h = b - a$$

the following result is reached:

$$I = I(a,b) = M(h) + \frac{(b-a)}{6}f''(d)h^2 \quad (5.57)$$

Proposition 5.5 *Let f be a function in $C^2[a,b]$, interpolating the set of data D_n where $n = 2m$. Then*

$$I = \int_a^b f(x)dx = M(h) + O(h^2)$$

where

$$M(h) = 2h\sum_{k=0}^{m-1} f(x_{2k+1})$$

The `MATLAB` code of the composite midpoint rule is as follows:

Algorithm 5.3 Midpoint Rule

```
function I = CompositeMidpoint(x, y)
% Inputs:
%     x = [x(1),...,x(n+1)] with x(1)=a, x(n+1)=b
%     y =[y(1),...,y(n+1)]
% Output: I = Approximation using composite midpoint rule
n=length(x)-1; m= length(y) - 1; a=x(1); b=x(n+1);
if n==m
    h= (b-a)/n ;
    % Test that n is an even integer
    if floor(n/2)  == ceil(n/2)
        Y = y(2:2:n);
        I = 2*h*sum(Y);
    end
end
```

5.6.4   Recurrence Relation for Trapezoid Rule

We prove now the following result.

Proposition 5.6 *For* $n = 2m$, $T(h) = \frac{1}{2}(T(2h) + M(h))$.

Proof. We start by writing:

$$I \approx T(2h) = \Sigma_{k=0}^{m-1} T_k' = \Sigma_{k=0}^{m-1} h(f(x_{2k}) + f(x_{2k+2}))$$

On the other hand:

$$M(h) = \Sigma_{k=0}^{m-1} M_k = 2h\Sigma_{k=0}^{m-1} f(x_{2k+1})$$

To prove the recurrence relation, note that:

$$T(2h) + M(h) = h \sum_{k=0}^{m-1} (f(x_{2k}) + f(x_{2k+2})) + 2f(x_{2k+1})$$

$$= h \sum_{k=0}^{m-1} [f(x_{2k}) + f(x_{2k+1})] + [f(x_{2k+1}) + f(x_{2k+2})]$$

$$= h \sum_{k=0}^{n-1} (f(x_k) + f(x_{k+1})) = 2T(h).$$

■

This directly leads to the required result, that is:

$$T(h) = \frac{1}{2}(T(2h) + M(h)) \tag{5.58}$$

Such formula is useful for example whenever one needs to compute a sequence of trapezoid rule values:

$$\mathcal{T}_k = \{T(h_0), T(h_0/2), T(h_0/4), ..., T(h_0/2^k)\}.$$

For such purpose, one starts with $T(h_0)$, then computes the sequence:

$$\mathcal{M}_k = \{M(h_0/2), M(h_0/4), ..., M(h_0/2^k)\}.$$

Use of (5.58) on $T(h_0)$ in addition to $\mathcal{M}_k$, allows one to obtain $\mathcal{T}_k$ by summing up fewer terms than in computing such sequence directly.

5.6.5 Simpson's Rule

Like the midpoint rule, Simpson's rule is applicable only if the number of subintervals is even $(n = 2m)$. Its higher accuracy than the trapezoid and midpoint rules requires more regularity conditions on f. Specifically, we assume that

$$f(x) \text{ is at least in } C^4(a,b).$$

We derive Simpson's rule as a $\frac{1}{3}$, $\frac{2}{3}$ linear combination of, respectively, the trapezoid and midpoint rules.
More precisely for $I'_k = \int_{x_{2k}}^{x_{2k+2}} f(x)dx$, one has by extending Taylor's series expansion of $f(x)$ to order 3, with $m_k = x_{2k+1}$ in (5.54):

$$f(x) = f(m_k) + f'(m_k)(x - m_k) + f''(m_k)\frac{(x - m_k)^2}{2} \quad (5.59)$$

$$+ f'''(m_k)\frac{(x - m_k)^3}{6} + f^{(4)}(c_k(x))\frac{(x - m_k)^4}{24},$$

with $c_k(x) \in (m_k, x)$ continuously depending on x. Integrating from x_{2k} to x_{2k+2} and using the second mean value theorem, one obtains :

$$I'_k = M_k + \frac{h^3}{3}f''(x_{2k+1}) + \frac{f^4(c_k)}{120}h^5 = M_k + \frac{h^3}{3}f''(x_{2k+1}) + O(h^5) \quad (5.60)$$

with $c_k \in (x_{2k}, x_{2k+2})$.(Note that this integration process annihilates all integral terms of odd powers in $(x - m_k)$). On the other hand, since:

$$T'_k = h(y_{2k} + y_{2k+2}),$$

then, by Taylor's expansion, one has successively:

$$y_{2k+2} = f(x_{2k+1}) + hf'(x_{2k+1}) + \frac{h^2}{2}f''(x_{2k+1}) + \frac{h^3}{6}f'''(x_{2k+1}) + O(h^5),$$

and:

$$y_{2k} = f(x_{2k+1}) - hf'(x_{2k+1}) + \frac{h^2}{2}f''(x_{2k+1}) - \frac{h^3}{6}f'''(x_{2k+1}) + O(h^5).$$

By adding the last 2 identities and multiplying by 2, one gets:

$$T_k' = h(y_{2k} + y_{2k+2}) = 2hf(x_{2k+1}) + h^3 f''(x_{2k+1}) + O(h^5).$$

Hence we obtain the relationship between T_k' and M_k:

$$T_k' = M_k + h^3 f''(x_{2k+1}) + O(h^5) \tag{5.61}$$

And therefore by combining algebraically (5.80) and (5.61), using:

$$3 \times (5.80) - (5.61),$$

one gets:

$$3I_k' - T_k' = 2M_k + O(h^5) \tag{5.62}$$

Define now the **simple integration Simpson's rule** as:

$$S_k = \frac{2}{3}M_k + \frac{1}{3}T_k' = \frac{h}{3}(f(x_{2k}) + 4f(x_{2k+1}) + f(x_{2k+2})) \tag{5.63}$$

then (5.62) is equivalent to:

$$I_k' = S_k + O(h^5). \tag{5.64}$$

Note that a more explicit expression of $I_k' - S_k$ can be found by first noting (see [21]):

$$I_k' - S_k = \int_{x_{2k}}^{x_{2k+2}} (f(x) - p_{2k\,2k+1\,2k+2}(x))dx, \tag{5.65}$$

where $p_{2k\,2k+1\,2k+2}(x)$ is the quadratic polynomial interpolating $f(x)$ at x_{2k}, x_{2k+1} and x_{2k+2}.
The right hand side in (5.65) can be handled in one of the following ways:

1. Given that from [21]:

$$f(x) - p_{2k\,2k+1\,2k+2}(x) = (x-x_{2k})(x-x_{2k+1})(x-x_{2k+2})[x_{2k}, x_{2k+1}, x_{2k+2}, x].$$

Then, letting $w(x) = \int_{x_{2k}}^{x} (t - x_{2k})(t - x_{2k+1})(t - x_{2k+2})dt$ and noting that $w(x_{2k}) = w(x_{2k+2}) = 0$ one has, using integration by parts:

$$I_k' - S_k = -\int_{x_{2k}}^{x_{2k+2}} w(x)\frac{d}{dx}([x_{2k}, x_{2k+1}, x_{2k+2}, x])dx,$$

given that if $f \in C^4$, $[x_{2k}, x_{2k+1}, x_{2k+2}, x] \in C^1$.
As $w(x) \geq 0$, using the second mean value theorem and the fact that:

$$\text{for } f \in C^4 : \frac{d}{dx}([x_{2k}, x_{2k+1}, x_{2k+2}, x]) = \frac{1}{4!}f^{(4)}(c_k(x)),$$

$(c_k(x)$ depending continuously on $x)$, one obtains:

$$\int_{x_{2k}}^{x_{2k+2}} w(x)\frac{d}{dx}([x_{2k}, x_{2k+1}, x_{2k+2}, x])dx$$

$$= \frac{1}{24}f^{(4)}(c_k)\int_{x_{2k}}^{x_{2k+2}} w(x)dx, \ c_k \in (x_{2k}, x_{2k+2}).$$

Since $\int_{x_{2k}}^{x_{2k+2}} w(x)dx = \frac{4}{15}$, then:

$$I'_k - S_k = -\frac{h^5}{90}f^{(4)}(c_k), \ c_k \in (x_{2k}, x_{2k+2}) \tag{5.66}$$

2. A second way to proceed is through the use of generalized divided differences in writing:

$$f(x) - p_{2k\,2k+1\,2k+2}(x) = (x - x_{2k})(x - x_{2k+1})(x - x_{2k+2})[x_{2k}, x_{2k+1}, x_{2k+1}, x_{2k+2}]$$

$$\ldots + (x - x_{2k})(x - x_{2k+1})^2(x - x_{2k+2})\frac{1}{4!}f^{(4)}(c_k(x)),$$

with $c_k(x)$ depending continuously on x. Through integration from x_{2k} to x_{2k+2}, use of:

$$\int_{x_{2k}}^{x_{2k+2}} (x - x_{2k})(x - x_{2k+1})(x - x_{2k+2})[x_{2k}, x_{2k+1}, x_{2k+1}, x_{2k+2}] = 0,$$

and of the second mean value theorem, one obtains as in (5.66):

$$I'_k - S_k = \frac{1}{4!}f^{(4)}(c_k)\int_{x_{2k}}^{x_{2k+2}} (x - x_{2k})(x - x_{2k+1})^2(x - x_{2k+2})dx$$

$$= -\frac{h^5}{90}f^{(4)}(c_k), \ c_k \in (x_{2k}, x_{2k+2}).$$

Summing up (5.66) over k, one derives the **composite Simpson's rule**, namely:

$$I = I(a, b) = \sum_{k=0}^{m-1}(S_k - \frac{h^5}{90}f^{(4)}(c_k)) = S(h) - \frac{(b-a)h^4}{180}f^{(4)}(c), \ c \in (a, b), \tag{5.67}$$

with

$$S(h) = \sum_{k=0}^{m-1}(\frac{2}{3}M_k + \frac{1}{3}T'_k) = \frac{2M(h) + T(2h)}{3}.$$

Thus, the following error estimate is obviously deduced:

Proposition 5.7 *Let f be a function in $C^4[a,b]$, interpolating the set of data D_n. Then:*

$$I = I(a,b) = S(h) + O(h^4),$$

where $S(h) = (y_0 + 4\sum_{k=0}^{m-1} y_{2k+1} + 2(\sum_{k=1}^{m-1} y_{2k}) + y_{2m})\frac{h}{3}$

The `MATLAB` code of the composite Simpson's rule is as follows:

Algorithm 5.4 Composite Simpson's Rule

```
function I = CompositeSimpson(x,y,N)
% Inputs: N, number of intervals
%             x = [a=x(1),...,x(N+1)=b] , y =[y(1),...,y(N+1)]
N=length(y)-1 ;h= (b-a)/N ;
%Verify that the components of x are equally spaced
%Test that N is an even integer
I = (y(1) + y(N+1)) ;
Y1 = y(3:2:N-1) ;
Y2 = y(2:2:N) ;
I1 = 2*sum(Y1) ;
I2= 4*sum(Y2) ;
I = (h/3)*(I+I1+I2) ;
```

5.7   Romberg Integration

Romberg integration is a Richardson extrapolation process applied to accelerate convergence of the composite midpoint or trapezoidal rules. It is based on the following facts (one of which is proved in the appendix of Section 5.8):

Proposition 5.8 *Let $h = \frac{b-a}{2^l}$, $l = 0, 1, 2, \ldots$ and $f(.)$ be an analytical function, i.e., with continuous derivatives up to any order, then:*

$$I = I(a,b) = T(h) + \tau_1 h^2 + \tau_2 h^4 + .. + \tau_j h^{2j} + ..., \qquad (5.68)$$

and

$$I = I(a,b) = M(h) + \mu_1 h^2 + \mu_2 h^4 + .. + \mu_j h^{2j} + ..., \qquad (5.69)$$

where the sequences $\{\mu_j\}$, $\{\tau_j\}$ are independent from h, and depend on the function f (and its derivatives) at a and b.

The Formulae

On the basis of (5.68), we can implement Richardson's extrapolation, by writing this equation simultaneously for h and $\frac{h}{2}$. Specifically, in that case we obtain:

(a) $I = T(h) + \tau_1 h^2 + \tau_2 h^4 + \ldots + \tau_j h^{2j} + \ldots$

(b) $I = T\left(\frac{h}{2}\right) + \tau_1 \left(\frac{h}{2}\right)^2 + \tau_2 \left(\frac{h}{2}\right)^4 + \ldots + \tau_j \left(\frac{h}{2}\right)^{2j} + \ldots$

In order to eliminate the dominant term of the error, by performing the algebraic operation $4(b) - (a)$, we obtain:

$$3I = 4T\left(\frac{h}{2}\right) - T(h) + O(h^4)$$

and therefore:

$$I = \frac{4T\left(\frac{h}{2}\right) - T(h)}{3} + t_2 h^4 + t_3 h^6 + \ldots \tag{5.70}$$

where the sequence $\{t_i\}$ is independent of h.

Defining the first Romberg integration operator as:

$$R^1(h/2) = \frac{4T\left(\frac{h}{2}\right) - T(h)}{3} \text{ or equivalently } R^1(h) = \frac{4T(h) - T(2h)}{3} \tag{5.71}$$

equation (5.70) provides then an approximation to the integral $I(a,b)$ of order $O(h^4)$ verifying:

$$I = R^1(h) + O(h^4) \tag{5.72}$$

In a similar way, we can derive a second Romberg integration formula by writing again the equation above simultaneously in terms of h and $\frac{h}{2}$:

(a) $I = R^1(h) + t_2 h^4 + t_3 h^6 + \ldots$

(b) $I = R^1\left(\frac{h}{2}\right) + t_2 \left(\frac{h}{2}\right)^4 + t_3 \left(\frac{h}{2}\right)^6 + \ldots$

Performing the algebraic operation $16(b) - (a)$ yields:

$$15I = 16R^1\left(\frac{h}{2}\right) - R^1(h) + O(h^6)$$

And therefore:

$$I = \frac{16R^1\left(\frac{h}{2}\right) - R^1(h)}{15} + t_3 h^6 + t_4 h^8 + \ldots \tag{5.73}$$

where the sequence $\{t_i\}$ is independent of h.

Defining the second Romberg integration operator as:

$$R^2(h/2) = \frac{16R^1(\frac{h}{2}) - R^1(h)}{15} \text{ or equivalently } R^2(h) = \frac{16R^1(h) - R^1(2h)}{15} \tag{5.74}$$

equation (5.52) is then equivalent to:

$$I = R^2(h) + O(h^6) \tag{5.75}$$

As for differentiation, this process can be repeated.

The first Romberg extrapolation formulae obtained based on the composite trapezoid rule are as follows:

h	$T(h)$	$R^1(h)$	$R^2(h)$	$R^3(h)$
h_0	$\times$			
$h_0/2$	$\times$	$\times$		
$\frac{h_0}{4}$	$\times$	$\times$	$\times$	
$\frac{h_0}{8}$	$\times$	$\times$	$\times$	$\times$

TABLE 5.8: A template to apply Romberg integration formulae

Proposition 5.9 *Let f belong to $C^\infty[a,b]$*

$$I = I(a,b) = \int_a^b f(x)dx = \begin{cases} R^1(h) + O(h^4) \\ R^2(h) + O(h^6) \\ R^3(h) + O(h^8) \\ ... \\ R^k(h) + O(h^{2k+2}) \end{cases}$$

with

$$R^1(h) = \frac{2^2 T(h) - T(2h)}{2^2 - 1}; \; R^2(h) = \frac{2^4 R^1(h) - R^1(2h)}{2^4 - 1}; \; R^3(h) = \frac{2^6 R^2(h) - R^2(2h)}{2^6 - 1},$$

and in general the kth-order Romberg operator:

$$R^k(h) = \frac{2^{2k} R^{k-1}(h) - R^{k-1}(2h)}{2^{2k} - 1}$$

with an error of h^{2k+2}. Table 5.8 provides a template for applying Romberg integration based on the **composite trapezoidal rule**.

Remark 5.5 *Referring to Proposition (5.6), since $M(h) = 2T(h) - T(2h)$, then one concludes that:*

$$R^1(h) = \frac{4T(h) - T(2h)}{3} = \frac{2M(h) + T(2h)}{3} = S(h)$$

meaning that Simpson's Composite Numerical Integration formula is equivalent to the first Romberg Trapezoidal Extrapolation formula.
In a consistent manner with the Composite Midpoint and Trapezoidal Rules, one has when $h = \frac{b-a}{2^l}, l = 0, 1, 2, ...$:

$$I = I(a,b) = S(h) + s_2 h^4 + s_6 h^6 + ... + s_j h^{2j} + ...$$

where all the coefficients $\{s_i\}$ are independent of h.
This allows starting a Romberg integration process beginning with composite Simpson's formula $S(h)$.

5.8 Appendix: Error Term for Midpoint Rule, $h = \frac{b-a}{2^l}$

For the purpose of applying Richardson's extrapolation (5.56) can be used in its infinite series expansion form. Let $h_0 = (b - a)$. Then one has for $m = (a + b)/2$, $M(h_0) = h_0 f(m)$,

$$I = M(h_0) + f^{(2)}(m)\frac{h_0^3}{24} + ... + f^{(2j)}(m)\frac{h_0^{2j+1}}{4^j(2j+1)!} + ...$$

which is equivalent to:

$$I = M(h_0) + h_0 \Sigma_{j \geq 1} \gamma_j f^{(2j)}(m) h_0^{2j}, \tag{5.76}$$

Similarly to (5.31), there exists a sequence of universal constants $\{a_i : i = 1, 2, ...\}$, such that:

$$f^{(2j)}(m) = \frac{f^{(2j-1)}(b) - f^{(2j-1)}(a)}{2h_0} + \sum_{i=1}^{\infty} a_i f^{(2j+2i)}(m) h_0^{2i}. \tag{5.77}$$

Combining (5.76) with (5.77), one deduces:

$$I = M(h_0) + \Sigma_{j=1}^{\infty} \mu_j h_0^{2j}, \tag{5.78}$$

where:

$$\mu_j = \left(\sum_i^j \gamma_i^{j-i}\right)[f^{(2j-1)}(b) - f^{(2j-1)}(a)],$$

and the sequence γ_i^l defined by the recurrence relations:

$$\begin{cases} \gamma_j^0 = \gamma_j, & (5.79.1) \\ \gamma_j^l = \Sigma_{i=1}^{j-1} \gamma_i^{l-1} a_{j-1}, \, l \geq 1. & (5.79.2) \end{cases} \tag{5.79}$$

Let

$$\nu_j = \sum_i^j \gamma_i^{j-i}.$$

Then (5.78) is equivalent to:

$$I = M(h_0) + \Sigma_{j=1}^{\infty} \nu_j (f^{(2j-1)}(b) - f^{(2j-1)}(a)) h_0^{2j}. \tag{5.80}$$

For $h = \frac{h_0}{2}$, let $I_1 = \int_a^m f(x)dx$ and $I_2 = \int_m^b f(x)dx$ with $M_1(h_0/2)$ and $M_2(h_0/2)$, respectively their approximations using the midpoint rule. Obviously, from (5.80), we have successively:

$$I_1 = M_1(h_0/2) + \Sigma_{j=1}^{\infty} \nu_j (f^{(2j-1)}(m) - f^{(2j-1)}(a))(h_0/2)^{2j}$$

and

$$I_2 = M_2(h_0/2) + \Sigma_{j=1}^{\infty} \nu_j (f^{(2j-1)}(b) - f^{(2j-1)}(a))(h_0/2)^{2j}.$$

Adding up these 2 equations leads to:

$$I = I_1 + I_2 = M_1(h_0/2) + M_2(h_0/2) + \Sigma_{j=1}^{\infty} \nu_j (f^{(2j-1)}(b) - f^{(2j-1)}(a))(h_0/2)^{2j},$$

which is equivalent to:

$$I = M(h_0/2) + \Sigma_{j=1}^{\infty} \nu_j (f^{(2j-1)}(b) - f^{(2j-1)}(a))(h_0/2)^{2j}, \qquad (5.81)$$

i.e., (5.78) with h_0, replaced by $h_0/2$. This argument can be repeated proving (5.78) with h_0, replaced by $h_0/2^l$, $l \geq 0$. This result can be generalized to both trapezoid and Simpson's rules and is of major importance for the implementation of **Romberg integration**.

5.9 Exercises

Numerical Differentiation

1. Use the most accurate of the forward, backward or central difference approximation formulae to determine the empty entries in the following table:

i	x_i	$f(x_i)$	$f'(x_i)$
0	0.0	5	.
1	0.1	4.960	.
2	0.2	4.842	.
3	0.3	4.651	.
4	0.4	4.393	.

2. Use the most accurate of the forward, backward or central difference approximation formulae to determine the empty entries in the following table:

i	x_i	$f(x_i)$	$f'(x_i)$
0	-0.9	0.097	.
1	-0.7	-0.122	.
2	-0.5	-0.387	.
3	-0.3	-0.655	.
4	-0.1	-0.895	.

3. Let $f(x) = e^{x^2} + 2x$. Fill in the empty entries in the table below to approximate $f'(0)$, using the forward difference approximation formula $\frac{\Delta_h f(0)}{h}$.

h	$\frac{\Delta_h f(0)}{h}$
0.125	.
0,250	.
0.375	.
0.500	.
0.625	.

4. Let $0 < h \leq 1$ and $D = \phi(h) + d_1 h + d_2 h^2 + d_3 h^3 - ...$ where the constants c_i, $\forall i = 1, 2, ...$ are independent from h. What combination of $\phi(h)$ and $\phi(h/2)$ should lead to a more accurate estimate of D?

5. Let $0 < h \leq 1$ and $D = \phi(h) + c_1 h^{1/2} + c_2 h^{2/2} + c_3 h^{3/2} - ...$ where the constants c_i, $\forall i = 1, 2, ...$ are independent from h. What combination of $\phi(h)$ and $\phi(h/2)$ should lead to a more accurate estimate of D?

6. Let $\phi(h) = L - O(h^p)$, where $0 < h \leq 1$. Show that Richardson's extrapolation can be carried out for any two values h_1 and h_2 of h.

7. Use Richardson extrapolation based on the central difference approximation formula to estimate the first derivative of $y = \cos x$ at $x = \pi/4$, with initial value of $h = \pi/3$. Compare with the actual value of $f'(\pi/4)$, by computing the absolute relative error in that case.

8. Use Richardson extrapolation based on the forward difference approximation formula to estimate the first derivative of $y = \ln x$ at $x = 4$, with initial value of $h = 0.5$. Compare with the actual value of $f'(4)$, by computing the absolute relative error in that case.

9. Consider the following table of data associated with some unknown function $y = f(x)$

i	x_i	y_i
0	0.00	1.000
1	0.25	2.122
2	0.50	3.233
3	0.75	4.455
4	1.00	5.566
5	1.25	−1.000
6	1.50	−1.255
7	1.75	−1.800
8	2.00	−2.000

(a) Find an approximation to $f'(0.25)$ using successively the forward, backward and central difference approximations if $h = 0.25$.

(b) Find approximations to $f'(1)$ using the central difference approximation with $h = 0.25$, $h = 0.50$ then $h = 1.00$. Improve these results by computing central difference Richardson's extrapolation approximations of the first and second order, $\psi^1_{0.25}(.)$ and $\psi^2_{0.25}(.)$ to approximate $f'(1)$.

(c) Approximate $f'(0)$ and $f'(2)$ with $h = 0.25$.

(d) Find approximations to $f''(1)$ and $f'''(1)$ using the forward difference approximations, with $h = 0.25$.

10. Consider the following table of data for the function $f(x)$

i	x_i	y_i
0	0.000	1.0000000
1	0.125	1.1108220
2	0.250	1.1979232
3	0.375	1.2663800
4	0.500	1.3196170
5	0.625	1.3600599
6	0.750	1.3895079
7	0.875	1.4093565
8	1.000	1.4207355

Use the central difference formula to approximate $f'(0.5)$, followed by Richardson's extrapolation of the 1st and 2nd orders to improve the results. Fill out the following table:

h	$\psi_h(.)$	$\psi_h^{(1)}(.)$	$\psi_h^{(2)}(.)$
0.5	×		
0.25	×	×	
0.125	×	×	×

11. Based on the set of data of exercise 10:

   (a) Calculate the second derivative $f''(0.5)$, using the central difference approximation with $h = 0.25$ and $h = 0.125$. Use Richardson's extrapolation operator of the first order, $\psi_{0.125}^1(f(0.5))$ to improve these results.

   (b) Calculate the third derivative $f'''(1.000)$, using the backward difference approximation.

12. Based on the set of data of Exercise 10, use the Forward Difference formula to approximate $f'(0)$, followed by Richardson's extrapolation of the first and second orders. Fill out the following table:

h	$\phi_h(.)$	$\phi_h^{(1)}(.)$	$\phi_h^{(2)}(.)$
0.5	×		
0.25	×	×	
0.125	×	×	×

13. Based on the set of data of the preceding exercise, use the backward difference formula to approximate $f'(1)$, followed by Richardson's extrapolation of the first and second orders. Fill out the following table:

h	$\chi_h(.)$	$\chi_h^{(1)}(.)$	$\chi_h^{(2)}(.)$
0.5	×		
0.25	×	×	
0.125	×	×	×

14. Consider the following set of data:

$$D_n = \{(x_i, y_i)|i = 0, ..., n \text{ with } y_i = f(x_i)\}$$

where the X-coordinates are **equally spaced**, that is $x_i = x_0 + ih$ for all i, with $0 < h \leq 1$ and $n \geq 4$. Based on D_n:

(a) Use **Newton's** quadratic interpolating polynomial $p_{012}(x)$ to determine its derivative $p'_{012}(x)$ and the value of $p'_{012}(\mathbf{x_0})$ in terms of the 3 points x_0, x_1 and x_2 ("The 3 points formula".) Express $D_h(x_0)$ in terms of x_0 and h. For notation purposes, let in that case, $p'_{012}(\mathbf{x_0}) = D_h(\mathbf{x_0})$.

(b) Given that $f \in C^3$, the polynomial interpolation error is estimated by the following identity:

$$f(x) = p_{012}(x) + \frac{1}{3!}(x - x_0)(x - x_1)(x - x_2)f^{(3)}(c(x))$$

where $c(x) \in (x_0, x_3)$ depends continuously on x.
<u>Through differentiation</u> of this identity, find the expression of the Error if $f'(x_0) \approx p'_{012}(x_0)$, and show that this Error is $O(h^2)$.

(c) Given that

$$f'(x_0) = D_h(x_0) + C_1 h^2 + C_2 h^3 + C_3 h^4 + \ldots + C_i h^{i+1} + \ldots.$$

where all the coefficients C_i are independent of h.
Apply Richardson's extrapolation procedure once to improve the approximation of $f'(x_0)$, then define the first-order Richardson's extrapolation operator $D_h^1(x_0)$. What is the order of the error if $f'(x_0) \approx D_h^1(x_0)$.

15. Consider the following set of data:

i	x_i	$f(x_i)$
0	0.000	1.0000
1	0.125	1.1108
2	0.250	1.1979
3	0.375	1.2663
4	0.500	1.3196
5	0.625	1.3600
6	0.750	1.3895
7	0.875	1.4093
8	1.000	1.4207

Use this set of data and the results derived in the preceding exercise to compute $D_{0.125}^1(0)$.

Numerical Integration

16. Derive the estimates on $I_k - A_k, I_k - B_k, I - A(h)$ and $I - B(h)$, in Proposition 5.3.

17. Approximate $I = \int_a^b f(x)\,dx$ based on the set of data given in Exercise 15, using the midpoint rectangular rule.

18. Use the composite midpoint rectangular rule to approximate $I = \int_0^2 e^{3x}\cos(2x)dx$, if 9 partition points are used.

19. (a) Estimate the value of $I = \int_0^1 (x^2 + 1)^{-1}dx$ by the composite midpoint rule if 7 partition points are used.

    (b) Find the absolute error in this approximation. Obtain also an upper bound on the absolute error, if 7 partition points are used.

20. The Bessel function of order 0 is defined by the equation

$$J_0(x) = \frac{1}{\pi}\int_0^\pi \cos(x\sin\theta)d\theta$$

Approximate $J_0(1)$ by the composite midpoint rectangular rule using 5 equally spaced partition points, then find an upper bound to the error in this approximation. (Let $\cos\sqrt{2}/2 = B$.)

21. How many equi-spaced partition points should be used in the approximation of $I = \int_0^1 e^{-x^2}\,dx$ by means of the composite midpoint rectangular rule, if the absolute error $|\epsilon| \leq \frac{10^{-4}}{2}$?

22. Determine the value of h required to approximate $I = \int_0^1 xe^x dx$ up to 3 decimal figures.

23. Establish "Composite Right" and "Left Rectangular Rules" that approximate the definite integral $I = \int_a^b f(x)dx$, in case the partition points are not equally spaced.

24. Let $I = \int_0^2 x^2 e^{-x^2}\,dx$.

    (a) Use the midpoint rectangular rule to approximate I with 3 equally spaced partition points.

    (b) Derive the formulae of the Romberg operators applied to the midpoint rectangular rule.

    (c) Fill in the empty slots of the following table adequately.

h	$M(h)$	$M^1(h)$	$M^2(h)$
$h_0 = 1$			
$\frac{h_0}{2} = 0.5$			
$\frac{h_0}{4} = 0.25$			

25. (a) Estimate the value of $I = \int_0^4 2^x dx$ by the composite trapezoidal rule if 9 partition points are used.

    (b) Find the absolute error in this approximation. Obtain also an upper bound on the absolute error in this case.

26. Determine the value of h if the composite trapezoid rule is to estimate $\int_0^\pi \sin x dx$ with error $\leq 10^{-7}$? Will the integral be over or under estimated?

27. Obtain an upper bound on the absolute error using 55 equally spaced points, when we compute $\int_0^6 \sin x^2 dx$ by means of:

    - the composite trapezoid rule

    - the composite midpoint rectangular rule

    - Simpson's rule

28. Let $f(x) = x^2 \cos x$. Approximate $I = \int_0^\pi f(x)dx$ by the composite trapezoid rule using the partition points $0, \pi/2, \pi$. Repeat by using partition points $0, \pi/4, \pi/2, 3\pi/4, \pi$. Use these results to apply Romberg extrapolation approximation $R^1(\pi/4)$ and obtain a better evaluation for I.

29. Consider the data given in Exercise 15. Fill in the following Table, using $h_0 = 1$.

h	$T(h)$	$R^1(h)$	$R^2(h)$	$R^3(h)$
$h_0 = 1$	×			
$h_0/2 = 0.5$	×	×		
$\frac{h_0}{4} = 0.25$	×	×	×	
$\frac{h_0}{8} = 0.125$	×	×	×	×

30. Consider the Bessel function $J_0(x)$ as defined in exercise 19.

    (a) Approximate $J_0(1)$ by the trapezoid rule using 3 equally spaced partition points, then find an upper bound to the absolute error in this approximation. (Let $\cos 1 = A$ and $\cos \sqrt{2}/2 = B$.

    (b) Apply Romberg extrapolation procedure once on the trapezoidal rule in (a), to obtain a better approximation to $J_0(1)$.

31. Let $a = x_0 < x_1 < \ldots < x_n = b$ be a set of partition points of the interval $[a, b]$, with $h_i = x_{i+1} - x_i$ leading to a non-uniform spacing. Establish the composite trapezoid rule formula to approximate $\int_a^b f(x)dx$, then find an upper bound for the error term in this approximation.
Hint: On the interval $[x_k, x_{k+1}]$, use:

$$\int_{x_k}^{x_{k+1}} f(x)dx = T_k + \frac{1}{2}\int_{x_k}^{x_{k+1}} (x - x_k)(x - x_{k+1})f''(c(x))dx, \ c(x) \in (x_k, x_{k+1}).$$

32. Compute $I = \int_0^2 x^2 \ln(x^2 + 1)dx$ by Simpson's rule using 5 partition points in 2 different ways.

33. Find an approximate value of $\int_1^2 x^{-1}dx$ using the composite Simpson's rule with $h = 0.25$. Give a bound on the absolute error.

34. Let $D_n = \{(x_i, y_i)|i = 0, 1, \ldots, n = 2m,$ where $y_i = f(x_i))\}$ be a given set of data, where the X-coordinates are equally spaced, and where n is an even integer.

    (a) Derive the first 2 Romberg approximation formulae: $S^1(h)$ and $S^2(h)$, **applied to the composite Simpson's rule**, given that:

    $$I = S(h) + s_1 h^4 + s_2 h^6 + \ldots + s_j h^{2j+2} + \ldots.$$

    (b) The next question deals with the following set of values for a function $f(x)$, arranged in a table as follows:

i	x_i	$f(x_i)$
0	0.000	1.0000
1	0.125	1.0157
2	0.250	1.0645
3	0.375	1.1510
4	0.500	1.2840
5	0.625	1.4779
6	0.750	1.7551
7	0.875	2.1503
8	1.000	2.7183

    In order to approximate $\int_0^1 f(x)\,dx$, based on Simpson's rule and the formulae obtained in (a), fill in the empty slots of the following table adequately, carrying 5 significant figures with rounding to the closest.

h	$M(h)$	$T(h)$	$S(h)$	$S^1(h)$	$S^2(h)$
$h_0 = 0.5$					
$\frac{h_0}{2} = 0.25$					
$\frac{h_0}{4} = 0.125$					

35. Consider the integral $I = erf(x) = \frac{2}{\sqrt{\pi}} \int_0^x e^{-t^2} dt$

   (a) Let $I = erf(1)$. Approximate I up to 2 decimal figures by means of the composite trapezoidal rule with equi-spaced partition points if the exact value of $I = 0.84$. (Use rounding to the closest.)

   (b) What is the number of partition points needed, if it is known that the composite trapezoid rule has to be followed by the Romberg process in order to improve the accuracy of the approximation in the preceding question.

36. Consider the **logarithmic integral** defined by the equation

$$I = li(x) = \int_2^x \frac{1}{\ln t} dt$$

   (a) Approximate $li(11)$ by means of the composite Simpson's rule using 9 equally spaced nodes, then apply Romberg extrapolation of 1st order to improve the result.

   (b) Compute the relative errors in both approximations, given that the exact value $li(11) = 5.5458$.

5.10   Computer Projects

Exercise 1: Numerical Differentiation

Let $x = [x_1, x_2, ..., x_n]$ and $y = [y_1, y_2, ..., y_n]$ be 2 vectors of equal length n, representing a set of n points in the plane:

$$D_n = \{(x_i, y_i) | x_1 < x_2 < ... < x_n \; ; i = 1, 2, ..., n\}$$

where $y_i = f(x_i)$ for some real valued function f.

1. Consider the set of points D_n as given above, where the x-components are equally spaced with $x_{i+1} - x_i = h$; $0 < h \leq 1$, and let $m = 1, 2$ or 3. Write a MATLAB
   function d1 = ApproxDerivative(x,y,xi,m)
   that approximates the first derivative of some unknown function f at a node x_i, using the backward difference formula $(m = 1)$, the forward difference formula $(m = 2)$ or the central difference formula $(m = 3)$. Your function should check first the validity of the input and display an error message if the derivative cannot be computed at node x_i, that is check the following:

   - The components of x should be equally spaced

   - $0 < h \leq 1$

   - The value of m should be consistent with the index of x_i. For example, if m=1 then to apply the backward difference formula, x_{i-1} should also be an element of x.

2. Write a MATLAB
   function R = Richardson(f,a,h,k)
   that takes as input: a function f, a real number a, $0 < h \leq 1$ being the smallest value of the increment, and a positive integer k. This function applies k iterations of Richardson's extrapolation procedure to improve the approximation of the first derivative $f'(a)$, using the **central difference** formula, and outputs the results in a square lower triangular matrix R of size $(k + 1 \times k + 1)$.
   N.B. The entries of the first column of the matrix are approximations to $f'(a)$ using the central difference approximation formulas for different values of h.

3. Test each one of the functions above for 2 different test cases, and save the results in a Word document.

Exercise 2: Numerical Integration

The purpose of this exercise is to find approximations to the following definite integrals using the composite trapezoidal rule followed by the Romberg process:

1. $I_1 = \int_0^1 \frac{1}{1+x^2}\,dx = \pi$

2. $I_2 = \int_1^2 \frac{1}{x}\,dx = \ln 2 = 0.693147180559945...$

3. The exact value of $\int_0^\infty e^{-x^2}\,dx = \frac{\sqrt{\pi}}{2} = 1.570796326794897....$, with

$$\left| \int_0^\infty e^{-x^2}\,dx - I_3 \right| < C \times 10^{-16}$$

where $I_3 = \int_0^6 e^{-x^2}\,dx$.

1. Write a MATLAB
   function I = CompositeTrapezoid(f,a,b,n,p)
   that takes as inputs:

   (a) a real valued function f

   (b) 2 real numbers a and b, with $a < b \in D_f$ (domain of f)

   (c) a positive integer n, representing the number of subintervals of equal length determined by the partition points

   $$\{x_i \; ; \; i = 0, 1, ..., n \,|\, a = x_0, b = x_n; \; h = x_{i+1} - x_i\}$$

   (i.e., $h = \frac{b-a}{n}$, with $0 \le h \le 1$.)

   (d) p, a positive integer representing some precision fixed by the user

   This function approximates the integral $\int_a^b f(x)\,dx \approx I$ using the composite trapezoidal rule, and outputs I, displayed up to p decimal figures.
   Hint: Use the MATLAB function num2str(I, p) to round the computed I to the closest, up to p decimal figures.

2. Write a MATLAB
   function R = RombergCompositeTrapezoid(f,a,b,n,p,tol)
   that takes as inputs f, a, b, n, p as defined in part 1 above with n of the form: $n = 2^k$,(k positive integer), in addition to some tolerance $tol = 0.5 * 10^{-10}$. This function applies j iterations of the Romberg process based on the composite trapezoidal rule with $j \le k$, where j is the first integer for which

   $$|R(k+1, j-1) - R(k+1, j)| < tol$$

   and outputs the results in a matrix R of size $k+1 \times j+1$

Hints:

- The successive values of h are $\{h = \frac{b-a}{2^i}, \text{ for } i = 0, 1, ..., k\}$
- The entries of the first column of the matrix are values of $T(h)$, $\forall\, h$
- The entries of the remaining columns of the matrix are values of $R^1(h), ..., R^j(h)$,

3. Test the 2 MATLAB functions above on I_1, I_2 and I_3. Save your results in a Word document.

Chapter 6

Advanced Numerical Integration

6.1   Numerical Integration for Non-Uniform Partitions

In what has preceded, we have considered numerical integration on a set of equidistant points $\{x_i\}$, whereas $x_{i+1} - x_i = h$, $\forall i$. The formulae derived were well suited for the cases when a function $f(x)$ is given through a table $D_n = \{(x_n, y_n = f(x_n))\}$.

Naturally, each of the formulae we have derived can be generalized to non-uniform partitions $\{x_i\}$.

6.1.1   Generalized Formulae and Error Analysis

1. In the case of the trapezoidal rule, the number of intervals is n and such partitions satisfy:

$$x_i - x_{i-1} = h_i, \text{ with } h_i \neq h_j, \text{ for at least one pair } (i,j), \, i \neq j$$

Let also:

$$h = \max_{1 \leq i \leq n} \{h_i\}.$$

Written in that context, the **composite trapezoidal rule** becomes:

$$T(h) = \sum_{i=1}^{n} T_i = \frac{1}{2} \sum_{i=1}^{n} h_i(y_{i-1} + y_i)$$

with the local error (for the simple trapezoid rule) expressed as previously:

$$I_i = T_i - \frac{h_i^3}{12} f''(c_i), \ i = 1, ..., n, \ c_i \in (x_{i-1}, x_i),$$

while, using the *Intermediate Value Theorem*, the global error for $T(h)$ would become:

$$I - T(h) = -\frac{h_1^3 + h_2^3 + ... + h_n^3}{12} f''(c), \text{ with } x_0 < c < x_n,$$

that leads to:

$$|I - T(h)| \leq \frac{h^2}{12}(h_1 + h_2 + ... + h_n) \max_{x \in (a,b)} |f''(x)| = \frac{h^2}{12}(b - a) \max_{x \in (a,b)} |f''(x)|.$$

2. Similar considerations may be carried out also for the **midpoint and Simpson's rules**. In that case, we maintain the constraint of partitioning (a, b) into an even number of subintervals ($n = 2m$) in the following way:

$$a = x_0 < x_1 < ... < x_{2n-1} < x_{2n} = b, \text{ with } x_{2i-1} = \frac{x_{2i-2} + x_{2i}}{2},$$

$i = 1, .., m$. Moreover:

$$x_{2i} - x_{2i-2} = 2h_i, \text{ with } h_i \neq h_j, \text{ for at least one pair } (i, j), \ i \neq j.$$

As above, let also:

$$h = \max_i \{h_i\}$$

The expressions of the composite midpoint and Simpson's formulae become respectively:

$$M(h) = 2 \sum_{i=1}^{m} h_i y_{2i-1}$$

and

$$S(h) = \frac{1}{3} \sum_{i=1}^{m} h_i(y_{2i-2} + 4y_{2i-1} + y_{2i}) = \frac{1}{3}(M(h) + T(2h)). \quad (6.1)$$

Furthermore, the local errors for the simple rules remain unchanged, specifically:

$$\int_{x_{2i-2}}^{x_{2i}} f(x)dx = 2hf(x_{2i}) + \frac{h_i^3}{12} f''(c_i),$$

and

$$\int_{x_{2i-2}}^{x_{2i}} f(x)dx = \frac{h_i}{3}(y_{2i-2} + 4y_{2i-1} + y_{2i}) - \frac{h_i^5}{90}f''(c_i). \qquad (6.2)$$

The bounds for the global errors can be easily derived, yielding respectively:

$$|I - M(h)| \leq \frac{h^2}{6}(b-a) \max_{x \in (a,b)} |f''(x)|, \qquad (6.3)$$

and

$$|I - S(h)| \leq \frac{h^4}{180}(b-a) \max_{x \in (a,b)} |f^{(4)}(x)|. \qquad (6.4)$$

6.1.2 The Case for Adaptive Numerical Integration

Adaptive numerical integration consists in *"adapting" the partition of the interval (a,b) to the behavior of the function $f(x)$.* To illustrate that point, consider applying the non-uniform global Simpson's formula (6.1) to approximate the integral $I = \int_a^b f(x)\,dx$ and let us assume that there exists some subinterval (d,b) of (a,b) wherein the behavior of $f(x) \approx p(x)$ with $p(x) \in \mathbb{P}_3$, i.e., a polynomial of degree at most 3. Since from (6.2), one has:

$$I - S(h) = -\frac{1}{90} \sum_{i=1}^{m} h_i^5 f^{(4)}(c_i), \ c_i \in (x_{2i-2}, x_{2i}),$$

then one can select the partition of $\{x_i\}_i$, so that $x_{2m-2} = d$ and $x_{2m} = b$, the remaining points $\{x_i | i = 0, 1 ... 2m-2\}$ subdividing uniformly or non-uniformly the interval (a,d).

To illustrate this situation, consider the function:

$$f(x) = x^5(x-5)^2 e^{-x},$$

which graph is given in Figure 6.1. Obviously, as indicated by the graph of f, the function is very close to zero on $[20, 40]$, but has significant variations over $[0, 20]$. Evaluating $I = \int_0^{40} x^5(x-5)^2 e^{-x} dx$ by placing a uniform mesh on $(0, 40)$ would not be an appropriate strategy, as one needs a highly refined mesh on $(0, 20)$ and a coarse grid on $(20, 40)$. For that purpose, it would be convenient to consider methods that would adapt the partitioning $[0, 40]$ according to the behavior of $f(x)$ as to be explained in the following section.

6.1.3 Adaptive Simpson's Integration

Thus, Adaptive Numerical Integration is motivated by the need to compute an accurate approximation to I, taking into account a user defined computational tolerance, ϵ_{tol}, in the sense that one seeks $I_c \approx I$, such that:

$$|I - I_c| \leq \epsilon_{tol}, \text{ (absolute error)}$$

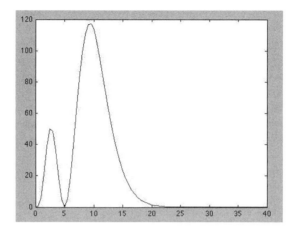

FIGURE 6.1: Graph of $f(x) = x^5(x-5)^2 e^{-x}$

or

$$\frac{|I - I_c|}{|I|} \leq \epsilon_{tol} \text{ (relative error)}.$$

Considering that the adaptive algorithm is one that must take into account whether $f(x)$ has sharp variations on subintervals of (a, b), and smooth ones on the remaining ones, then if the approximation error is to be evenly distributed, the partition points should be generated adaptively so that fine partitions, with small step sizes, are chosen in the first case, and coarse partitions, with larger step sizes, are used in the second one.

One of the methods used for such purpose is the recursive "Adaptive Simpson's Rule" that evaluates I in view of reaching the user's fixed computational tolerance (absolute or relative) *tol*. In case of relative computational tolerance, a "rough" estimate I_{est} of I would also be needed. One choice for I_{est} would be for example, $\frac{b-a}{6}(f(a) + 4f(\frac{a+b}{2}) + f(b))$.

In the adaptive process, the steps of the recursive algorithm are as follows:

a. Divide the initial interval $[a, b]$ into 2 subintervals of equal length, with $h = (b-a)/2$.

b. Compute $S(h)$ and $S(h/2)$.

c. As $h = \frac{b-a}{2^k}$, then using a similar argument to that of Proposition 5.8 (Section 5.8 in Chapter 5), one proves for $f \in C^6$:

$$I = S(h) + \alpha h^4 + O(h^6), \tag{6.5}$$

where α is independent from h. Using Richardson extrapolation on (6.5), one proves that:

$$I - S(h/2) = \frac{S(h/2) - S(h)}{15} + O(h^6). \tag{6.6}$$

Thus $\frac{S(h/2)-S(h)}{15}$ provides the $O(h^4)$, principal part of the error in $I - S(h/2)$. Hence, one of 2 following situations may occur, using absolute (or relative) errors:

c.1 $|\frac{S(h/2)-S(h)}{15}| \le tol$, (or $|\frac{S(h/2)-S(h)}{15I_{est}}| \le tol$).

c.2 $|\frac{S(h/2)-S(h)}{15}| > tol$, (or $|\frac{S(h/2)-S(h)}{15I_{est}}| > tol$).

In the first occurrence, we approximate I by $S = S(h/2)$ and stop the process.

Otherwise, letting $m = \frac{a+b}{2}$, we proceed with recurring the process by writing:

$$I = I_1 + I_2 = \int_a^m f(x)dx + \int_m^b f(x)dx,$$

and then apply **a. - b. - c.** in parallel on I_1 and I_2, in view of reaching respectively S_1 and S_2, such that (in case of absolute errors):

$$|I_k - S_k| \le tol/2 \quad (|\frac{I_k - S_k}{I_{est}}| \le tol/2), \; k = 1, 2.$$

Clearly, $|I_k - S_k| \le tol/2, \; k = 1, 2$ implies in case of use of absolute errors:

$$|I - (S_1 + S_2)| \le |I_1 - S_1| + |I_2 - S_2| \le tol,$$

or when using relative errors:

$$|\frac{I - (S_1 + S_2)}{I_{est}}| \le |\frac{I_1 - S_1}{I_{est}}| + |\frac{I_2 - S_2}{I_{est}}| \le tol,$$

A pseudo code for such procedure that uses relative errors would be as follows:

```
function S=RecurAdaptSimp(f,a,b,tol,Est)
h=(b-a)/2
Evaluate S(h) and S(h/2)
If |S(h)-S(h/2)|/|Est|>15*tol
 > m=(a+b)/2
 > S1=RecurAdaptSimp(f,a,m,tol/2,Est)
 > S2=RecurAdaptSimp(f,m,b,tol/2,Est)
 > S=S1+S2
else
 > S=S(h/2)
end
```

A detailed MATLAB implementation is as follows.

Algorithm 6.1 Adaptive Simpson's Integration (Recursive Version)

```
function [S,x]=RecurAdaptSimp(f,a,b,tol,i,Est)
% Input: Est, an estimate of the value of I
%          the Integral of f(x) from a to b
%          tol: Relative tolerance
%          i: level of recurrence
% Output: S: approximation of I, such that: |(I-S)/Iest|<=tol
%          x: the partition points
% Initialize parameters
x=[];%No partition of the interval (a,b)
h=(b-a);% Initial value of h
% Get S(h) and S(h/2)
m=(a+b)/2;
m1=(a+m)/2;
m2=(b+m)/2;
T1=h*(f(a)+f(b))/2;%Evaluate T(h)
M1=h*f(m);%Evaluate M(h/2)
T=(T1+M1)/2;%Evaluate T(h/2)
S1=(T1+2*M1)/3;%Evaluate S(h/2)
M=h*(f(m1)+f(m2))/2;%Evaluate M(h/4)
S=(T+2*M)/3;%Evaluate S(h/4)
x=[x m1 m2];%Update x with m1 and m2
if abs((S-S1)/Est)>15*tol% if |(S(h)-S(h/2))/Iest|>tol
    i=i+1;%raise recurrence level by 1
    %Apply AdaptSimp on (a,m) and (m,b)
    [S1,x1]=RecurAdaptSimp(f,a,m,tol/2,i,Est);
    [S2,x2]=RecurAdaptSimp(f,m,b,tol/2,i,Est);
    S=S1+S2;x=[x1 x2];%Updare S and x
    i=i-1;%decrease recurrence level by 1
end
if i==1 % Final update at recurrence level 1
    x=[x m];% Update x with m
    x=sort(x);% sort x
    x=[a x b];% Update x with a and b
end
```

Example 6.1 *Consider approximating* $I = \int_0^{40} 100xe^{-x}dx$ *which exact value can be verified to be* $100 - 4100 * e^{-40} \approx 100.$

Proceeding by a standard Simpson's rule on a uniform mesh h can prove to be catastrophic! For relative computational tolerances ϵ_{tol}, one computes from (5.67) the corresponding minimum number of uniform intervals. The results are indicated in Table 6.1. Obviously, such summations with high number of

ϵ_{tol}	$n_{\min}(\epsilon_{tol})$
0.5×10^{-4}	1229
0.5×10^{-5}	2185
0.5×10^{-6}	3884
0.5×10^{-7}	6907
0.5×10^{-8}	12283
0.5×10^{-9}	21841
0.5×10^{-10}	38840

TABLE 6.1: Minimum number of intervals for uniform partitions using Simpson's rule to compute $I = \int_0^{40} 100xe^{-x}dx$ up to a relative tolerance ϵ_{tol}

ϵ_{tol}	$n(\epsilon_{tol})$	$\min h$	$\max h$
0.5×10^{-4}	34	7.8125000×10^{-2}	10
0.5×10^{-5}	58	3.9062500×10^{-2}	10
0.5×10^{-6}	94	1.9531250×10^{-2}	10
0.5×10^{-7}	166	9.765625×10^{-3}	10
0.5×10^{-8}	286	4.8828125×10^{-3}	5
0.5×10^{-9}	496	4.8828125×10^{-3}	5
0.5×10^{-10}	912	2.4414062×10^{-3}	5

TABLE 6.2: Number of intervals as a function of the user's tolerance ϵ_{tol} in adaptive Simpson's rule

elements would lead to an excessive round-off error propagation. On the other hand, adaptive numerical integration using the MATLAB program:

$$\text{RecurAdaptSimp}(a, b, \text{tol}, i, \text{Est}) \text{ with tol} = \epsilon_{tol}$$

would provide a comparably moderate number of intervals $n(\epsilon_{tol})$ as shown in Table 6.2, where we also provide in the third and fourth columns $\min_{1 \le i \le n} \{h_i\}$ and $\max_{1 \le i \le n} \{h_i\}$. This asserts the strength of the method to automatically generate a highly non-uniform partition of the interval (a, b). Such features speak in favor of Adaptive Numerical Integration in terms of flexibility and high accuracy for minimal costs. On the other hand, the method introduces the feasibility of approximating $I = \int_a^b f(x)dx$, based on a set of points that are not uniformly distributed over (a, b).

Remark 6.1 *The MATLAB command* quad *is a notorious implementation of adaptive Simpson's approximation for definite integrals.*

6.2   Numerical Integration of Functions of Two Variables

Consider the double integral:

$$I = \int \int_{\Omega} f(x,y)dxdy,$$

where $\Omega \subset \mathbb{R}^2$ with boundary $\Gamma = \partial\Omega$. Assume also that $f(x,y)$ is at least continuous on Ω.

The methods derived so far are difficult to generalize in a systematic way to double integration approximations over all domains $\Omega \subset \mathbb{R}^2$. The simplest case would be when Ω is a rectangular region.

6.2.1   Double Integrals over Rectangular Domains

Let $\Omega = (a,b) \times (c,d)$, in which case, if we define the rectangle corners by:

$$M = (a,c), \ N = (b,c), \ P = (b,d), \ Q = (a,d).$$

Then:

$$\Gamma = \Gamma_1 \cup \Gamma_2 \cup \Gamma_3 \cup \Gamma_4,$$

with

$$\Gamma_1 = \overrightarrow{MN}, \Gamma_2 = \overrightarrow{NP}, \Gamma_3 = \overrightarrow{PQ} \text{ and } \Gamma_4 = \overrightarrow{QM}.$$

In this case, I may be written as:

$$I = \int_a^b \int_c^d f(x,y)dydx = \int_a^b [\int_c^d f(x,y)dy]\,dx \tag{6.7}$$

or equivalently:

$$I = \int_a^b F(x)dx \tag{6.8}$$

with:

$$F(x) = \int_c^d f(x,y)dy. \tag{6.9}$$

We have thus reduced the initial double integral into two simple integrals (6.8) and (6.9). To obtain approximations formulae for I we start by partitioning each of (a,b) and (c,d), using respectively n and m subintervals, as follows:

$$a = x_0 < x_1 < ... < x_n = b; c = y_0 < y_1 < ... < y_m = d,$$

with $x_{i+1} - x_i = h_{i+1}, \forall i = 0, ..., n-1$ and $y_{j+1} - y_j = k_{j+1}, \forall j = 0, ..., m-1$. A display of such partitions can be found in Figure 6.2.

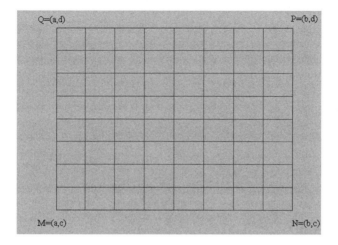

FIGURE 6.2: A partition of the rectangle $MNPQ$ with $m = n = 8$

6.2.2 Double Rectangular Rule

We generalize first the composite rectangular rule (5.41) as follows:

1. $I = \int_a^b F(x)dx \approx \sum_{i=1}^{n} h_i F(x_{i-1})$.

2. $F(x_{i-1}) = \int_c^d f(x_{i-1}, y)dy \approx \sum_{j=1}^{m} k_j f(x_{i-1}, y_{j-1})$.

Combining both approximations yields:

$$\int_a^b \int_c^d f(x, y)dxdy \approx \sum_{i=1}^{n} \sum_{j=1}^{m} h_i k_j f(x_{i-1}, y_{j-1}). \tag{6.10}$$

In case the partitioning points of (a, b) and (c, d) are equally spaced, i.e.,

$$h_i = h = \frac{b - a}{n}, \forall i \text{ and } k_j = k = \frac{d - c}{m}, \forall j,$$

then:

$$\int_a^b \int_c^d f(x, y)dxdy \approx hk \sum_{i=1}^{n} \sum_{j=1}^{m} f(x_{i-1}, y_{j-1}).$$

Under these conditions, the two-dimensional composite rectangular rule would be given by:

$$A(h, k) = hk \sum_{i=1}^{n} \sum_{j=1}^{m} f(x_{i-1}, y_{j-1}) \tag{6.11}$$

6.2.3 Double Trapezoidal and Midpoint Rules

A similar derivation may be also carried out for the composite trapezoidal rule. We leave out the details of the derivation and write directly the numerical integration formula::
$T(h, k) =$

$$\frac{hk}{4} \sum_{i=1}^{n} \sum_{j=1}^{m} (f(x_{i-1}, y_{j-1}) + f(x_{i-1}, y_j) + f(x_i, y_{j-1}) + f(x_i, y_j)). \quad (6.12)$$

In case n and m are even integers, the composite midpoint rectangular rule formula can be easily derived and given by:

$$M(h, k) = 4hk \sum_{i=1}^{n/2} \sum_{j=1}^{m/2} f(x_{2i-1}, y_{2j-1}) \quad (6.13)$$

6.2.4 Double Simpson's Rule

As for the composite Simpson's rule, **with n and m being even integers,** the double integration formula is derived as follows. In a first step, we use composite Simpson's integration with respect to y. This gives:

$$\int_c^d f(x, y)\, dy \approx \frac{k}{3} \sum_{j=1}^{m/2} f(x, y_{2j-2}) + 4f(x, y_{2j-1}) + f(x, y_{2j}).$$

Letting now:

$$F_k(x) = \frac{k}{3} \sum_{j=1}^{m/2} f(x, y_{2j-2}) + 4f(x, y_{2j-1}) + f(x, y_{2j}),$$

then proceeding with a composite Simpson's integration with respect to x on $F_k(x)$, we obtain:

$$\int_a^b F_k(x)dx \approx \frac{h}{3} \sum_{i=1}^{n/2} F_k(x_{2i-2}) + 4F_k(x_{2i-1}) + F_k(x_{2i}).$$

Thus we can write:

$$I \approx \int_a^b F_k(x)dx \approx S(h, k),$$

with:
$S(h, k) =$

$$S(h, k) = S_1(h, k) + S_2(h, k) + S_3(h, k). \quad (6.14)$$

with:

$$S_1(h,k) = \frac{hk}{9} \sum_{i=1}^{n/2} [f(x_{2i-2}, y_{2j-2}) + f(x_{2i-2}, y_{2j}) + f(x_{2i-2}, y_{2j-2}) + f(x_{2i}, y_{2j})],$$

$$S_2(h,k) = \frac{4hk}{9} \sum_{i=1}^{n/2} [f(x_{2i-2}, y_{2j-1}) + f(x_{2i-1}, y_{2j}) + f(x_{2i}, y_{2j-1}) + f(x_{2i-1}, y_{2j-2})],$$

and:

$$S_3(h,k) = \frac{16hk}{9} \sum_{j=1}^{m/2} f(x_{2i-1}, y_{2j-1}).$$

6.2.5 Error Estimates

Error estimates can also be easily derived for the approximating formulae (6.11), (6.12), (6.13) and (6.14). We start with an error analysis for the rectangular rule (6.11).

Theorem 6.1 *For f_x and $f_y \in C(\Omega)$, i.e., $f \in C^1(\Omega)$, the composite rectangular approximation is $O(h+k)$:*

$$I = A(h,k) + (b-a)(d-c)(h\frac{\partial f}{\partial x}(\xi,\eta) + k\frac{\partial f}{\partial y}(\xi_1,\eta_1)) \qquad (6.15)$$

where (ξ,η) and (ξ_1,η_1) are in the rectangle $(a,b) \times (c,d)$.

Proof. Starting with:

$$I = \int_a^b F(x)dx = \sum_{i=1}^n hF(x_{i-1}) + h(b-a)F'(\xi), \ \xi \in (a,b),$$

where $F'(\xi) = \int_c^d \frac{\partial f}{\partial x}(\xi,y)dy$, then using the mean-value theorem, we obtain:

$$F'(\xi) = \int_c^d \frac{\partial f}{\partial x}(\xi,y)dy = (d-c)\frac{\partial f}{\partial x}(\xi,\eta_i).$$

Hence:

$$I = \int_a^b F(x)dx = \sum_{i=1}^n hF(x_{i-1}) + h(b-a)\sum_{i=1}^n \frac{\partial f}{\partial x}(\xi,\eta_i), \ (\xi,\eta_i) \in (a,b) \times (c,d).$$

One concludes, using the intermediate value theorem, that:

$$I = \sum_{i=1}^n hF(x_{i-1}) + h(b-a)(d-c)\frac{\partial f}{\partial x}(\xi,\eta), \ (\xi,\eta) \in (a,b) \times (c,d). \qquad (6.16)$$

Furthermore, as:
$$F(x_{i-1}) = \int_c^d f(x_{i-1}, y)dy = ..$$

$$..k \sum_{j=1}^{n} f(x_{i-1}, y_{j-1}) + k(d-c)\frac{\partial f}{\partial y}(x_{i-1}, \zeta_i), \ \zeta_i \in (c, d), \tag{6.17}$$

then using the definition of $A(h, k)$ and combining (6.16) and (6.17), we reach:

$$I = A(h, k) + h(b-a)(d-c)\frac{\partial f}{\partial x}(\xi, \eta) + hk(d-c)\sum_{i=1}^{n}\frac{\partial f}{\partial y}(x_{i-1}, \zeta_i).$$

By applying a second time the intermediate value theorem on $\sum_{i=1}^{n}\frac{\partial f}{\partial y}(x_{i-1}, \zeta_i)$, one gets (6.15).   ∎

Similar procedures can be conducted to the other integration formulae: (6.12), (6.13) and (6.14). In what follows we give the results of such analyses.

1. For $f \in C^2(\Omega)$, the composite trapezoid and midpoint rectangular approximations satisfy the following estimates:

$$I = T(h, k) - \frac{(b-a)(d-c)}{12}(h^2\frac{\partial^2 f}{\partial x^2}(\xi, \eta) + k^2\frac{\partial^2 f}{\partial y^2}(\xi_1, \eta_1)), \tag{6.18}$$

and similarly we obtain:

$$I = M(h, k) + \frac{(b-a)(d-c)}{6}(h^2\frac{\partial^2 f}{\partial x^2}(\xi, \eta) + k^2\frac{\partial^2 f}{\partial y^2}(\xi_1, \eta_1)) \tag{6.19}$$

i.e., $I = T(h, k) + O(h^2) + O(k^2)$ and $I = M(h, k) + O(h^2) + O(k^2)$

2. Also, for $f \in C^4(\Omega)$, the composite double Simpson's rule satisfies the following estimate:

$$I = S(h, k) - \frac{(b-a)(d-c)}{180}(h^4\frac{\partial^4 f}{\partial x^4}(\xi, \eta) + k^4\frac{\partial^4 f}{\partial y^4}(\xi_1, \eta_1)) \tag{6.20}$$

i.e., $I = S(h, k) + O(h^4 + k^4)$.

Note that in (6.18), (6.19) and (6.20), the pairs (ξ, η) and (ξ_1, η_1) refer to generic points in Ω.
We illustrate through a case that uses the composite Simpson's rule.

Example 6.2 *Compute $I = \int_0^{2.5}\int_0^{1.4} x^4 y^4 \, dy \, dx$, using the composite Simpson's rule with $(m, n) \in \{(4, 4), (8, 8), (16, 16), (64, 64)\}$.*

Note that the exact value of I is $\frac{1}{25}(1.4)^5(2.5)^5 = 21.00875$. The results are summarized in Table 6.3.

(m, n)	$S(h, k)$	$|I - S(h, k)|$
$(4, 4)$	21.118313	5.215115×10^{-3}
$(8, 8)$	21.015589	3.255473×10^{-4}
$(16, 16)$	21.009177	2.03452×10^{-5}
$(64, 64)$	21.008752	7.94729×10^{-8}

TABLE 6.3: Results of use of double Simpson's rule to approximate $\int_0^{2.5} \int_0^{1.4} x^4 y^4 \, dy \, dx = 21.00875$

6.2.6 Double Integrals over Convex Polygonal Domains

Delaunay Meshing

For the general case of $\int_\Omega f(x, y) dx dy$ where Ω is a connected domain with a boundary $\partial\Omega$ that consists of a continuous finite sequence of smooth arcs, current practices start by "meshing" the domain Ω into triangles. More specifically, "meshing" Ω, consists in subdividing it into a set $\mathcal{T}$ of "triangles." In this chapter, we restrict our presentation to **convex polygonal domains** which can be easily "meshed" using the MATLAB delaunay command (for more details see [20], [13].)

A **Delaunay triangulation** starts with a set of nodes $\mathcal{P} = \{P_1, P_2, ..., P_N\}$ in Ω and its boundary $\partial\Omega$. It then generates a set $\mathcal{T} = \mathcal{T}(\mathcal{P})$, such that, if we let $\mathcal{C}(T)$ be the circumcircle associated with each $T \in \mathcal{T}$ which vertices are M, N, $P \in \mathcal{P}$, then:

$$\mathcal{C}(T) \text{ contains no node of } \mathcal{P} \text{ in its interior.} \tag{6.21}$$

In that way, Delaunay triangulations tend to maximize the minimum angle of all the angles of the triangles in $\mathcal{T}$ and therefore avoid "skinny" or "flat" triangles. In addition $\mathcal{T} = \{T_i \,|\, 1 \leq i \leq M\}$ satisfies the following properties:

$$\forall i, j \in \{1, 2, ..., M\} \ \overline{T_i} \cap \overline{T_j} = \begin{cases} \text{triangle itself when } i = j \\ \text{vertex} \\ \text{one side} \\ \phi \text{ empty set} \end{cases} \tag{6.22}$$

Note that the 2 triangles in Figure 6.3 do not conform to such meshing constraint. In addition to (6.21) and (6.22), $\mathcal{T}$ satisfy:

$$\bigcup_i \overline{T_i} = \overline{\Omega}. \tag{6.23}$$

Thus $\mathcal{T}$ covers Ω and one can write:

$$\int_\Omega f(x, y) dx dy = \sum_{T \in \mathcal{T}} \int_T f(x, y) dx dy.$$

FIGURE 6.3: Nonconforming triangles in meshing a domain

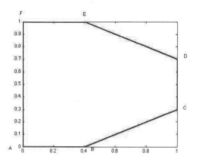

FIGURE 6.4: Plot of a two dimensional domain with a polygonal boundary
$\partial\Omega = \{(0,0),(0.4,0),(1,0.3),(1,0.7),(0.4,1),(0,1)\}$

We consider the following example of a hexagonal domain Ω with:

$$\partial\Omega = \{(0,0),(0.4,0),(1,0.3),(1,0.7),(0.4,1),(0,1)\}.$$

Figure 6.4 plots the boundary of this hexagonal domain: We mesh this domain
using a recursive procedure that starts with a "coarse mesh" based on the set
of nodes

$$\mathcal{P}_1 = \{(0,0),(0.4,0),(1,0.3),(1,0.7),(0.4,1),(0,1),G_1 = (0.4,0.5)\},$$

consisting of the vertices of Ω in addition to G_1, its barycenter (center of gravity). Applying MATLAB `delaunay` command on $\mathcal{P}_1$ followed by the `triplot`
command leads to $\mathcal{T}_1$, a triangulation consisting of 6 triangles, using a G_1
as a common vertex. The resulting meshing of Ω is shown in Figure 6.5. To
refine $\mathcal{T}_1$ we introduce the edges midpoints of each of its triangles, then proceeding again with the `delaunay` command followed by `triplet` which gives
a new mesh $\mathcal{T}_2$, consisting of 26 triangles, as shown in Figure 6.6. The previous steps are the core of our recursive procedure that allows **refinement
up to higher orders**. For instance, Figure 6.7 provides a mesh of 100 triangles resulting from subdividing the sides of the triangles in $\mathcal{T}_2$. The following

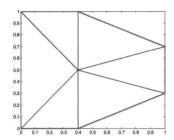

FIGURE 6.5: A coarse mesh for the polygonal domain with boundary $\partial\Omega = \{(0,0), (0.4,0), (1,0.3), (1,0.7), (0.4,1), (0,1)\}$

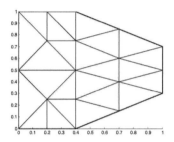

FIGURE 6.6: A 26 triangles mesh for the polygonal domain with boundary $\partial\Omega = \{(0,0), (0.4,0), (1,0.3), (1,0.7), (0.4,1), (0,1)\}$

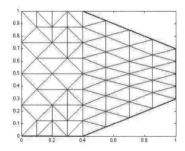

FIGURE 6.7: A 100 triangles mesh for the polygonal domain with boundary $\partial\Omega = \{(0,0), (0.4,0), (1,0.3), (1,0.7), (0.4,1), (0,1)\}$

algorithms generate such recursive processes starting with the initial coarse meshing:

Algorithm 6.2 Recursive Meshing of a Polygon

```
function [tri,x,y]=RecurProcess(x0,y0,reorder)
% Inputs: Vertices coordinates (x0,y0) of polygonal domain
% Outputs: Nodes [x,y] resulting from mid-edges refinement
%                 tri: Delaunay triangulation based on [x,y]
%Seek center of gravity of the polygon
m=length(x0); %equal to length(y0)
xg=0;yg=0;
for k=1:m
    xg=xg+x0(k);
    yg=yg+y0(k);
end
xg=xg/m;
yg=yg/m;
x=[x0 xg];y=[y0 yg];
tri=delaunay(x,y);
for k=1:reorder
    [tri,x,y]=MidEdges(tri,x,y);
end
triplot(tri)
```

Algorithm 6.2 uses the "Mid-edging" procedure 6.3 described as follows.

Algorithm 6.3 Mid-edging a Triangulation

```
function [tri1,x1,y1]=MidEdges(tri,x,y)
% Input: Triangulation tri based on the set of nodes (x,y)
% Outputs: Delaunay triangulation tri1 based on (x,y) in addition
%                 to mid-edges [x1,y1]
[m,n]=size(tri);
% Get the midpoints of all edges
mdx=zeros(3*m,1);mdy=zeros(3*m,1);
for k=1:m
  p=tri(k,1);q=tri(k,2);r=tri(k,3);
  mdx(3*k-2)=(x(p)+x(q))/2;mdy(3*k-2)=(y(p)+y(q))/2;
  mdx(3*k-1)=(x(q)+x(r))/2;mdy(3*k-1)=(y(q)+y(r))/2;
  mdx(3*k)=(x(r)+x(p))/2;mdy(3*k)=(y(r)+y(p))/2;
end
x=[x;mdx];y=[y;mdy];
% Eliminate any redundancy in the set of nodes [x,y]
Mdp=unique([x y],'rows');
x1=Mdp(:,1);
y1=Mdp(:,2);
tri1=delaunay(x1,y1);
```

Approximation of double integrals using triangular meshing

Since:

$$\int_{\Omega} f(x,y)dxdy = \sum_{T \in \mathcal{T}} \int_{T} f(x,y)dxdy,$$

the problem reduces to an approximation of a double integral over each of the triangles.

In case, M, N, P are the vertices of a triangle T, then a simple generalization of the one-dimensional trapezoidal rule uses an average of the value of $f(x,y)$ at these vertices, as follows:

$$\int_{T} f(x,y)dxdy \approx \frac{1}{3}\text{Area}(T)(f(M) + f(N) + f(P)). \qquad (6.24)$$

Moreover, the one-dimensional midpoint rule can be generalized using the center of gravity of the triangle T:

$$\int_{T} f(x,y)dxdy \approx Area(T)f(G). \qquad (6.25)$$

Both formulae (6.24) and (6.25) are exact for polynomials of the form $ax + by + c$.

Remark 6.2 *Note that to obtain more accurate approximations, a higher order formula that uses simultaneously the center of gravity and the vertices of T, can be used too. Specifically:*

$$\int_{T} f(x,y)dxdy \approx \frac{1}{12}\text{Area}(T)(f(M) + f(N) + f(P) + 9f(G)). \qquad (6.26)$$

is exact for polynomials of the form $axy + bx + cy + d$. Note that:

$$(6.26) = \frac{1}{4}(6.24) + \frac{3}{4}(6.25).$$

In Exercise 10, one proves that the formula:

$$\int_{T} f(x,y)dxdy \approx \frac{1}{6}\text{Area}(T)(f(m) + f(n) + f(p)), \qquad (6.27)$$

is exact for quadratic polynomials, i.e., polynomials of the form: $ax^2y^2 + bx^2 + cy^2 + dx + ey + f$, where m, n and p are respectively midpoints of the sides NP, PM and MN. Consequently, (6.27) is more accurate than (6.26).

In case **the domain Ω is not polygonal**, then its boundary is approached by a polygonal one: Ω_p, such that the area of $\Omega \cap \Omega_p$ is small, so that:

$$\int_{\Omega} f(x,y)dxdy \approx \int_{\Omega_p} f(x,y)dxdy.$$

Consequently, Ω_p is meshed using triangles, followed by applying (6.24) or (6.25) to approximate $\int_{\Omega_p} f(x,y)dy$.

6.3   Monte Carlo Simulations for Numerical Quadrature

In this section, we explore non-deterministic procedures for estimating definite integrals using **random numbers generation**.

6.3.1   On Random Number Generation

A sequence of numbers $\mathcal{R} = \{x_1, x_2, ..., x_n\}$ where $x_i \in (0,1) \; \forall \; i$ is said to be **random** if no correlation exists between successive numbers of this sequence. The elements $\{x_i\}$ are distributed throughout the interval $(0,1)$, with no pattern or rule linking the values of these elements. For example, if the numbers are monotonically increasing, they are not random; also if each $x_i = f(x_{i-1})$ where f is a simple continuous function, then the numbers are not randomly distributed. The integer n, which is the total number of elements in this sequence, is also called the number of **trials**.

In practice, the random sequence $\mathcal{R}_n$ is obtained using special **random-number generators** software procedures such as MATLAB, rand function, based on mathematical methods that can be extensively found in the literature, such as in [6], [17] and [24]. These procedures produce arrays of uniformly distributed "pseudo-random" numbers in the unit interval $(0,1)$ with each call of the random generation function (for example rand in MATLAB). More precisely such functions generate:

1. A sequence of numbers that is **uniformly distributed** in the interval $(0,1)$, i.e., with no subset of $(0,1)$ containing a share (of numbers) that is proportional to its size. For example, the probability that an element x of the sequence falls in the subinterval $[a, a+h]$ is h, and is independent from the number a. Similarly, if $p_i = (x_i, y_i)$ are uniformly distributed random points in some rectangle in the plane, then the number of these points that fall inside a square of area k should depend only on k and not on the location of the square inside the rectangle.

2. Moreover, the numbers produced by a computer code are **not completely random** since a "deterministic" mathematical algorithm is used to select these numbers. However, for practical purposes, these numbers are "sufficiently random" and for that reason, we refer to these as **pseudo- random** numbers.

Procedures to generate a sequence $\mathcal{R}_n = \{x_1, x_2, ..., x_n\}$ of pseudo-random numbers are usually based on an initial integer I_0 called the **seed** of the sequence. It is a number that controls whether the procedure repeats the same particular sequence after n reaches N, i.e.,

$$x_{N+i} = x_i, \; i = 1, 2, ..., N-1,$$

as theoretically, for a fixed value of the seed I_0, the random number generator can produce hundreds of thousands of pseudo random numbers before repeating itself. One example, [9], of an algorithm that generates a sequence of n pseudo-random numbers in single precision that are uniformly distributed in the interval $(0, 1)$ is as follows. Choose l_0 to be any integer between 1 and the *Mersenne prime number* $M = 2^{31} - 1 = 2147483647$. A MATLAB implementation of a pseudo-random numbers generator is as follows.

Algorithm 6.4 Pseudo-Random Generator

```
function x=myrand(n)
%Input: n is the length of the sequence
% Output:
%   Random numbers [x_1, x_2,...,x_n], x_i in (0, 1)
M=2^31-1;I0=M;
% I0 is the seed of the sequence
% 1<=l0, integer <=M=2^31-1 (Mersenne prime number).
% for example I0=M
x=ones(n,1);
y=I0;
 for i=1: n
    y=rem(7^5*y,M);
     x(i)=y / M;
end
```

Note that all the computed $x(i)$'s are numbers such that: $0 < x(i) < 1$.

6.3.2   Estimation of Integrals through Areas and Volumes

Consider the integral $I = \int_a^b f(x)dx$, which we identify with the area $\mathcal{A}$, located in a two-dimensional cartesian plane between:

$$x = a, \ x = b, \ y = 0 \text{ and } y = f(x).$$

Define now:

$$m = \min_{a \leq x \leq b} f(x) \text{ and } M = \max_{a \leq x \leq b} f(x).$$

Then one has:

$$m(b - a) \leq \mathcal{A} \leq M(b - a) \tag{6.28}$$

We assume now the existence of a procedure that **generates at random** any number N of ordered pairs $\{(x_i, y_i) | i = 1, ..., N\}$, where:

$$\forall i : a \leq x_i \leq b, \text{ and } m \leq y_i \leq M.$$

Of that N "throws," let us count n, as the number of hits, i.e., n is set initially to 0 and at each "throw," if:

$$y_i \times f(x_i) \geq 0 \text{ and } |y_i| \in [0, |f(x_i)|],$$

then n is incremented by 1. As a result and according to the law of large numbers:

$$(M - m) \times (b - a) \lim_{N \to \infty} \frac{n}{N} = \mathcal{A}.$$

Generation of random numbers is a rather difficult task. MATLAB rand function does generate **"pseudo-random"** numbers in the interval (0,1). The sequence generated through calling such functions is not perfectly random, but is reasonable for use in estimating integrals through Monte Carlo simulations. The following simple MATLAB procedure implements this type of method.

Algorithm 6.5 A Monte Carlo Simulation by "Hits"

```
function I=MonteCarlo1D(a,b,m,M,N)
%Input: The interval of integration (a,b)
%        The number of throws N
%        m and M, where m<=f(x)<=M
%Output: The Monte Carlo approximation I
n=0;% Initialize the number of hits
A=(M-m)*(b-a);%Area of rectangle in which area under f(x) lies
for i=1:N
    x=a+(b-a)*rand(1);
    y=m+(M-m)*rand(1);
    z=f(x);
    if y*z>=0
        if abs(y)<=abs(z)
            n=n+1;
        end
    end
end
I=A*n/N;
```

We give 2 examples resulting from this implementation.

Example 6.3 *The first deals with the integral* $I = \int_{-1}^{2} f(x)dx$, *with* $f(x) = 3(x - 1)^2 + 2(x - 1)$, *which exact value is* $I = 6$.

Two consecutive runs have been conducted, the first with 1,000 throws and the second with 10,000, obtaining respectively 238 and 2486 hits, leading to approximating I by respectively 5.95 and 6.215. These are illustrated in Figure 6.8.

Example 6.4 *The second considers the integral* $I = \int_{0}^{1} f(x)dx = \pi$, *with* $f(x) = 4\sqrt{1 - x^2}$.

Experiments were conducted for 100, 1,000, 5,000 and 10,000 throws. Figure 6.9 displays the results for the first 2 cases. Table 6.4 gives the results of these tests, while the graphs in Figure 6.9 illustrate the experiments for $N = 1000$ and 10,000. Thus, there is no indication that an increase in the number of

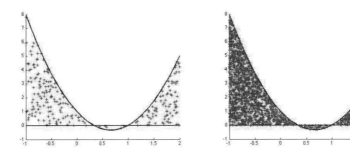

FIGURE 6.8: Application of Monte Carlo to $I = \int_{-1}^{2} 3(x-1)^2 + 2(x-1)dx$

N	n	I_n	$|I - I_n|/I$
100	81	3.24	3.132403×10^{-2}
1000	782	3.128	4.326676×10^{-2}
5000	782	3.124	5.6000×10^{-3}
10000	7,867	3.1468	1.657550×10^{-3}

TABLE 6.4: Application of Monte Carlo method to $I = \int_{0}^{1} 4\sqrt{1 - x^2}dx$

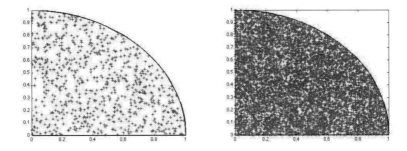

FIGURE 6.9: Application of Monte Carlo method to $I = \int_{0}^{1} 4\sqrt{1 - x^2}dx$

throws would allow a better approximation to π. With 15,000 throws, we reach only 3.1325 as approximation to π. This is caused by two factors:

- The pseudo-randomness of the numbers being generated, and

- The slowness of the convergence of a Monte Carlo simulation.

This implies that when opting to use this type of stochastic approximation, one must insure "almost perfect" randomness and at the same time expect long computation times.

6.3.3   Estimating Mean Values

One consequence of (6.28) is the mean value theorem that allows one to write:

For a simple integral:

$$I = \int_a^b f(x)dx = (b-a)f(c), \ c \in (a,b). \tag{6.29}$$

For a double and triple integral:

$$I = \int_\Omega f(x,y)dxdy = |\Omega|f(c), \ c \in \Omega, \tag{6.30}$$

$$I = \int_\Omega f(x,y,z)dxdydz = |\Omega|f(c), \ c \in \Omega, \tag{6.31}$$

with $|\Omega|$ being respectively the area and volume of the 2 (respectively 3) dimensions domain: Ω.

In either of these cases and regardless of the dimension of the domain Ω, the mean-value formula reduces the finding of I to:

1. Finding $|\Omega|$ ($|\Omega| = (b-a)$ in 1 dimension).

2. Estimating $f(c)$, with c being undetermined.

In case $|\Omega|$ is not known, then two tasks need to be carried out simultaneously. Noting first that:

- $\Omega = (a,b)$ in 1 dimension.

- In 2 dimensions, there exists $\{(a_i,b_i)|i=1,2\}$ such that $\Omega \subset (a_1,b_1) \times (a_2,b_2)$.

- In 3 dimensions, there exists $\{(a_i,b_i)|i=1,3\}$ such that $\Omega \subset (a_1,b_1) \times (a_2,b_2) \times (a_3,b_3)$.

Then, through a stochastic process of a Monte Carlo simulation, one generates a sequence of n hits out of N throws respectively on:

- (a, b) in 1 dimension. In such case $n = N$ with $|\Omega|_c = |\Omega| = b - a$

- $(a_1, b_1) \times (a_2, b_2)$, in 2 dimensions, leading to:

$$|\Omega|_c = (b_1 - a_1)(b_2 - a_2)\frac{n}{N}.$$

- And $(a_1, b_1) \times (a_2, b_2) \times (a_3, b_3)$, in 3 dimensions, giving:

$$|\Omega|_c = (b_1 - a_1)(b_2 - a_2)(b_3 - a_3)\frac{n}{N}.$$

The procedure that determines $|\Omega|_c \approx |\Omega|$, saves simultaneously the sequence:

$$\{P_1, P_2, ..., P_n\},$$

of the random points generated from these n hits. Consequently, the undetermined mean value $f(c)$ in (6.29), (6.30), (6.31) is estimated by the random sum:

$$f(c) \approx \frac{1}{n}(f(P_1) + f(P_2) + ... + f(P_n)), \tag{6.32}$$

which allows a final estimate of:

$$I \approx \frac{|\Omega|_c}{n}(f(P_1) + f(P_2) + ... + f(P_n)).$$

A possible measure of the error in approximating $f(c)$ by

$$\frac{1}{n}(f(P_1) + f(P_2) + ... + f(P_n)),$$

is given by the variance σ^2 of f, where

$$\sigma^2 = \overline{f^2} - (\overline{f})^2; \quad \text{with } \overline{f} = \frac{1}{n}\sum_{i=1}^{n} f(P_i) \text{ and } \overline{f^2} = \frac{1}{n}\sum_{i=1}^{n} f(P_i)^2$$

It is proved in [30] that the error incurred is of order $O(1/\sqrt{n})$.

We give examples of integrals over domains Ω, with a known value of $|\Omega|$, starting with a one-dimensional case on the computation of π computed previously using areas Monte Carlo simulation.

Example 6.5 *Compute $I = 4\int_0^1 \sqrt{1 - x^2}dx$ using mean-value Monte Carlo simulations.*

The results are given in Table 6.5. As in Example 6.4, one reaches the same conclusions regarding the slowness of the method and its dependence on "perfect" random generation.

We consider now two-dimensional examples, the first being an integral of an integral over a square.

Example 6.6 *Let $V = \int_0^{5/4} \int_0^{5/4}(4 - x^2 - y^2)\,dydx$.*

The analytical value of V can be found. Its exact value is $V = 4.622395833$. Using the Monte Carlo simulation with successively $n = 10^2, 10^3, 10^4$, the results are provided in Table 6.6.

n	I_n	$|I - I_n|/I$
100	3.245724	3.314608×10^{-2}
500	3.094700	1.492644×10^{-2}
1000	3.142022	1.367608×10^{-4}
2000	3.167981	8.39975×10^{-3}
5000	3.122143	6.190950×10^{-3}
10000	3.138454	9.990669×10^{-4}

TABLE 6.5: Results of Monte Carlo mean-value simulations to $I = 4 \int_0^1 \sqrt{1 - x^2} dx$

n	V_n	$\epsilon = |V - V_n|$
100	4.578791308	0.0436045
1000	4.581294418	0.0411014
10000	4.622980842	0.0005850

TABLE 6.6: Results for Monte Carlo approximations to $V = \int_0^{5/4} \int_0^{5/4} (4 - x^2 - y^2) \, dy dx$

6.4 Exercises

1. Derive estimate (6.3) for non-uniform meshes in the composite midpoint rule.

2. Derive estimate (6.4) for non-uniform meshes in the composite Simpson's rule.

3. Derive the identity (6.6) which estimates the $O(h^4)$ term in the composite Simpson's rule.

4. Derive the identity (6.19) that provides the error term in the composite double integration midpoint rule on rectangular domains.

5. Derive the identity (6.18) that explicits the error term in the composite double integration trapezoidal rule on rectangular domains.

6. With $m = n = 4$, approximate the following double integrals using the composite double trapezoid, and Simpson's rules.

   (a) $\int_{1.4}^{2} \int_{1}^{1.5} \ln(2xy) \, dy \, dx$

   (b) $\int_{2}^{2.2} \int_{2}^{2.6} (x^2 + y^3) \, dy \, dx$

7. With $m = n = 4$, approximate the following double integrals using the midpoint rule.

(a) $\int_2^4 \int_1^2 \ln(2xy)\, dy\, dx$

(b) $\int_2^3 \int_2^4 (x^2 + y^3)\, dy\, dx$

8. With $m = n = 2$, approximate the following double integrals using successively the composite double midpoint and Simpson's rules.

(a) $\int_0^1 \int_0^1 e^{y-x}\, dy\, dx$

(b) $\int_0^\pi \int_0^\pi \cos x\, dy\, dx$

9. With $m = n = 2$, approximate the following double integrals using successively the composite trapezoid rule.

(a) $\int_0^1 \int_0^1 e^{y-x}\, dy\, dx$

(b) $\int_0^\pi \int_0^\pi \cos x\, dy\, dx$

10. In reference to (6.27), let M, N, P be the vertices of a triangle T and m, n, p, respectively the midpoints of the sides of T: MN, NP, PM. Find the coefficients a, b, c, a_1, a_2, a_3 such that the approximation formula:

$$a_1 f(M) + a_2 f(N) + a_3 f(P) + b_1 f(m) + b_2 f(n) + b_3 f(p)$$

to $\int_T f(x, y)\, dx\, dy$ is **exact** for $f(x) = p(x)$, $p(x)$ a polynomial of degree 2, i.e., $p(x) = 1$, x, y, x^2, y^2, xy.

11. With $m = n = p = 2$, approximate the following triple integrals using successively the composite triple midpoint and trapezoid rules.

(a) $\int_{-1}^1 \int_1^2 \int_0^1 y\, dz\, dy\, dx$

(b) $\int_{-1}^1 \int_0^1 \int_1^2 xyz1\, dz\, dx\, dy$

6.5   Computer Exercises

1. Test MATLAB quad against this chapter function RecurAdaptSimp (Refer to Algorithm 6.1) for the following known integrals:

   - $\int_0^{100} (x^2 - 1)e^{-x}dx$, with absolute tolerance 0.5×10^{-7}.
   - $\int_0^{100} (x^3 - x)e^{-2x}dx$, with absolute tolerance 0.5×10^{-10}.

2. Write a MATLAB program that generates the results in Table 6.3 for $f(x, y) = x^4y^4$, then test your program for the following double integrals:

   - $\int_0^2 \int_0^1 (x^2 + y^2)e^{x+y}dxdy$.
   - $\int_0^2 \int_0^1 (x + y)(\sin^2(x) + \sin^2(y))dxdy$.

3. Consider the polygonal domain Ω shown in Figure 6.4. Test Algorithms 6.2 and 6.3 to approximate:

$$\int_\Omega e^{x+y}dxdy,$$

   on the meshes shown in Figures 6.5, 6.6 and 6.7.

4. Use Algorithm 6.5 to compute approximations I_n of $I = \int_a^b f(x)\,dx$ and as well $\frac{|I-I_n|}{|I|}$, for $n = 2^p$, $p = 4, 5, 6, 7, 8, 9, 10$, in the following cases:

   (a) $f(x) = \sqrt{x}$, $a = 0$, $b = 5$.
   (b) $f(x) = \sqrt{x + \sqrt{x}}$, $a = 0$, $b = 8$.

5. Extend Algorithm 6.5 to double integrals and apply it to find approximations I_n to $I = \int_0^{5/4} \int_0^{5/4} (\sqrt{4 - x^2 - y^2})\,dydx$ and simultaneously $\frac{|I-I_n|}{|I|}$ for $n = 10^p$, $p = 3, 4, 5, 6$, using the exact value $I = 2.66905414$.

6. Write a MATLAB program to approximate $\int \int_\Omega f(x, y)\,dxdy$ using a Monte Carlo method based on (6.30) that uses the approximation (6.32) in the following cases:

   (a) $f(x, y) = \sin(x)\cos(y)$, $\Omega = \{(x, y) : (x - 1)^2 + (y - 1)^2 \leq \frac{1}{4}\}$.
   (b) $f(x, y) = e^{x+y}$, Ω the polygonal domain shown in Figure 6.4.

Chapter 7

Numerical Solutions of Ordinary Differential Equations (ODEs)

7.1 Introduction

Differential equations involve the dependence of some variable $y(t)$ with respect to an independent time variable t. They are often used to model physical problems in engineering economics and natural and social sciences. There is a large number of references on the topics of analysis of ordinary differential equations and as well on numerical solutions to approximate solutions of differential equations. For that purpose, we cite [2], [5], [11], [18], [22], [21] [25], [28] and [30]. Note also that all standard textbooks on Scientific Computing include at least one chapter on Numerical Ordinary Differential Equations ([4], [7], [9], [29], [26] etc.).

We start this chapter by giving some specific ODEs models, with each describing a phenomenon for which one seeks a solution $y(t)$ over the time interval $[0, T]$.

Example 7.1 *The first one is that of a linear first-order ordinary differential equation that models a diffusive process of decay, for example that of a radioactive rate or that of a temperature with time.*

The modeling function $y(t)$ satisfies:

$$y'(t) + Ky(t) = s(t),\ 0 < t \leq T,\ y(0) = a. \tag{7.1}$$

where K is the rate of decay and $s(t)$ the "source" function of radioactivity or of heat.

Other well known models find their origin in dynamics and are based on the classical laws of motion. Examples are as follows:

Example 7.2 *The first-order rocket equation, where one seeks its velocity $y(t)$ that verifies:*

$$M(t)y' = K - F(y)y,\ 0 < t \leq T,\ y(0) = 0, \tag{7.2}$$

where K is the resulting rocket propulsive force, $M(t)$ is its time varying mass and $F(y)y$, a resistance force caused by friction with $F(y)y$ "smoothly" increasing with y, for example $F(y) = \frac{y^{1/2}}{\ln(2+y)}$.

Example 7.3 *The first-order population logistics equation:*

$$y' = a\left(1 - \frac{y}{b}\right)y,\ 0 < t,\ y(0) = y_0, \tag{7.3}$$

Example 7.4 *The second-order equation of the pendulum where $y(t)$ is its position verifies the following:*

$$y''(t) + A\sin(y(t)) = v(t),\ 0 < t \leq T,\ y(0) = a,\ y'(0) = b, \tag{7.4}$$

where A is a constant depending on the pendulum physical characteristics and $v(t)$ an external force depending on the time t; a and b are respectively the initial position and velocity of the pendulum.

Example 7.5 *The second-order **Van der Pol** equation associated with an oscillator subject to a non-linear damping force satisfies:*

$$y'' - \mu(1 - y^2)y' + y = 0,\ 0 < t \leq T,\ y(0) = a,\ y'(0) = b, \tag{7.5}$$

where $y(t)$ is the oscillator's position and μ a positive constant.

Although the pendulum and Van der Pol equations (7.4) and (7.5) are of the second-order, both can be reduced to a first-order system of two first-order differential equations. This can be done by introducing the variables:

$$\begin{cases} y_1(t) = y \\ y_2(t) = y' \end{cases}$$

One verifies in the case of (7.4), for example, that y_1 and y_2 satisfy:

$$\begin{cases} y_1' = y_2 \\ y_2' = -A\sin(y_1(t)) + s(t),\ 0 < t \leq T. \\ y_1(0) = a,\ y_2(0) = b. \end{cases} \tag{7.6}$$

By introducing vector notations, specifically:

$$Y(t) = \begin{pmatrix} y_1(t) \\ y_2(t) \end{pmatrix}$$

and

$$f(t, Y(t)) = \begin{pmatrix} y_2 \\ -A\sin(y_1(t)) + s(t) \end{pmatrix},$$

then the pendulum problem can be written as follows:

$$Y'(t) = f(t, Y(t)), \text{ with } 0 < t \leq T, \; Y(0) = Y_0 \tag{7.7}$$

given that:

$$Y_0 = \begin{pmatrix} a \\ b \end{pmatrix}.$$

More generally, an n-order initial value ordinary differential equation with $n \geq 1$ and written as:

$$y^{(n)}(t) = g(t, y, y', ..., y^{(n-1)}), \text{ with } t_0 < t \leq T, \; y^{(k)}(t_0) \text{ given } \forall \; 0 \leq k \leq n - 1$$

is amenable to a system of n first-order differential equations of the form (7.7) with an n-dimensional initial value vector $Y(t_0) = Y_0$ and $f : [t_0, T] \times \mathbb{R}^n \to \mathbb{R}^n$.

Although the computational methods considered in this chapter are applicable to (7.7), we will restrict our presentation to the general initial-value problem of a **first-order scalar ordinary differential equation**:

$$(IVP) \begin{cases} y'(t) = f(t, y(t)), \; t \in [t_0, T] \\ y(t_0) = y_0 \text{ given.} \end{cases}$$

where $y_0 \in \mathbb{R}$ and the function $f(.,.) : [t_0, T] \times \mathbb{R} \to \mathbb{R}$ is at least continuous over its domain. The interval $[t_0, T]$ (that could be finite or infinite) is also called the existence interval of the solution.

In the remaining part of this chapter, we start in Section 7.2 by presenting specific ODEs systems, for which analytical solutions can be found and intervals of existence are clearly specified. In the sequel, we give a general theorem on existence and uniqueness of solutions to ODEs. Then in Section 7.3 we provide the reader with general mathematical settings in which numerical methods for solving ODEs can be defined. Section 7.4 is dedicated to explicit Runge-Kutta methods while Section 7.5 presents Adams-Bashforth explicit and Adams-Moulton implicit methods. Section 7.6 gives a brief discussion on Multi-step Backward Difference Formulae while the last section handles a two-point boundary value problem using a finite-difference discretization.

7.2   Analytic Solutions to ODEs

Analytical Solutions In all of the above examples, only equation (7.1) leads to an expression of $y(t)$ in terms of t and the problem parameters. Specifically, one has:

$$y(t) = y(0)e^{-Kt} + \int_0^t e^{-K(t-s)}v(s)ds.$$

In case the integral $\int_0^t e^{-K(t-s)}v(s)ds$ can be formally found, then $y(t)$ can be obtained from this formula for all $t \in [0, T]$. Otherwise we can resort, using the techniques of the previous chapter, to a numerical computation of such integral.

Consider now the following simple initial value problem for which an analytic solution can be easily found:

Example 7.6 *Let*

$$y' = ay^p, \, a > 0, \, y(0) = 1, \tag{7.8}$$

Using the method of separation of variables, the solution of this initial value problem satisfies the formulae:

$$y(t) = \begin{cases} (1 + a(1-p)t)^{\frac{1}{1-p}}, \, p \neq 1, \\ e^{at}, \, p = 1. \end{cases} \tag{7.9}$$

The existence and properties of the solution depend on the values of the parameters a and p. The following results can be easily derived through standard separation of variables techniques to obtain analytic solutions. Specifically:

1. **Case 1: $a > 0$**
   If $p > 1$: the existence interval is finite with $[t_0, T) = [0, \frac{1}{a(p-1)})$ and $y(t) \to \infty$ as t increases. Note that the growth to ∞ of the solution can be fast (highly "steep")
   If $p \leq 1$: the existence interval is infinite with $[t_0, T) = [0, \infty)$. As above $y(t) \to \infty$ as t increases, but the growth to ∞ of the solution is rather slow.
   Figure 7.1 illustrates these results for $a = 1$.

2. **Case 2: $a < 0$**
   If $p < 1$: the existence interval is $[0, \frac{1}{a(p-1)})$, and the decay to 0 as t increases can be fast (highly "steep")
   If $p \geq 1$: the existence interval is $[0, \infty)$, and the decay to 0 as t increases is rather slow.
   For $a = -1$, these results are illustrated in Figure 7.2.

Existence Results
In general, analytical or formal solutions cannot be computed for (IVP).

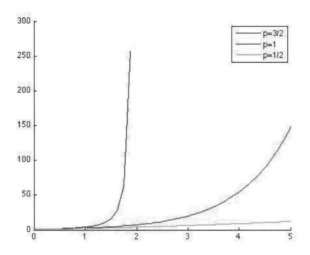

FIGURE 7.1: Graph of the solution to $y' = ay^p$, $a > 0$, $y(0) = 1$, $a = 1$

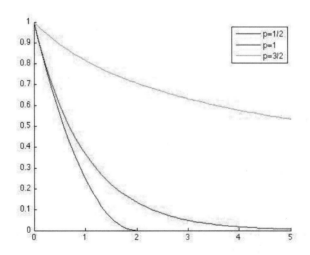

FIGURE 7.2: Graph of the solution to $y' = ay^p$, $a > 0$, $y(0) = 1$, $a = -1$

However, some results on existence and behavior of the solutions can be stated. For that purpose the initial value problem (IVP) is first written equivalently as an integral equation:

$$y(t) = y(t_0) + \int_{t_0}^{t} f(s, y(s))ds. \tag{7.10}$$

Such a problem could be handled through the study of an operator-function:

$$\mathcal{T} : v(t) \rightarrow z(t),$$

given by:

$$z(t) = \mathcal{T}(v(t)) = y(t_0) + \int_{t_0}^{t} f(s, v(s))ds, \tag{7.11}$$

and by proving that $\mathcal{T}$ has a unique solution that solves the (IVP) problem. This is usually obtained by imposing assumptions on the function f. Specifically, let $z_1(t) = \mathcal{T}(v_1(t))$ and $z_2(t) = \mathcal{T}(v_2(t))$, for 2 distinct functions v_1 and v_2. One writes then:

$$z_1(t) - z_2(t) = \int_{t_0}^{t} (f(s, v_1(s)) - f(s, v_2(s)))ds,$$

and introduces the following definition:

Definition 7.1 *A function $f(t, v) : D \subset \mathbb{R}^2 \rightarrow \mathbb{R}$ satisfies a **Lipschitz** condition in the variable v on the set D, if there exists a positive constant L_0, with:*

$$(\mathcal{L}_0) \quad |f(t, v_1) - f(t, v_2)| \leq L_0|v_1 - v_2|,$$

for all (t, v_1) and $(t, v_2) \in D$. The constant L_0 is called a Lipschitz constant for f on D.

Example 7.7 *Let $f(t, y) = t^2|y|$. Show that $f(.,.)$ satisfies a Lipschitz condition on the set $D = \{(t, y) \,|\, 1 \leq t \leq 5; -3 \leq y \leq 4\}$.*

Let (t, v_1) and $(t, v_2) \in D$, then

$$|f(t, v_1) - f(t, v_2)| = t^2||v_1| - |v_2|| \leq 55|v_1 - v_2|$$

Obviously, the Lipschitz constant is here, $L_0 = 25$.

■

Based on the Lipschitz condition $(\mathcal{L}_0)$, a general result of existence and uniqueness of the solution to (IVP) can be proved ([2]), by showing the operator $\mathcal{T}$ is "contracting" in the sense that:

$$|z_1(t) - z_2(t)| = |\mathcal{T}(v_1)(t) - \mathcal{T}(v_2)(t)| \leq \gamma \max_{0 \leq s \leq (t-t))} |v_1(s) - v_2(s)|.$$

where $\gamma < 1$ and $L_0(t - t_0) = \gamma < 1$. As γ and L_0 are independent from y_0, then one obtains:

Theorem 7.1 *Let $D = [t_0, T] \times \mathbb{R}$. If $f(t, y)$ is continuous and satisfies a Lipschitz condition ($\mathcal{L}_0$) in the variable y on D, then the initial value problem:*

$$(IVP) \ y'(t) = f(t, y(t)), \ t \in [t_0, T], \ y(t_0) = y_0$$

has a unique solution $y(t)$, $\forall t \in [t_0, T]$.

Remark 7.1 *Note in case the set D is given by $D = [t_0, T] \times \mathcal{I}_0$, where $\mathcal{I}_0 \subset \mathbb{R}$ contains the initial condition y_0, then $\mathcal{L}_0$ depends on y_0 and the existence interval (t_0, t_1) of Theorem 7.1 depends on y_0.*

Thus, by induction, one reaches a sequence of existence intervals, $[t_0, t_1]$, $[t_1, t_2]$,... that yields the final interval for the solution $[t_0, T_f]$, $T_f \leq \infty$.

Remark 7.2 *Note that the solution to (IVP) can be computed using an iterative scheme called **Picard's** iteration applied on (7.11), where a sequence of functions $\{y^{(k)}\}$ defined over the interval $[t_0, T]$ is generated, following the iterative process:*

$$\begin{cases} y^{(0)}(t) = y_0, \ \forall t \in [t_0, T] \\ y^{(k)}(t) = y_0 + \int_{t_0}^t f(s, y^{(k-1)}(s))ds, \ k \geq 1, \ \forall t \in [t_0, T]. \end{cases} \tag{7.12}$$

*(7.12) is a **Predictor-Corrector** type process leading to a sequence $\{y^{(k)}\}$ that converges to $y(t)$ on $[t_0, T]$.*

Example 7.8 *Consider the following initial value problem:*

$$y'(t) = -y + t + 1, \ 0 \leq t \leq 1, \ y(0) = 1$$

Use Picard's method to generate $y^{(i)}(t)$ for $0 \leq i \leq 3$

Using the iterative process (7.12), the following functions are obtained:

1. $y^{(0)}(t) = 1$, $\forall t \in [0, 1]$

2. $y^{(1)}(t) = 1 + \int_0^t s \, ds = 1 + \frac{t^2}{2}$, $\forall t \in [0, 1]$

3. $y^{(2)}(t) = 1 + \int_0^t (-1 - \frac{s^2}{2} + s + 1)ds = 1 + \frac{t^2}{2} - \frac{t^3}{6}$, $\forall t \in [0, 1]$

4. $y^{(3)}(t) = 1 + \frac{t^2}{2} - \frac{t^3}{6} + \frac{t^4}{24}$, $\forall t \in [0, 1]$

Note that the actual solution to this problem is $y(t) = t + e^{-t}$, while the first few terms of the Picard iteration correspond to the Mac Laurin series of $y(t)$, i.e.,

$$1 + \frac{t^2}{2} - \frac{t^3}{6} + \frac{t^4}{24} - \frac{t^5}{120} + ...,$$

with for $0 < t \leq T$,

$$|y(t) - y^{(k)}(t)| \leq \frac{t^{k+2}}{(k+2)!} \leq \frac{T^{k+2}}{(k+2)!},$$

indicating the rapid convergence of the Picard iteration to the solution $y(t)$. In particular, if one is working on an interval $(t_0, t_0 + h)$, then Picard's iteration gives:

$$|y^{(1)}(t) - y(t)| = |\int_{t_0}^{t} (f(s, y^{(0)}(s)) - f(s, y(s)))ds| \leq Lt \max_{t_0 \leq s \leq t_0 + h} |y^{(0)}(s) - y(s)|,$$

$\forall t \in (t_0, t_0 + h)$ and therefore:

$$|y^{(2)}(t) - y(t)| = |\int_{t_0}^{t} (f(s, y^{(1)}(s)) - f(s, y(s)))ds|$$

$$\leq L^2 \max_{t_0 \leq s \leq t_0 + h} |y^{(0)}(s) - y(s)| \int_{t_0}^{t} sds,$$

i.e.,

$$|y^{(2)}(t) - y(t)| \leq \frac{(Lh)^2}{2} \max_{t_0 \leq s \leq t_0 + h} |y^{(0)}(s) - y(s)|.$$

More generally, one has by induction:

$$|y^{(k)}(t) - y(t)| \leq \frac{(Lh)^k}{k!} \max_{t_0 \leq s \leq t_0 + h} |y^{(0)}(s) - y(s)|. \tag{7.13}$$

This indicates that Picard's iteration oder's of convergence is $O(\frac{h^k}{k!})$.

Remark 7.3 *Solutions to some ordinary differential equations may also exhibit an oscillatory behavior over long time intervals. Such is the case for example of the second-order ODEs (7.4) and (7.5) that are respectively the pendulum and Van der Pol equations.*

7.3   Mathematical Settings for Numerical Solutions to ODEs

We consider now some computational aspects related to the initial value problem

$$(IVP) \begin{cases} y'(t) = f(t, y(t)), \ t \in [t_0, T] \\ y(0) = y_0 \end{cases}$$

Numerical methods are devised to produce **discrete solutions** that are approximations to the exact solution $y(t)$ of (IVP) on a set of discrete points. Specifically, a discrete solution is usually a solution of a **difference equation** on a discrete set of time values $\{t_i | i = 0, 1, ..., N\}$, that partition the interval $[t_0, T]$ such that:

$$t_0 < t_1 < ... < t_{N-1} < t_N = T,$$

and that are usually equally spaced, i.e.,

$$t_i = t_0 + ih, \; \forall i = 1, ..., N$$

with $t_N = T = t_0 + Nh$ and $h = t_{i+1} - t_i = \frac{T - t_0}{N}$ being the **time step**. The interval $[t_0, T]$ is thus subdivided into N subintervals

$$\{[t_i, t_{i+1}] \,|\, i = 0, 1, ..., N - 1\}.$$

of equal lengths. However, there are (IVP) problems for which a uniform partition of $[t_0, T]$ is not convenient. In such case "adaptive" methods are designed that adapt the discrete time distribution according to the behavior of the solution. This topic is analyzed in later sections of this chapter. In all cases, one seeks a discrete solution in the form of a finite sequence:

$$Y_N = \{y_0, y_1, ..., y_N\}$$

that approximates the set of exact values of the solution $y(t)$:

$$\mathcal{Y}_N = \{y(0), y(t_1), ..., y(t_N)\}.$$

The elements of Y_N are such that:

$$y_0 = y(0) \text{ and } y_i \approx y(t_i), \; 0 < i \le N$$

Moreover, the sequence $Y_N = \{y_i\}_{i=0}^N$ satisfies a **difference equation**, which fits one of the following categories:

1. **One-step explicit difference equation** for $i \ge 1$, obtained from expressions of the form:

$$y_i = F^E(t_i, t_{i-1}, y_{i-1}) \Leftrightarrow \frac{y_i - y_{i-1}}{h} = G^E(t_i, t_{i-1}, y_{i-1}). \qquad (7.14)$$

requiring 1 initial value: $y_0 = y(0)$.

2. **One-step implicit difference equation** where for $i \ge 1$:

$$y_i = F^I(t_i, t_{i-1}, y_i, y_{i-1}) \Leftrightarrow \frac{y_i - y_{i-1}}{h} = G^I(t_i, t_{i-1}, y_i, y_{i-1}), \qquad (7.15)$$

requiring 1 initial value: $y_0 = y(0)$. Unlike (7.14), this last equation is generally nonlinear, requiring use of roots finding methods as described in Chapter 2 or a Picard iteration that would start with one application of an explicit scheme (7.14). Implicit methods may in some cases provide better discrete solutions than explicit methods, but require more computational effort at each step.

3. **k-Multi-steps explicit difference equation** for $k > 1$ and $i \geq k$, where:

$$y_i = F^{E,k}(t_i, ..., t_{i-k}, y_{i-1}, ..., y_{i-k}) \Leftrightarrow$$

$$\frac{y_i - y_{i-1}}{h} = G^{E,k}(t_i, ..., t_{i-k}, y_{i-1}, ..., y_{i-k}), \qquad (7.16)$$

requiring k initial values: $y_0 = y(t_0)$ and y_1, ..., y_{k-1}, usually obtained using one-step methods.

4. **k-Multi-steps implicit difference equation** , for $k > 1$ and $i \geq k$, where:

$$y_i = F^{I,k}(t_i, ..., t_{i-k}, y_i, ..., y_{i-k}) \Leftrightarrow$$

$$\frac{y_i - y_{i-1}}{h} = G^{I,k}(t_i, ..., t_{i-k}, y_i, ..., y_{i-k}) \qquad (7.17)$$

which also require k initial values in addition to solving at each time step some nonlinear equation.

Remark 7.4 *Combined use of explicit and implicit difference equations lead to a* **Picard's iteration predictor-corrector** *process, as indicated in Remark 7.2 with a rapid convergence as expressed by the estimate (7.13).*

For example in considering the one-steps methods (7.14) and (7.15), the explicit scheme gives a prediction y_i^P:

$$y_i^P = F^E(t_i, t_{i-1}, y_{i-1}),$$

and y_i^P is in turn corrected once through:

$$y_i^C = F^I(t_i, t_{i-1}, y_i^P, y_{i-1}),$$

leading to the final suggested approximation $y_i = y_i^C$. Note that several corrections can be applied to improve the first approximation y_i^P.

For the purpose of analyzing convergence of a numerical method solving (IVP), we start by introducing the error vector:

$$\mathcal{E} = \{e_0, e_1, ..., e_n\},$$

where $e_i = y(t_i) - y_i$, $i = 0, 1, ..., n$ with $e_0 = 0$. We may now define convergence of the discrete scheme as follows.

Definition 7.2 *A numerical method of the form (7.14), (7.15), (7.16) or (7.17), solving (IVP) is* **convergent** *on $[t_0, T]$, if:*

$$\lim_{h \to 0} \max_{1 \leq i \leq N} |e_i| = 0.$$

Furthermore, the convergence of the numerical method is of **order** *p, if $\max_{1 \leq i \leq N} |e_i| = O(h^p)$.*

Convergence and **order of convergence** results are usually determined from the analysis of the **local truncation error** of a method. Specifically:

Definition 7.3 *For all* $i = 1, 2, ..., N$, *the* **local truncation error** *of the difference equations (7.14), (7.15), (7.16) and (7.17), with respect to the exact solution* $y(t)$ *are respectively given by:*

$$E_i = E(y(t_i)) = \begin{cases} y(t_i) - F^E(t_i, t_{i-1}, y(t_{i-1})), \ i = 1, ..., N, \\ y(t_i) - F^I(t_i, t_{i-1}, y(t_i), y(t_{i-1})), \ i = 1, ..., N, \\ y(t_i) - F^{E,k}(t_i, ..., t_{i-k}, y(t_{i-1}), ..., y(t_{i-k})), \ i = k, ..., N, \\ y(t_i) - F^{I,k}(t_i, ..., t_{i-k}, y(t_i), y(t_{i-1}), ..., y(t_{i-k})), \ i = k, ..., N. \end{cases}$$
$$(7.18)$$

Furthermore, the difference method is said to be of order p, *if* $\max_i |E_i| = O(h^{p+1})$.

To obtain convergence results for a numerical method, an additional assumption on the difference method being used is needed. In this chapter, we **only illustrate this concept on a one-step explicit method**, (7.14). For this purpose, assume that the function $G \equiv G^E(t_i, t_{i-1}, w)$ satisfies a Lipschitz property with respect to w over a domain $\mathcal{D}_y \subset \mathbb{R}$ that includes the range $\mathcal{R}$ of the exact solution $y(t)$, i.e.,

$$\mathcal{R} = \{y(t) : t \in [t_0, T]\} \subset \mathcal{D}_y.$$

Then, $\forall t_i, t_{i-1} \in [t_0, T]$, and $\forall w, z \in \mathcal{D}_y$,

$$|G(t_i, t_{i-1}, w) - G(t_i, t_{i-1}, z)| \leq K|w - z|, \qquad (7.19)$$

where K is a function of $y(.)$ and T, but is independent from i and h. On that basis, we may prove the following convergence result:

Theorem 7.2 Convergence *If the local truncation error of the difference method (7.14) solving (IVP) is* $O(h^{p+1})$ *(p > 0), and the function* $G(.)$ *satisfies the Lipschitz property (7.19), then the sequence* $Y_N = \{y_0, y_1, ..., y_N\}$ *that solves (7.14) is such that:*

$$\max_{1 \leq i \leq N} |y_i - y(t_i)| = \max_{1 \leq i \leq N} |e_i| = O(h^p).$$

Proof. For simplicity and with no loss of generality, we prove this result for the case $t_0 = 0$. Proceeding by induction and given that:

$$\frac{y(t_1) - y(0)}{h} = G(t_1, 0, y(0)) + \frac{1}{h}E_1, \qquad (7.20)$$

using then (7.14):

$$\frac{y_1 - y_0}{h} = G(t_1, 0, y_0). \qquad (7.21)$$

then subtracting (7.20) and (7.21) leads to:

$$e_1 = e_0 + h(G(t_1, 0, y(0)) - G(t_1, 0, y_0)) + E_1.$$

Since $y(0) = y_0$, then $e_0 = 0$ and one has:

$$|e_1| = |E_1| = O(h^{p+1}).$$

Thus, $y_1 \approx y(t_1)$ and $y_1 \in \mathcal{D}_y$. Taking the procedure one step further, one has:

$$e_2 = e_1 + h(G(t_2, t_1, y(t_1)) - G(t_2, t_1, y_1)) + E_2.$$

Hence:

$$|e_2| \le |e_1| + h|G(t_2, t_1, y(t_1)) - G(t_2, t_1, y_1)| + |E_2|,$$

and therefore:

$$|e_2| \le (1 + hK)|e_1| + |E_2| = (1 + hK)|E_1| + |E_2|.$$

This implies that $y_2 \approx y(t_2)$, i.e., $y_2 \in \mathcal{D}_y$, allowing pursuing of the recurrence. Thus, more generally, one has:

$$|e_i| \le (1 + hK)^{i-1}|E_1| \ldots + (1 + hK)|E_{i-1}| + |E_i|, \ i \ge 1.$$

Hence:

$$|e_i| \le ((1 + hK)^{i-1} + \ldots + (1 + hK) + 1) \max_{1 \le k \le i} |E_k|, \ i \ge 1,$$

i.e.,

$$|e_i| \le \frac{(1 + hK)^i}{hK} \max_{1 \le k \le i} |E_k| \equiv \frac{(1 + hK)^i}{K} O(h^p).$$

Let $\epsilon_0 << 1$ be a small number. Then for $h \le h_0 = \frac{\epsilon_0}{K}$, one has:

$$|e_i| \le \frac{(1 + \epsilon_0)^i}{K} O(h^p),$$

which indicates simultaneously that $y_i \approx y(t_i)$ and $|y(t_i) - y_i| = O(h^p)$. Consequently:

$$\max_{1 \le i \le N} |y_i - y(t_i)| \le \frac{(1 + \epsilon_0)^N}{K} \max_{1 \le k \le N} |E_k|/h \equiv C_N h^p. \qquad (7.22)$$

with $C_N = \frac{1}{K} e^{\epsilon_0 N}$ ∎

Remark 7.5 *Note that the error estimate (7.22) depends on a constant C_N that grows exponentially like $e^{\epsilon_0 N}$. Reducing the effect of such growth implies using higher order methods in which the term $O(h^p)$ would damp large values taken by C_N.*

Remark 7.6 *It is also important to note that given the estimate:*

$$|e_i| \leq (1 + hk)^{i-1}|E_1|... + (1 + hK)|E_{i-1}| + |E_i|, \ i \geq 1,$$

then for "starting values of i," $i = 1, 2, 3, 4$, one has $|e_i| \leq (1 + hK)^4 \times O(h^{p+1}) \leq C_4 h^{p+1}$. Thus, the convergence order at the beginning of the numerical quadrature has the same order as the order of the truncation error.

Theorem 7.3 Stability *Let:*

$$\mathcal{Z}_N = \{z_0, z_1, ..., z_N\} \text{ and } \mathcal{W}_N = \{w_0, w_1, ..., w_N\},$$

*be two sets of solutions to (7.14), with respective initial conditions z_0 and w_0. Then under (7.19) and as $h \to 0$, the numerical scheme (7.14) is **stable** in the sense that:*

$$\forall i, \ 1 \leq i \leq N, \ |w_i - z_i| \leq c_N |w_0 - z_0|, \tag{7.23}$$

with $c_N = e^{\epsilon_0 N}$ defined in the previous theorem.

Proof. Given that:

$$w_i - z_i = w_{i-1} - z_{i-1} + h(G(t_{i-1}, t_i, w_{i-1}) - G(t_{i-1}, t_i, z_{i-1})),$$

one concludes using (7.19), that:

$$|w_i - z_i| \leq (1 + Kh)|w_{i-1} - z_{i-1}|.$$

Hence, by induction, one gets:

$$\forall i : 0 \leq i \leq N, \ |w_i - z_i| \leq (1 + Kh)^i |w_0 - z_0|,$$

and therefore:

$$\forall i : 0 \leq i \leq N, \ |w_i - z_i| \leq (1 + Kh)^N |w_0 - z_0|.$$

As in the previous theorem, using similar considerations for the choice of h, leads to the estimate (7.23).   ■

In what follows we will present the most widely used numerical methods starting with one-step explicit Runge-Kutta methods up to multi-step Adams methods.

7.4   Explicit Runge-Kutta Schemes

In numerical integration of ODEs, explicit Runge-Kutta methods (RK methods) form an important family of explicit one-step methods. These techniques were developed around 1900 by the German mathematicians C. Runge and M.W. Kutta.

One simple procedure that leads to a relation between $y(t_i)$ and $y(t_{i-1})$ is based on the numerical integration methods developed in Chapter 5. For that purpose, we start by transforming the initial value problem (IVP) into a sequence of integral equations obtained by integrating $y'(t) = f(t, y(t))$ from t_{i-1} to t_i, yielding:

$$y(t_i) - y(t_{i-1}) = \int_{t_{i-1}}^{t_i} f(t, y(t))dt, \ i = 1, ..., N. \tag{7.24}$$

7.4.1   Euler Explicit Method

The first and simplest formula is the rectangular rule (5.41) that gives for $f \in C^1[t_0, T]$ and $y \in C^2[t_0, T]$:

$$\int_{t_{i-1}}^{t_i} f(t, y(t)) \, dt = h \, f(t_{i-1}, y(t_{i-1})) + O(h^2)$$

thus yielding for all i, $1 \le i \le N$:

$$y(t_i) = y(t_{i-1}) + hf(t_{i-1}, y(t_{i-1})) + O(h^2) \tag{7.25}$$

Discretizing this last equation by replacing simultaneously $y(t_j)$ by y_j, $(j = i-1, i)$ and dropping the $O(h^2)$ truncation term, the classical **Euler explicit** scheme is obtained. This scheme consists of finding a discrete sequence $Y_N = \{y_i | i = 0, 1, ..., N\}$ such that:

$$\begin{cases} y_i = y_{i-1} + hf(t_{i-1}, y_{i-1}), \Leftrightarrow \frac{y_i - y_{i-1}}{h} = f(t_{i-1}, y_{i-1}), \ i = 0, 1, ..., N-1, \\ y_0 = y(t_0), \end{cases}$$
$$\tag{7.26}$$

Obviously, the local truncation error of $O(h^2)$. In the notations of (7.14):

$$F(t_i, t_{i-1}, y_{i-1}) \equiv y_{i-1} + hf(t_{i-1}, y_{i-1}) \text{ and } G(t_i, t_{i-1}, y_{i-1}) \equiv f(t_{i-1}, y_{i-1}).$$

Thus, if $f(., .)$ satisfies a Lipschitz condition as in (7.19), Theorems 7.2 and 7.3 are applicable and yield for Euler's method the following result:

Theorem 7.4 *If* $|f(t_{i-1}, w) - f(t_{i-1}, z)| \le K|w - z|$, $\forall \ i = 1, ..., N$, *and* $\forall \ w, z \in \mathcal{D}_y \subset \mathbb{R}$, *with* $\mathcal{D}_y$ *containing the range of* $y(t)$, *then for* h *sufficiently small:*

$$\max_{1 \le i \le N} |y_i - y(t_i)| \le C_N h,$$

with C_N *as defined in Theorem 7.2.*

Thus **Euler's method is of order 1**. For practical purposes, we express (7.26) in the format of a **one-stage Runge-Kutta** method. Specifically:

$$(RK1) \begin{cases} k_1 = f(t_{i-1}, y_{i-1}) \\ y_i = y_{i-1} + hk_1 \end{cases}$$

i	t_i	k_1	y_i	$y(t_i)$	$|y_i - y(t_i)|$
0	0.00	0.0000E+00	1.0000E+00	1.000000E+00	0.0000E+00
1	0.25	1.5625E-02	1.0000E+00	1.000977E+00	9.7704E-04
2	0.50	1.2549E-01	1.0039E+00	1.015748E+00	1.1841E-02
3	0.75	4.3676E-01	1.0353E+00	1.082314E+00	4.7036E-02
4	1.00	1.1445E+00	1.1445E+00	1.284025E+00	1.3956E-01
5	1.25	2.7941E+00	1.4306E+00	1.841079E+00	4.1049E-01
6	1.50	7.1858E+00	2.1291E+00	3.545308E+00	1.4162E+00
7	1.75	2.1039E+01	3.9256E+00	1.043042E+01	6.5049E+00
8	2.00	7.3481E+01	9.1852E+00	5.459815E+01	4.5413E+01

TABLE 7.1: Results of Euler's method for $y'(t) = t^3 y$, $t \in [0, 2]$, $y(0) = 1$

Computationally, implementing Euler's method would require **one function evaluation** $f(.,.)$, at each time step as shown in the following algorithm.

Algorithm 7.1 Euler's Method

```
% Input:  function f,  interval of existence [t0, T],  initial
%condition y0, and  time step h
% Output: sequence of approximations to the exact solution
% {y1, y2, ..., yn}
function y = Euler(f, 0, T, y0, h)
for  i=0:n-1
        k1 = f(t(i), y(i)) ;
        y(i+1) = y(i) + h*k1  ;
end
```

Example 7.9 *Use Euler's explicit scheme to solve the following initial value problem with time step $h = 0.25$:*

$$\begin{cases} y'(t) = t^3 y & t \in [0, 2] \\ y(0) = 1 \end{cases}$$

The corresponding discrete scheme with one-stage is given by:

$$(RK1) \begin{cases} k_1 = t_i^3 y_i \\ y_{i+1} = y_i + h k_1 \end{cases}$$

Since the analytical or exact solution is given by $y(t) = e^{\frac{t^4}{4}}$, we can therefore compute the absolute and relative errors at each t_i. These are provided in the last 2 columns of Table 7.1. Note the deterioration of the absolute error as t_i increases; $\max_i |y_i - y(t_i)| = O(h)$ for $t_i \leq 1$. This is compatible with the estimate found in Theorem 7.4, motivating the search for more accurate methods to approximate the solution for larger times.

7.4.2   Second-Order Explicit Runge-Kutta Methods

Second-order Runge-Kutta methods can be derived by approximating successively in (7.24), the integral $\int_{t_{i-1}}^{t_i} f(t, y(t))dt$ by the midpoint then the trapezoidal rules.

In the sequel, we will be using extensively the following consequence of the mean value theorem.

Proposition 7.1 *If* $f(.,.) : \mathbb{R}^2 \to \mathbb{R}$ *is a function of 2 variables and is of class* C^1, *then:*

$$f(t, z + O(\epsilon)) = f(t, z) + O(\epsilon)$$

∎

a. Use of the Midpoint Rule

Based on the midpoint rule, (7.24) can be written as:

$$y(t_i) = y(t_{i-1}) + h f(t_{i-1} + \frac{h}{2}, y(t_{i-1} + \frac{h}{2})) + O(h^3) \qquad (7.27)$$

Using Taylor's expansion on $y(t)$ yields:

$$y(t_{i-1} + \frac{h}{2}) = y(t_{i-1}) + \frac{h}{2}y'(t_{i-1}) + O(h^2) = y(t_{i-1}) + \frac{h}{2}f(t_{i-1}, y(t_{i-1})) + O(h^2).$$

Equation (7.27) becomes then:

$$y(t_i) = y(t_{i-1}) + h f(t_{i-1} + \frac{h}{2}, y(t_{i-1}) + \frac{h}{2}f(t_{i-1}, y(t_{i-1})) + O(h^2)) + O(h^3) \qquad (7.28)$$

Using Proposition 7.1 yields:

$$y(t_i) = y(t_{i-1}) + h f(t_{i-1} + \frac{h}{2}, y(t_{i-1}) + \frac{h}{2}f(t_{i-1}, y(t_{i-1})) + O(h^3) \qquad (7.29)$$

Dropping the $O(h^3)$ truncation error term and replacing $y(t_i)$ by y_i for all i leads to a second-order explicit method given by:

$$y_i = y_{i-1} + h f(t_{i-1} + \frac{h}{2}, y_{i-1} + \frac{h}{2}f(t_{i-1}, y_{i-1})), \; i = 1, 2, ..., N, \qquad (7.30)$$

or equivalently:

$$\frac{y_i - y_{i-1}}{h} = f(t_{i-1} + \frac{h}{2}, y_{i-1} + \frac{h}{2}f(t_{i-1}, y_{i-1})), \; i = 1, 2, ..., N. \qquad (7.31)$$

Using the notations in (7.14), we note that:

$$G(t_{i-1}, t_i, y_{i-1}) \equiv f(t_{i-1} + \frac{h}{2}, y_{i-1} + \frac{h}{2}f(t_{i-1}, y_{i-1})).$$

In that case, if $f(.,.)$ satisfies the Lipschitz condition:

$$|f(t, w) - f(t, z)| \le c|w - z|, \; \forall w, z \in \mathcal{D}_y, \forall t \in [0, T],$$

then for h sufficiently small:

$$|G(t_{i-1}, t_i, w) - G(t_{i-1}, t_i, z)| \leq c|w - z| + c\frac{h}{2}|w - z|,$$

i.e.,

$$|G(t_{i-1}, t_i, w) - G(t_{i-1}, t_i, z)| \leq K|w - z|, \forall w, z \in \mathcal{D}_y, \forall i = 1, ..., N$$

Thus, Theorems 7.2 and 7.3 are applicable and yield for this "modified" Euler's method the following result:

Theorem 7.5 *Under the assumptions of Theorem 7.4 and for h sufficiently small, the sequence $Y_N = \{y_0, y_1, ..., y_n\}$ obtained from the modified Euler equation (7.30) satisfies:*

$$\max_{1 \leq i \leq N} |y_i - y(t_i)| \leq C_N h^2,$$

with $y_0 = y(t_0)$ and C_N as defined in Theorem 7.2.

Thus the **modified Euler's method is of order 2**. For practical purposes, we express (7.30) in the format of a **two-stage Runge-Kutta** method. Specifically:

$$(RK2) \begin{cases} k_1 = f(t_{i-1}, y_{i-1}) \\ k_2 = f(t_{i-1} + \frac{h}{2}, y_{i-1} + \frac{h}{2}k_1) \\ y_i = y_{i-1} + hk_2 \end{cases}$$

with a local truncation error of $O(h^3)$.

Computationally, the implementation of $(RK2)$ requires **two function evaluations** $f(.,.)$ at each time step as shown in the following algorithm.

Algorithm 7.2 Modified Euler's Method

```
% Input:  function f,  interval of existence [t0, T],  initial
%condition y0, and  time step h
% Output: sequence of approximations to the exact solution
% {y1, y2, ..., yn}
function y = ModifiedEuler(f, 0, T, y0, h)
for  i=0:n-1
        k1 = f(t(i), y(i)) ;
        k2 = f(t(i)+h/2, y(i)+h*k1/2) ;
        y(i+1) = y(i) + h*k2 ;
end
```

b. Use of the Trapezoidal Rule Method: Heun's Method

Another second-order Runge Kutta method of order 2 (referred to as Heun's method) is obtained based on the trapezoidal rule applied to (7.24). One then obtains:

$$y(t_i) = y(t_{i-1}) + \frac{h}{2}\left[f(t_{i-1}, y(t_{i-1})) + f(t_i, y(t_i))\right] + O(h^3). \qquad (7.32)$$

Using Taylor's formula, one has

$$y(t_i) = y(t_{i-1}) + hy'(t_{i-1}) + O(h^2) = y(t_{i-1}) + h\,f(t_{i-1}, y(t_{i-1})) + O(h^2),$$

implying that:

$$f(t_i, y(t_i)) = f(t_i, y(t_{i-1}) + h\,f(t_{i-1}, y(t_{i-1})) + O(h^2))$$

Using Proposition 7.1, equation (7.32) becomes:

$$y(t_i) = y(t_{i-1}) + \frac{h}{2}\left[f(t_{i-1}, y(t_{i-1})) + f(t_i, y(t_{i-1}) + hf(t_{i-1}, y(t_{i-1})))\right] + O(h^3)$$
$$(7.33)$$

Again, by dropping the $O(h^3)$ term and replacing $y(t_i)$ by y_i, for all i, yields according to the notations in (7.14):

$$y_i = y_{i-1} + \frac{h}{2}\left[f(t_{i-1}, y_{i-1}) + f(t_i, y_{i-1} + hf(t_{i-1}, y_{i-1}))\right] \equiv F(t_{i-1}, t_i, y_i),$$
$$(7.34)$$

or equivalently:

$$\frac{y_i - y_{i-1}}{h} = \frac{1}{2}\left[f(t_{i-1}, y_{i-1}) + f(t_i, y_{i-1} + hf(t_{i-1}, y_{i-1}))\right] \equiv G(t_{i-1}, t_i, y_{i-1}).$$

As for the previous second-order Runge-Kutta method, Theorems 7.2 and 7.3 are applicable in case the function $f(.,.)$ satisfies a Lipschitz condition, thus yielding the second-order property of the method. Specifically:

Theorem 7.6 *Under the assumptions of Theorem 7.4, then for h sufficiently small, the sequence $Y_N = \{y_0, y_1, ..., y_n\}$ obtained from (7.34) satisfies:*

$$\max_{1 \le i \le N} |y_i - y(t_i)| \le C_N h^2,$$

with $y_0 = y(t_0)$ and C_N as defined in Theorem 7.2.

(7.34) can be also expressed in the format of a 2-stage Runge-Kutta method:

$$(RK2.H) \begin{cases} k_1 = f(t_{i-1}, y_{i-1}) \\ k_2 = f(t_i, y_{i-1} + hk_1) \\ y_i = y_{i-1} + \frac{h}{2}(k_1 + k_2), \end{cases} \qquad (7.35)$$

which has a local truncation error of $O(h^3)$ and a convergence order of $O(h^2)$. As a straightforward application, we consider now the following example.

t_i	k_1	k_2	y_i	$y(t_i)$	$|y_i - y(t_i)|$
0.00	0.00000E+00	1.56250E-02	1.00000E+00	1.00000E+00	0.00000E+00
0.25	1.56555E-02	1.25489E-01	1.00195E+00	1.00098E+00	9.76086E-04
0.50	1.27450E-01	4.36863E-01	1.01960E+00	1.01575E+00	3.84845E-03
0.75	4.59901E-01	1.14762E+00	1.09014E+00	1.08231E+00	7.82100E-03
1.00	1.29108E+00	2.83684E+00	1.29108E+00	1.28403E+00	7.05028E-03
1.25	3.52942E+00	7.58782E+00	1.80706E+00	1.84108E+00	3.40139E-02
1.50	1.07889E+01	2.43602E+01	3.19672E+00	3.54531E+00	3.48588E-01
1.75	4.06796E+01	1.01402E+02	7.59036E+00	1.04304E+01	2.84006E+00
2.00	2.02805E+02	5.77518E+02	2.53506E+01	5.45982E+01	2.92475E+01

TABLE 7.2: Results of Heun's method for $y'(t) = t^3 y$, $t \in [0, 2]$, $y(0) = 1$

Example 7.10 *Use the second-order Runge-Kutta method (Heun's form) to solve the initial value problem of the preceding example.*

The corresponding discrete scheme resulting from $(RK2.H)$ gives:

$$\begin{cases} k_1 = t_i^3 y_i \\ k_2 = (t_i + h)^3 (y_i + h k_1) \\ y_{i+1} = y_i + \frac{h}{2}[k_1 + k_2] \end{cases}$$

The numerical results are presented in Table 7.2. Note that $\max_i |e_i| = 0.135$ is compatible with the $O(h^2)$ order of the method.

Remark 7.7 An implicit second-order Runge-Kutta method
Note that if we discretize directly (7.32), we get the implicit second-order method:

$$y_i = y_{i-1} + \frac{h}{2} \left[f(t_{i-1}, y_{i-1}) + f(t_i, y_i) \right], \ i = 1, 2, ..., N \qquad (7.36)$$

that can be put in the form (7.15):

$$y_i = F(t_{i-1}, t_i, y_{i-1}, y_i) \equiv y_{i-1} + \frac{h}{2} \left[f(t_{i-1}, y_{i-1}) + f(t_i, y_i) \right].$$

Equation (7.36) is non-linear in y_i and may be solved through a predictor-corrector process. Several choices are available:

- $\begin{cases} y_i^{(P)} = y_{i-1} \\ y_i^{(C)} = y_{i-1} + \frac{h}{2} \left[f(t_{i-1}, y_{i-1}) + f(t_i, y_i^{(P)}) \right]. \end{cases}$

- $\begin{cases} y_i^{(P)} = y_{i-1} + h f(t_{i-1}, y_{i-1}), \ y_i^{(P)} \ \textit{is obtained using Euler's method} \\ y_i^{(C)} = y_{i-1} + \frac{h}{2} \left[f(t_{i-1}, y_{i-1}) + f(t_i, y_i^{(P)}) \right]. \end{cases}$

$a_1 = 0$	
a_2	b_{21}
a_3	$b_{31}\ b_{32}$
a_4	$b_{41}\ b_{42}\ b_{43}$
......	....
a_s	$b_{s1}\ b_{s2} ....b_{s,s-1}$
	$w_1\ w_2\ w_3\ ....w_{s-1}\ w_s$

TABLE 7.3: Coefficients of an s-stage Runge-Kutta method

Note also that the second alternative is precisely Heun's method, therefore asserting that the predicted estimate is a good choice. As for the first alternative, the predicted value being inaccurate, a second correction would be necessary to reach an acceptable approximation for y_i, specifically:

$$\begin{cases} y_i^{(P)} = y_{i-1} \\ y_i^{(C),1} = y_{i-1} + \frac{h}{2}\ [f(t_{i-1}, y_{i-1}) + f(t_i, y_i^{(P)})] \\ y_i^{(C),2} = y_{i-1} + \frac{h}{2}\ [f(t_{i-1}, y_{i-1}) + f(t_i, y_i^{(C),1})]. \end{cases}$$

7.4.3 General Explicit Runge-Kutta Methods

The three methods introduced above: Euler explicit ($RK1$), modified Euler ($RK2$) and Heun's ($RK2.H$) methods belong in fact to the more general family of Runge-Kutta methods whose order of convergence is greater than zero and with general form given by:

$$(RK_s) \begin{cases} k_1 = f(t_{i-1} + a_1 h, y_{i-1}), (a_1 \text{ usually } 0) \\ k_2 = f(t_{i-1} + a_2 h, y_{i-1} + b_{21} h k_1) \\ k_3 = f(t_{i-1} + a_3 h, y_{i-1} + b_{31} h k_1 + b_{32} h k_2) \\ .... \\ k_s = f(t_{i-1} + a_s h, y_{i-1} + b_{s,1} h k_1 + b_{s,2} h k_2 + ... + b_{s,s-1} h k_{s-1}) \\ y_i = y_{i-1} + h(w_1 k_1 + w_2 k_2 + ... + w_s k_s), \end{cases}$$

$$(7.37)$$

All the coefficients of an (RK_s) method are usually put in a tabular form as in 7.3 implying that an (RK_s) method can be described by a column vector $\{a_i|\ i = 1, ..., s\}$, an $s \times s$ strictly lower triangular matrix for the coefficients $\{b_{ij}\}$ and a row vector for the weights $\{w_i|\ i = 1, ..., s\}$.

The basic criteria for the selection of the coefficients is to reach an $O(h^{s+1})$ **truncation error**, i.e., given that $y \in C^{s+1}$, $f \in C^s$,

$$y(t_i+h)-y(t_i)-h(w_1 k_1(y(t_{i-1}))+w_2 k_2(y(t_{i-1}))+....+w_s k_s(y(t_{i-1}))) = O(h^{s+1}),$$

$s \geq 1$, which in turn practically implies that:
1. $\forall i = 2, ..., s,\ \sum_{j=1}^{i-1} b_{ij} = a_i$
2. $\sum_{i=1}^{s} w_i$

We proceed with general RK methods of order 2, 3 and 4.

1. Methods of order 2.
This class is described by the formulae:

$$(RK_2) \begin{cases} k_1 = f(t_{i-1}, y_{i-1}) \\ k_2 = f(t_{i-1} + ah, y_{i-1} + bhk_1) \\ y_i = y_{i-1} + h(w_1 k_1 + w_2 k_2), \end{cases} \tag{7.38}$$

There are four coefficients a, b, w_1 and w_2 to be determined on the assumption that $y \in C^3$ in view of having:

$$y(t + h) - y(t) - w_1 k_1(y) - w_2 k_2(y) = O(h^3).$$

For that purpose, we proceed with a Taylor's expansion to write:

$$y(t + h) = y(t) + hy'(t) + \frac{h^2}{2} y''(t) + O(h^3).$$

Equivalently:

$$y(t + h) = y(t) + hf(t, y(t)) + \frac{h^2}{2}(f_t(t, y(t)) + f_y(t, y(t))f(t, y(t)) + O(h^3).$$

If $y(t+h) - y(t) - w_1 k_1(y) - w_2 k_2(y) = O(h^3)$, it means that after expanding $k_1(y)$ and $k_2(y)$, we would select the four coefficients of the method in view of canceling the three terms $f(t, y(t))$, $f_t(t, y(t))$ and $f_y(t, y(t))$. Obviously, this would lead to three equations in four unknowns and hence a family of method that depends on one parameter.

On the basis that $f \in C^2$ (since $y \in C^3$), a two-variable Taylor's expansion for:

$$\phi(h) = f(t + ah, y(t) + bhf(t, y(t)),$$

gives :

$$\phi(h) = \phi(0) + h\phi'(0) + O(h^2),$$

i.e.,

$$\phi(h) = f(t, y(t)) + ahf_t(t, y(t)) + bhf_y(t, y(t))f(t, y(t)) + O(h^2)$$

Consequently,

$$y(t+h) - y(t) - w_1 k_1(y) - w_2 k_2(y) = (1 - w_1 - w_2)f(t, y(t)) + h^2((\frac{1}{2} - w_2 a)f_t(t, y(t)) +$$

$$(\frac{1}{2} - w_2 b)f_y(t, y(t))f(t, y(t)) + O(h^3).$$

This leads to the equations:

$$w_1 + w_2 = 1, \quad aw_2 = \frac{1}{2}, \quad bw_2 = \frac{1}{2}. \tag{7.39}$$

0	
$\frac{1}{2w}$	$\frac{1}{2w}$
	$1 - w$ w

TABLE 7.4: Coefficients of a general two-stage Runge-Kutta method

Hence, the solution can be written in terms of one parameter $w = w_2 > 0$, the other three being:

$$w_1 = 1 - w \, ; a = b = \frac{1}{2w}.$$

Consequently, we obtain a second-order Runge-Kutta family that depends on one parameter w, $\frac{1}{2} \leq w \leq 1$:

$$(RK_2(w)) \begin{cases} k_1 = f(t_{i-1}, y_{i-1}) \\ k_2 = f(t_{i-1} + \frac{1}{2w}h, y_{i-1} + \frac{h}{2w}k_1) \\ k_3 = f(t_{i-1} + a_3h, y_{i-1} + b_{31}hk_1 + b_{32}hk_2) \\ y_i = y_{i-1} + h((1 - w)k_1 + wk_2). \end{cases} \qquad (7.40)$$

In a tabular form, a general second-order Runge Kutta is given in Table 7.4. The previous schemes of modified Euler and Heun, obtained by numerical integration, are particular cases of this family $(RK_2(w))$, respectively for $w = 1$ and $w = \frac{1}{2}$.

2. Runge-Kutta methods of order higher than 2.

As we proceeded for second-order Runge-Kutta methods, third-order ones are also established on the basis of Taylor's expansions. On the basis of (RK_s), a general third-order Runge Kutta method has the following form:

$$(RK_3) \begin{cases} k_1 = f(t_{i-1}, y_{i-1}) \\ k_2 = f(t_{i-1} + a_2h, y_{i-1} + b_{21}hk_1) \\ k_3 = f(t_{i-1} + a_3h, y_{i-1} + b_{31}hk_1 + b_{32}hk_2) \\ y_i = y_{i-1} + h(w_1k_1 + w_2k_2 + w_3k_3). \end{cases} \qquad (7.41)$$

The eight coefficients $\{w_i\}$, $\{a_i\}$, et $\{b_{ij}\}$ are determined on the basis that for $y \in C^4$ $(f \in C^3)$, one has:

$$y(t + h) - y(t) - w_1k_1(y) - w_2k_2(y) - w_3k_3(y) = O(h^4). \qquad (7.42)$$

Writing the method in tabular form, gives Table 7.5. The (7.42) would imply canceling in the expansion of $y(t + h) - y(t) - w_1k_1(y) - w_2k_2(y) - w_3k_3(y)$ the six terms:

$$f(t, y(t)), \ f_t(t, y(t)) \ f_y(t, y(t)), \ f_{tt}(t, y(t)), \ f_{ty}(t, y(t)) \text{ and } f_{yy}(t, y(t)),$$

$$
\begin{array}{c|ccc}
0 & & & \\
a_2 & b_{21} & & \\
a_3 & b_{31} & b_{32} & \\
\hline
 & w_1 & w_2 & w_3
\end{array}
$$

TABLE 7.5: Coefficients of a general three-stage Runge-Kutta method

thus leading to six equations in eight unknowns and therefore a family of third-order Runge-Kutta methods depending on two variables. This will not be done here. For that purpose, we refer the reader to [18].

Instead, we give a third-order Runge-Kutta method (Heun of order 3) that can be obtained using numerical quadrature on the integral equation (7.24). Specifically, one uses the numerical integration formula:

$$
\int_{t_{i-1}}^{t_i} f(t, y(t))dt = \frac{h}{4}(f(t_{i-1}, y(t_{i-1})) + 3f(t_{i-1} + \frac{2h}{3}, y(t_{i-1} + \frac{2h}{3})) + O(h^4).
$$

Thus:

$$
y(t_i) = y(t_{i-1}) + \frac{h}{4}(f(t_{i-1}, y(t_{i-1})) + 3f(t_{i-1} + \frac{2h}{3}, y(t_{i-1} + \frac{2h}{3})) + O(h^4).
$$

Combined with the formula for the modified Euler on $[t_{i-1}, t_{i-1} + \frac{2h}{3}]$, one has:

$$
y(t_{i-1} + \frac{2h}{3}) = y(t_{i-1}) + \frac{2h}{3}f(t_{i-1} + \frac{h}{3}, y(t_{i-1}) + \frac{h}{3}f(t_{i-1}, y(t_{i-1})) + O(h^3),
$$

to conclude, combining the last identities, with:

$$
y(t_i) = y(t_{i-1}) + \frac{h}{4}(f(t_{i-1}, y(t_{i-1})) +
$$

$$
\frac{h}{4}(3f(t_{i-1} + \frac{2h}{3}, y(t_{i-1}) + \frac{2h}{3}f(t_{i-1} + \frac{h}{3}, y(t_{i-1}) + \frac{h}{3}f(t_{i-1}, y(t_{i-1}))) + O(h^4).
$$

Discretizing this equation by dropping the $O(h^4)$ term and replacing the $y(t_i)$ by y_i for all i gives the three-stage Runge-Kutta Heun of order 3:

$$
(RK3.H) \quad
\begin{cases}
k_1 = f(t_{i-1}, y_{i-1}) \\
k_2 = f(t_{i-1} + \frac{h}{3}, y_{i-1} + \frac{h}{3}k_1) \\
k_3 = f(t_{i-1} + \frac{2h}{3}, y_{i-1} + \frac{2h}{3}k_2) \\
y_i = y_{i-1} + h(\frac{1}{4}k_1 + \frac{3}{4}k_3).
\end{cases}
\quad (7.43)
$$

This method is summarized in Table 7.6. The same analysis can be carried

0			
$\frac{1}{3}$	$\frac{1}{3}$		
$\frac{2}{3}$	0	$\frac{2}{3}$	
	$\frac{1}{4}$	0	$\frac{3}{4}$

TABLE 7.6: Coefficients of a three-stage Runge-Kutta Heun method

0			
$\frac{1}{2}$	$\frac{1}{2}$		
$\frac{1}{2}$	0	$\frac{1}{2}$	
1	0	0	1
	$\frac{1}{6}$	$\frac{2}{6}$	$\frac{2}{6}$	$\frac{1}{6}$

TABLE 7.7: Coefficients of the classical fourth-order Runge-Kutta method

out for fourth-order Runge-Kutta, defined by:

$$(RK_4) \begin{cases} k_1 = f(t_{i-1}, y_{i-1}) \\ k_2 = f(t_{i-1} + a_2 h, y_{i-1} + b_{21} h k_1) \\ k_3 = f(t_{i-1} + a_3 h, y_{i-1} + b_{31} h k_1 + b_{32} h k_2) \\ k_4 = f(t_{i-1} + a_4 h, y_{i-1} + b_{41} h k_1 + b_{42} h k_2 + b_{43} h k_3) \\ y_i = y_{i-1} + h(w_1 k_1 + w_2 k_2 + w_3 k_3 + w_4 k_4). \end{cases} \tag{7.44}$$

Seeking the thirteen unknown coefficients in order to have for $y \in C^5$ $f \in C^4$:

$$y(t+h) - Y(t) - a_1 k_1(y) - a_2 k_2(y) - a_3 k_3(y) - a_4 k_4(y) = O(h^5),$$

gives rise to a system of ten equations in thirteen unknowns and therefore a family of methods that depend on three parameters.

We choose to give some of the mostly used fourth-order Runge-Kutta methods:

1. First fourth-order Runge-Kutta summarized in Table 7.7 expressed in formulae as:

$$(RK4.1) \begin{cases} k_1 = f(t_{i-1}, y_{i-1}) \\ k_2 = f(t_{i-1} + \frac{h}{2}, y_{i-1} + \frac{1}{2} k_1) \\ k_3 = f(t_{i-1} + \frac{h}{2}, y_{i-1} + \frac{1}{2} k_2) \\ k_4 = f(t_{i-1} + h, y_{i-1} + k_3) \\ y_i = y_{i-1} + h(\frac{1}{6} k_1 + \frac{2}{6} k_2 + \frac{2}{6} k_3 + \frac{1}{6} k_4). \end{cases} \tag{7.45}$$

2. Second fourth-order Runge Kutta

Uses the "3/8 rule," which is given in Table 7.8. and the consequent formulae:

$$
\begin{array}{c|cccc}
0 & & & & \\
\frac{1}{3} & \frac{1}{3} & & & \\
\frac{2}{3} & -\frac{1}{3} & 1 & & \\
1 & 1 & -1 & 1 & \\
\hline
& \frac{1}{8} & \frac{3}{8} & \frac{3}{8} & \frac{1}{8}
\end{array}
$$

TABLE 7.8: Coefficients of the "$\frac{3}{8}$" fourth-order Runge-Kutta method

$$
(RK4.2) \begin{cases}
k_1 = f(t_{i-1}, y_{i-1}) \\
k_2 = f(t_{i-1} + \frac{h}{3}, y_{i-1} + \frac{1}{3}k_1) \\
k_3 = f(t_{i-1} + \frac{h}{3}, y_{i-1} - \frac{1}{3}k_1 + k_2) \\
k_4 = f(t_{i-1} + h, y_{i-1} + k_1 - k_2 + k_3) \\
y_i = y_{i-1} + h(\frac{1}{8}k_1 + \frac{3}{8}k_2 + \frac{3}{8}k_3 + \frac{1}{8}k_4).
\end{cases}
\tag{7.46}
$$

7.4.4 Control of the Time-Step Size

When using one-step methods, there are two ways to handle the time step control: Richardson extrapolation and embedded Runge-Kutta methods. In what follows, we summarize the methods using the arguments given in [18].

1. Richardson Extrapolation

Let *tol* be a user's computational tolerance.
For a given one-step method of order p, that has yielded the approximate solution $y_0, y_1, ..., y_{n-1}$ at times $0, t_1, ...t_{n-1}$, such that:

$$
\max_{0 \le i \le n-1} |y(x_i) - y_i| \le tol,
$$

where $d_i = \max\{1, |y_i|\}$. Then, based on $h = t_{n-1} - t_{n-2}$, we perform the following:
a. Compute successively, $y_n(h)$ and $y_{n+1}(h)$ based on y_{n-1}, such that $|y(x_{n-1}) - y_{n-1}| \le tol$.
b. Compute with a big step $2h$, $y_{n+1}^1(2h)$.
It is shown in [18] that:

$$
y(x_{n+1}) - y_{n+1}(h) = \frac{y_{n+1}(h) - y_{n+1}^1(2h)}{2^p - 1} + O(h^{p+2}) + O(|y(x_{n-1}) - y_{n-1}|).
\tag{7.47}
$$

Let $Err = \frac{y_{n+1}(h) - y_{n+1}^1(2h)}{2^p - 1}$. Since such term estimates an error expression of the form Ch^{p+1} and given that $O(|y(x_{n-1}) - y_{n-1}|) = O(tol)$ with h satisfying $O(h^{p+2}) = O(h^2 \times tol)$, then two situations may occur:

Case 1 If $|Err| \leq h^{\epsilon_0} \times tol$. In that case then we continue the computation with the same h.

Case 2 Otherwise, if $|Err| > h^{\epsilon_0} \times tol$, then we repeat the computation with $h/2$.

Whenever we reach **case 1**, we end up with:

$$y(x_{n+1}) - y_{n+1}(h) = O(h^{\epsilon_0} \times tol) + O(tol) = O(tol),$$

and continue hereon the adaptive process with the most recent value of h. The above arguments using absolute errors can also be done using instead relative errors. This is specifically done in the following MATLAB program, in which we have selected $\epsilon_0 = tol$. The consequent adaptive process is implemented using the fourth-order Runge-Kutta (7.45).

Algorithm 7.3 Adaptive Runge-Kutta Algorithm

```
function [t,Y]=myodeRK4Adaptive(T,h0,y0,tol)
% Input: T defines the interval [0,T];
%            h0 defines the initial mesh size
%    y0 the initial condition;
%            tol sets the user's relative tolerance
% Output: t is the set of discrete times: t(i) (t(1)=0);
%            Y the set of approximations Y(i) at t(i)
h=h0;% set the initial value of h.
t=zeros(50000,1);Y=zeros(50000,1);%Initialize the vectors t and Y
t(1)=0;Y(1)=y0;i=1;
% Start the process
while t(i)<=T
     Err=1;  % Insure we go in the loop at least once
     while Err>(h^(tol))*tol
          Yim1=Y(i);
          tim1=t(i);
          %  Evaluate with 2 steps h with 4th order RK
          Y1=RK4step(tim1,Yim1,h);Y2=RK4step(tim1+h,Y1,h);
          %  Evaluate with step 2h with same 4th order RK
          Y21=RK4step(tim1,Yim1,2*h,a,b);
          % Get relative error and conduct test
          Err=abs(Y2-Y21)/max(abs(Y2),1);
          if Err>h^(tol)*tol
               h=h/2;% Divide h by 2
          end
     end % End of computation at t(i)... Update i, t, Y
     i=i+1;t(i)=t(i-1)+h;Y(i)=Y1;i=i+1;t(i)=t(i-1)+h;Y(i)=Y2;
     if h/h0<10^(-6)*tol,  break,  end % Test against small h
     if abs(Y2)> realmax/2, break, end % Test against overflows
end
t=t(1:i);Y=Y(1:i);% End of process: extract t and Y
```

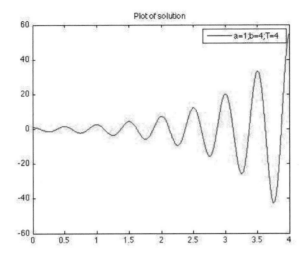

FIGURE 7.3: Graph of the solution to $y'(t) = ay(t) - be^{at}\sin(bt)$, $t > 0$; $y(0) = 1$

ϵ_{tol}	$\max_i \frac{|y_i - y(t_i)|}{max(1,|y_i|)}$	$\min h$	$\max h$
0.5×10^{-4}	1.542402×10^{-5}	9.765625×10^{-4}	0.0156
0.5×10^{-5}	7.914848×10^{-5}	4.882813×10^{-4}	0.0156
0.5×10^{-6}	1.245236×10^{-6}	2.441406×10^{-4}	0.0078
0.5×10^{-7}	1.923099×10^{-7}	1.220703×10^{-4}	0.0078
0.5×10^{-8}	4.543017×10^{-8}	6.103516×10^{-5}	0.0039

TABLE 7.9: Results of applying Algorithm 7.3 to solve $y'(t) = a * y(t) - be^{at}\sin(bt)$, $t > 0$; $y(0) = 1$

As an example, consider the linear non-homogeneous initial value problem:

$$y'(t) = a * y(t) - be^{at}\sin(bt), \ t > 0; \ y(0) = 1,$$

where a and b are constants. The solution of such problem is given by $y(t) = e^{at}\cos(bt)$. The solution exhibits simultaneously an "explosive" behavior (particularly for large values of a) in addition to its oscillatory character due to the presence of the trigonometric term. The plot of the solution for the case when $a = 1$; $b = 4\pi$; $T = 4$ is given in Figure 7.3. The results obtained when applying Algorithm 7.3 are given in Table 7.9. One drawback of this adaptive fourth-order Runge Kutta method is its cost in terms of evaluations of the function $f(.,.)$. Specifically, it requires eight evaluations of $f(.,.)$ with two steps of size h to obtain $y_{n+1}(h)$, followed by three evaluations of $f(.,.)$, with a step of size $2h$ to obtain $y_{n+1}^1(2h)$; hence a total of eleven $f(.,.)$ eval-

t		
0	0	
$\frac{1}{3}$	$\frac{1}{3}$	
	$-\frac{1}{2}$	$\frac{3}{2}$

TABLE 7.10: A second-order RK method embedded in third-order RK Heun method

uations to decide about the admissibility of h in pursuing the computation. A remedy to such excess of function evaluations is reached through the use of *embedded Runge-Kutta methods*.

2. Embedded Runge-Kutta Schemes
Given that (the adaptive) Algorithm 7.3 may be severely costly, we circumvent the problem of time step control by using one pair of embedded Runge-Kutta methods.

Definition 7.4 *A Runge-Kutta method of order p is said to be* **embedded** *in a Runge-Kutta method of order q with $p < q$, if the implementation of the order q method uses the same f function evaluations as those of the order p method.*

Here are examples giving pairs of embedded Runge-Kutta methods.

1. **An Embedded (1,2)**: Euler's explicit scheme is embedded in both the modified Euler and Heun methods.

2. **An Embedded (2,3)**: Another interesting case of a second-order method embedded in a third-order one is obtained by taking $w = \frac{3}{2}$ in $(RK_2(w))$. We get then a second-order Runge-Kutta method which is given in Table 7.10. One can check then that this table is embedded in the third-order Heun scheme $(RK3.H)$ since both use the same values of k_1 and k_2. A similar approach is used in MATLAB ode23 solver on the basis of the Runge-Kutta (2,3) pair of Bogacki and Shampine [3].

3. **An Embedded (2,4)**: The second-order modified Euler is embedded in the fourth-order Runge-Kutta method $(RK4_1)$, since both use the same values k_1 and k_2.

4. **An Embedded (4,5)**: This pair (referred to as the **Dormand-Prince** pair [12]), uses two embedded Runge-Kutta schemes of order 4 and 5, which coefficients are shown respectively in Table 7.11 and Table 7.12. Note that both methods use the same values of k_1, k_2, k_3 and k_4. The MATLAB ode45 solver is based on a similar pair of embedded Runge-Kutta methods.

t					
0	0				
$\frac{1}{4}$	$\frac{1}{4}$				
$\frac{3}{8}$	$\frac{3}{32}$	$\frac{9}{32}$			
$\frac{12}{13}$	$\frac{1932}{2197}$	$-\frac{7200}{2197}$	$\frac{7296}{2197}$		
1	$\frac{439}{216}$	-8	$\frac{3680}{513}$	$-\frac{845}{4104}$	
	$\frac{25}{216}$		$\frac{1408}{2565}$	$\frac{2197}{4104}$	$-\frac{1}{5}$

TABLE 7.11: Coefficients of the fourth-order RK used in MATLAB ode45 solver

t					
0	0				
$\frac{1}{4}$	$\frac{1}{4}$				
$\frac{3}{8}$	$\frac{3}{32}$	$\frac{9}{32}$			
$\frac{12}{13}$	$\frac{1932}{2197}$	$-\frac{7200}{2197}$	$\frac{7296}{2197}$		
1	$\frac{439}{216}$	-8	$\frac{3680}{513}$	$-\frac{845}{4104}$	
$\frac{1}{2}$	$-\frac{8}{27}$	2	$-\frac{3544}{2565}$	$\frac{1859}{4104}$	$-\frac{11}{40}$
	$\frac{16}{135}$		$\frac{6656}{12825}$	$\frac{28561}{56430}$	$-\frac{9}{50}$	$\frac{2}{55}$

TABLE 7.12: Coefficients of the fifth-order RK used in MATLAB ode45 solver

We consider now an alteration of Algorithm 7.3 (based exclusively on a one-step Runge-Kutta method), by implementing a pair of embedded Runge-Kutta methods: $\mathcal{M}_1$ and $\mathcal{M}_2$, of respective orders p and $p+1$.
Based on $h = t_n - t_{n-1}$ and on y_n, such that $|y(x_n) - y_n| \leq tol$, we perform the following:

a. Compute $y_{n+1}^1(h)$ with step h, using $\mathcal{M}_1$, as a substitute to $y_{n+1}^1(2h)$ in (7.47). This is done at the cost of four functions evaluations.
b. Compute $y_{n+1}(h)$, using $\mathcal{M}_2$ at the cost of one additional function evaluation.
As a total, **a.** and **b.** would then require five function evaluations, instead of eleven as was the case in Algorithm 7.3, i.e., an economy of six evaluations of $f(.,.)$! Thus, one writes:

$$y(x_{n+1}) - y_{n+1}(h) = \frac{y_{n+1}(h) - y_{n+1}^1(h)}{2^p - 1} + O(h^{p+2}) + O(|y(x_n) - y_n|).$$

Let $Err = \frac{y_{n+1}(h) - y_{n+1}^1(h)}{2^p - 1}$. Given that such term estimates an error expression of the form Ch^{p+1} and as $O(|y(x_n) - y_n|) = O(tol)$, then, with h satisfying $O(h^{p+2}) = O(h^2 \times tol)$, two situations may occur:
Case 1 If $|Err| \leq h^{\epsilon_0} \times tol$, we continue the computation with the same h.
Case 2 Otherwise, ($|Err| > h^{\epsilon_0} \times tol$), we repeat the computation with $h/2$. Whenever we reach **case 1**, we end up with:

$$y(x_{n+1}) - y_{n+1}(h) = h^{\epsilon_0} \times tol + O(tol) = O(tol),$$

and continue the adaptive process based on the 2 embedded Runge-Kutta methods $\mathcal{M}_1$ and $\mathcal{M}_2$.

7.5   Adams Multistep Methods

When using higher order one-step Runge-Kutta methods, the number of function evaluations increase significantly. This is particularly so when the function $f(.,.)$ is vector-valued. For example in case $f : \mathbb{R}^n \times [0, T] \to \mathbb{R}^n$, then using a fourth-order Runge-Kutta method would require $4n$ scalar function evaluations at each step. Such necessity may be too time consuming.
Thus, use of multistep methods is precisely to avoid such issue of multiple function evaluations when using one-step methods. In this chapter, we give an overview of three types of multistep methods:

1. Adams-Bashforth **multistep explicit schemes**

2. Adams-Moulton **multistep implicit schemes**

3. Backward difference (BDF) methods, used to solve special "stiff" systems of ODEs. These are ODEs systems that have solutions with sharp variations in short times.

A major reference on multistep methods is [18]. In this section, we focus mainly on Adams type methods. To present these methods, our starting point is the sequence of integral equations (7.24) introduced above:

$$y_i - y_{i-1} = \int_{t_{i-1}}^{t_i} f(t, y(t))dt, \ i = 1, ..., N.$$

Letting $g(t) = f(t, y(t))$, then obtaining an Adams type method consists in replacing $\int_{t_{i-1}}^{t_i} g(t)dt$ by a numerical integration formula derived through the replacement of $g(t)$ with a Lagrange interpolation polynomial $p(t)$ (as introduced in Chapter 3) based on a specific set of points $\{t_j\}$. Thus:
- To obtain a k-multistep explicit Adams-Bashforth $p(t)$ is of degree $k-1$ and based on the data set of k pairs:

$$\{(t_{i-1}, g(t_{i-1})), \ ..., \ (t_{i-k}, g(t_{i-k})), \ i - k \geq 0,$$

while:
- Obtaining a k-multistep implicit Adams-Moulton $p(t)$ is also of degree $k-1$ and is based on the k pairs:

$$\{(t_i, g(t_i)), \ (t_{i-1}, g(t_{i-1})), \ ..., \ (t_{i-k}, g(t_{i-k+1})), \ i - k + 1 \geq 0,$$

As a result, we obtain the following schemes, using the notation $f_j \equiv f(t_j, y_j)$.

7.5.1   Adams Schemes of Order 1

One obtains successively:

$$y_i = y_{i-1} + h f_{i-1} \ i \geq 1, \tag{7.48}$$

for Adams-Bashforth and

$$y_i = y_{i-1} + h f_i \ i \geq 1, \tag{7.49}$$

for Adams-Moulton. These are respectively Euler's explicit and implicit one-step methods. The first was discussed earlier and the second requires solving the (usually) nonlinear equation:

$$y_i - h f(t_i, y_i) = y_{i-1}. \tag{7.50}$$

Solving (7.50) is considered in the last section of this chapter, within the context of Backward Difference Formulae (BDF) methods.

7.5.2   Adams Schemes of Order 2

$$y_i = y_{i-1} + h[\frac{3}{2}f_{i-1} - \frac{1}{2}f_{i-2}], \; i \geq 2, \tag{7.51}$$

is the Adams-Bashforth version, and:

$$y_i = y_{i-1} + \frac{1}{2}h[f_i + f_{i-1}], \; i \geq 1. \tag{7.52}$$

is the Adams-Moulton one. It is precisely the trapezoidal rule formula obtained earlier in (7.32). As (7.49), (7.52) requires also solving a nonlinear equation:

$$y_i - \frac{1}{2}hf(t_i, y_i) = y_{i-1} + \frac{1}{2}hf_{i-1}. \tag{7.53}$$

However, an important point about Adams methods can be noted here. Through a predictor-corrector approach that uses Euler's explicit as a predictor scheme, followed for correction by the second-order Adams-Moulton, one gets the following method:

$$\begin{cases} i \geq 1 : \\ y_i^{(P)} = y_{i-1} + hf_{i-1} \\ f_i^{(P)} = f(t_i, y_i^{(P)}), \\ y_i = y_i^{(C)} = y_{i-1} + \frac{1}{2}h[f_i^{(P)} + f_{i-1}]. \end{cases} \tag{7.54}$$

It is easily checked that (7.54) is precisely Heun's second-order Runge-Kutta method $(RK_2.H)$. This reveals the following points in the use of Adams methods:

a- The combination of Adams-Bashforth's method of order 1, as predictor with Adams-Moulton's method of order 2 as corrector gives an explicit method of order 2.

b- A first-order Adams-Bashforth method is thus **embedded** in a second-order Adams-Moulton scheme, suggesting embedding the second-order (7.51) in a third-order Adams-Moulton as is done in what follows.

c- This embedded predictor-corrector pair can be obviously used for controlling the step size h as explained in Section 7.4.4.

7.5.3   Adams Schemes of Order 3

On that basis, we couple (7.51) as a predictor scheme with the third-order Adams-Moulton implicit formula:

$$y_i = y_{i-1} + h[\frac{5}{12}f(t_i, y_i) + \frac{2}{3}f(t_{i-1}, y_{i-1}) - \frac{1}{12}f(t_{i-2}, y_{i-2})], \tag{7.55}$$

for $i \geq 2$.

As a result of the pair (7.51) - (7.55), one gets a two step third-order predictor-corrector scheme:

$$
\begin{cases}
i \geq 2 : \\
y_i^{(P)} = y_{i-1} + h\left[\frac{3}{2} f_{i-1} - \frac{1}{2} f_{i-2}\right] \\
f_i^{(P)} = f(t_i, y_i^{(P)}), \\
y_i = y_i^{(C)} = y_{i-1} + h\left[\frac{5}{12} f_i^{(P)} + \frac{2}{3} f_{i-1} - \frac{1}{12} f_{i-2}\right].
\end{cases}
\tag{7.56}
$$

One advantage of this two step method of order 3 over a third-order Runge-Kutta procedure is in terms of function f evaluations when computing y_i, $i \geq 2$, which requires computing:
$$f_{i-1} = f(t_{i-1}, y_{i-1}) \text{ and } f_i^{(P)} = f(t_i, y_i^{(P)}).$$
Note at the same time some disadvantages of this method. Since the implementation of (7.56) begins at $i = 2$ and given that $y_0 = y(0)$, it is necessary to obtain y_1 by using a one-step method of order 3. For that purpose, we can use a Runge-Kutta method of order 3 such as $(RK3.H)$ or even simply a Runge-Kutta method of order 2, given that the error coincides with the local truncation of the method at $t = 0$ (see Remark 7.6). On the other hand, (7.56) requires that after computing y_i, $i \geq 2$, one saves f_{i-1} for use when computing at the next step, y_{i+1}.

7.5.4   Adams Methods of Order 4

On the same basis, we obtain higher order Adams method. We restrict ourselves to a fourth -order Adams multistep method:

$$
\begin{cases}
i \geq 3 : \\
y_i^{(P)} = y_{i-1} + h\left[\frac{23}{12} f_{i-1} - \frac{4}{3} f_{i-2} + \frac{5}{12} f_{i-3}\right] \\
f_i^{(P)} = f(t_i, y_i^{(P)}), \\
y_i = y_i^{(C)} = y_{i-1} + h\left[\frac{3}{8} f_i^{(P)} + \frac{19}{24} f_{i-1} - \frac{5}{24} f_{i-2} + \frac{1}{24} f_{i-3}\right].
\end{cases}
\tag{7.57}
$$

Similarly to the third-order Adams, (7.56), this Adams fourth-order predictor-corrector, (7.57), requires also two f function evaluations, in addition to storing simultaneously f_i and f_{i-1} to compute later on y_{i+1}.

On the other hand, starting the method requires in addition to $y_0 = y(0)$, y_1 and y_2. These can be computed using a Runge-Kutta Heun method of order 3 or even 2 (as noted in Remark 7.6).

7.6   Multistep Backward Difference Formulae

Consider the Euler implicit scheme (7.49):

$$i \geq 1 : y_i = y_{i-1} + hf(t_i, y_i) \Leftrightarrow y_i - hf(t_i, y_i) = y_{i-1}.$$

At this point, there are two distinct cases that arise:

a. f linear in y:
In this case when $f(t, y) = -a(t)y + b(t)$, and (7.49) becomes explicit in the sense that Euler implicit formula becomes:

$$(1 + ha(t_{i-1}))y_i = y_{i-1} + b(t_{i-1})$$

and y_i can be found explicitly, provided:

$$1 + ha(t_{i-1}) \neq 0, \ \forall i,$$

in which case one has:

$$y_i = \frac{y_{i-1} + b(t_{i-1})}{1 + ha(t_{i-1})}.$$

Such is the case:

1. For all h, whenever $a(t) \geq 0$, $f(.,.)$ being then **monotone decreasing** with respect to y.

2. Otherwise, one must put a restriction on h:

$$h \leq h_0 = \frac{c_0}{\max_{t \in [0,T]} |a(t)|}, \ c_0 < 1. \tag{7.58}$$

Such a condition is similar to that found in Theorem 7.2 for explicit schemes.

b. f nonlinear in y:
In that case, we let $r(y) = y - hf(t_i, y) - y_{i-1}$ and obtaining y_i, reduces to solving:

$$r(y) = 0.$$

Out of the methods studied in Chapter 2, we retain Newton's method, on the basis that it can be straightforwardly generalized when f is a vector function. Newton's iterative formula to solve $r(y_i) = 0$ is given by:

$$\begin{cases} r'(y_i^{(k)})(y_i^{(k+1)} - y_i^{(k)}) = -r(y_i^{(k)}), \\ y_i^{(0)} = y_{i-1} \text{ or using Euler's explicit: } y_i^{(0)} = y_{i-1} + hf_{i-1}. \end{cases} \tag{7.59}$$

Now $r'(y) = 1 - hf_y(t_i, y)$. As when $f(t, y)$ is linear in y, we also distinguish here two cases:

1. $f(.,.)$ is **monotone decreasing** with respect to y.

2. If not, one must put a restriction on h:

$$h \leq h_0 = \frac{c_0}{\max_{(t,y)\in[0,T]\times\mathcal{D}_y} |f_y(t,y)|}, \ c_0 < 1. \tag{7.60}$$

In either case Newton's iteration becomes:

$$y_i^{(k+1)} = y_i^{(k)} - \frac{r(y_i^{(k)})}{1 - hf_y(t_i, y_i^{(k)})}.$$

The interesting property of the Euler implicit scheme is its strong stability property when $f(.,.)$ is monotone decreasing with respect to y. Specifically, consider the distinct solutions $\{w_i\}$ and $\{z_i\}$, obtained from:

$$i \geq 1 : \ w_i - hf(t_i, w_i) = w_{i-1},$$

and

$$i \geq 1 : \ z_i - hf(t_i, z_i) = z_{i-1},$$

$w_0, z_0 \in \mathcal{D}_y$. Subtracting the second equation from the first yields:

$$i \geq 1 : \ w_i - z_i - h[f(t_i, w_i) - f(t_i, z_i)] = w_{i-1} - z_{i-1}.$$

Multiplying this equation by $w_i - z_i$ and using the monotony of f, yields:

$$i \geq 1 : \ (w_i - z_i)^2 - h(w_i - z_i)[(f(t_i, w_i) - f(t_i, z_i)] = (w_{i-1} - z_{i-1})(w_i - z_i).$$

Given the positiveness of the left hand side of this identity, one obtains the inequality:

$$i \geq 1 : \ (w_i - z_i)^2 - h(w_i - z_i)[(f(t_i, w_i) - f(t_i, z_i)] \leq |w_{i-1} - z_{i-1}|.|(w_i - z_i|.$$

This gives the following stability result:

Theorem 7.7 *If:*

$$(f(t, w) - f(t, z))(w - z) \leq 0, \ \forall t \in [0, T], \ w, z \in \mathcal{D}_y,$$

then:

$$|w_i - z_i| \leq |w_0 - z_0|, \ \forall i.$$

This stability property applies when we consider multistep generalizations of the Euler implicit scheme in the form of the Backward Difference Formulae (BDF). Here are up to fourth oder BDF formulae:

- Order 2:

$$i \geq 2 : \ y_i - \frac{2}{3}hf(t_i, y_i) = \frac{4}{3}y_{i-1} - \frac{1}{3}y_{i-2}.$$

- Order 3:

$$i \geq 3 : y_i - \frac{6}{11} h f(t_i, y_i) = \frac{18}{11} y_{i-1} - \frac{9}{11} y_{i-2} + \frac{2}{11} y_{i-3}.$$

- Order 4:

$$i \geq 4 : y_i - \frac{12}{25} h f(t_i, y_i) = \frac{48}{25} y_{i-1} - \frac{36}{25} y_{i-2} + \frac{16}{25} y_{i-3} - \frac{3}{25} y_{i-4}.$$

7.7 Approximation of a Two-Points Boundary Value Problem

Let b be a real-valued function on an interval $\Omega = (0, L)$, with $b(x) \geq 0$. Consider the one-dimensional boundary-value problem: Find $u : \overline{\Omega} \longrightarrow \mathbb{R}$, such that:

$$-u''(x) + b(x)u = f(x), \forall x \in (0, L), \ (1) \ u(0) = \alpha, \ u(1) = \beta \ (2) \qquad (7.61)$$

A finite-difference discretization consists in replacing the differential equation 7.61 (1) with a difference equation. Specifically, consider the discrete domain:

$$\overline{\Omega}_h = \{x_i = ih | 0 = x_0 < x_1 < \dots < x_N = 1\}, \ Nh = 1,$$

that uniformly partitioned Ω. Le $b_i = b(x_i)$. The discrete system corresponding to (7.61) is defined as follows:

$$-\delta_h^2 U_i + b_i U_i = f_i = f_i = f(x_i), \ \forall i, \ 0 < i < N, \ (1) \ U_0 = \alpha, \ U_1 = \beta \ (2)$$
$$(7.62)$$

For this one-dimensional model, note that the solution to (7.62) depends on $N + 1$ parameters, of which $M = N - 1$ $[U_1 \, U_2 .... U_M]^T$ are unknowns, since $[U_0 \, U_N]$ are given. Thus, the resulting system obtained from (7.62) takes the following matrix form:

$$AU = F, \qquad (7.63)$$

the matrix $A \in \mathbb{R}^{M,M}$ being tri-diagonal. In case, $a(x) = 1$ and $b(x) = 0$, A is the well-known "central difference matrix":

$$A = \frac{1}{h^2} \begin{pmatrix} 2 & -1 & 0 & \dots & 0 \\ -1 & 2 & -1 & 0 & 0 \\ \dots & \dots & \dots & \dots & \dots \\ \dots & 0 & -1 & 2 & -1 \\ 0 & 0 & \dots & -1 & 2 \end{pmatrix}, \ F = \begin{pmatrix} f_1 + \alpha/h^2 \\ f_2 \\ \dots \\ f_{M-1} \\ f_M + \beta/h^2 \end{pmatrix}.$$

It can be checked that:

- A is a sum of a tri-diagonal matrix and a diagonal matrix.

- A is symmetric.

Remark 7.8 *Note that the solution of the continuous problem is in a (fixed) vector space of the type $C^k(\Omega)$ while that of the discrete problem is in a (variable) finite-dimension space $\mathbb{R}^N$, with $\lim_{h\to 0} N = \infty$.*

To provide a coherent framework for analyzing the finite-difference discretization, one introduces the concepts of the "restriction" and "prolongation (extension)" operators.

Definition 7.5 *Given a function $v : \overline{\Omega} \to \mathbb{R}$, such that $v \in C(\overline{\Omega})$, the restriction $r_{h,N}(v)$ of v on $\overline{\Omega}_h$ is such that $V = r_{h,N}(v) \in \mathbb{R}^{N+1}$, with:*

$$V_i = v(x_i), \ \forall i = 0, ...N.$$

Similarly, one defines the restriction of v on Ω_h, $r_{h,N}(v) \in \mathbb{R}^{N+1}$. By convention we take: $r_h \equiv r_{h,N}$.

We now define the prolongation:

Definition 7.6 *Given $V \in \mathbb{R}^{N+1}$, $V = [V_0 \ V_1 .... V_N]^T$, a prolongation $p_{h,N} V$ of V in $C(\overline{\Omega})$ is a function $v \in C(\overline{\Omega})$, such that $r_{h,N}(p_{h,N} V) = V$.*

Note that there exist several prolongation operators for a vector $V \in \mathbb{R}^{N+1}$. For example, one may use linear, quadratic or cubic spline interpolations. In the case of finite-differences, it is sufficient to use piecewise linear splines:

$$v(x) = r_{h,M}(V)(x) = (V_i(x_{i+1} - x) + V_{i+1}(x - x_i))/h, \ i = 0, ..., N.$$

We consider now the convergence of the discrete solution $r_{h,N}(U)$ of (7.62) to the solution u of Poissons's equation (7.61). A preliminary result shall be first stated.

Theorem 7.8 *If the solution u to (7.61) is in $C^2(\Omega) \cap C(\overline{\Omega})$, then one has:*

$$\max_x |u(x) - p_{h,N}(r_{h,N}(u))(x)| \le ch^2 \max_x |u''(x)|.$$

On the basis of this result and the inequality:

$$\max_x |u(x) - p_{h,N} U(x)|$$

$$\le \max_x ||u(x) - p_{h,N}(r_{h,N} u)(x)| + \max_x |p_{h,N}(r_{h,N} u(x)) - p_{h,N} U(x)|,$$

then, it is sufficient to study the convergence of $p_{h,N} U$ to $p_{h,N}(r_{h,N} u)$ to obtain convergence of $p_{h,N} U$ to u. Since:

$$\max_x |p_{h,N}(r_{h,N} u)(x) - p_{h,N} U(x)| = |\max_i |u(x_i) - U_i|$$

one needs to estimate $\max_i |u_i - U_i|$ to obtain the convergence of the discrete solution $p_{h,N} U$ to the exact solution u. This requires first a **stability** result (found in [21]):

Theorem 7.9 *The matrix A in (7.63) is such that:*

$$\max_i |U_i| \le C \max_i |F_i|,$$

where C is independent from h.

Using the truncation error related to the second-order central difference formula, we can then prove:

Theorem 7.10 *If the solution u to (7.61) is such that, $u \in C^4(\Omega) \cap C(\overline{\Omega})$. Then the approximation $U_h = \{U_i\}$ to $u_h = \{u_i = u(x_i)\}$ that solves (7.63) satisfies the estimate:*

$$\max_i |u(x_i) - U_i| \le Ch^2,$$

C independent from h.

Proof. The proof is a classical procedure in numerical mathematics. It uses Theorem 7.9 and the estimate associated with the second-order central difference formula:

$$u''(x_i) = \delta_h^2 u(x_i) + h^2 \epsilon_i(u), \ v \in C^4, \ 1 \le i \le N.$$

where $\epsilon_i(u) = c \frac{d^4 u}{dx^4}(\eta_i)$, $x_{i-1} < \eta_i < x_{i+1}$. One checks $\epsilon = \{\epsilon_i(u)\}$ satisfies:

$$\max_i |\epsilon_i| \le C_1,$$

with C_1 independent from h and function of $\max_x |u^{(4)}(x)|$. To complete the proof, one uses:

$$A(u - U) = h^2 \epsilon,$$

Using the stability concept of Theorem 7.9, one directly obtains the estimates of the theorem. ∎

7.8   Exercises

1. Show that each of the following (IVP) has a unique solution:

   (a) $y' = y \sin(t)$, $0 \le t \le 1$, $y(0) = 1$

   (b) $y' = e^{(t-y)/2}$, $0 \le t \le 1$, $y(0) = 1$

   (c) $y' = \frac{2t^2 y}{1+t^4}$, $0 \le t \le 1$, $y(0) = 1$

2. Verify that each of the following functions $f(t, y(t))$ satisfies a Lipschitz condition on the set:

$$D = \{(t, y)|0 \le t \le 1, -\infty < y < +\infty\}$$

   and determine the corresponding Lipschitz constant in each case.

   (a) $f(t, y(t)) = t^3 y + 1$

   (b) $f(t, y(t)) = 1 - y^2$

   (c) $f(t, y(t)) = e^{(t-y)/2}$

   (d) $f(t, y(t)) = -ty + 3\frac{t}{y^2}$

3. Consider the following (IVP):

$$y' = -4y + t, \ 0 \le t \le 1, \ y(0) = 1$$

   Use Picard's method to generate the functions $y^{(i)}(t)$ for $i = 0, 1, 2$.

4. Use Euler's method to solve the following (IVP)

   (a) $y'(t) = e^{(t-y)/2}$, $0 \le t \le 1$, $y(0) = 1$, $h = 0.25$

   (b) $y'(t) = -y + ty^{3/2}$, $2 \le t \le 4$, $y(2) = 0$, $h = 0.25$

   (c) $y'(t) = 1 + y/t^2$, $1 \le t \le 2$, $y(1) = 1$, $h = 0.25$

5. Use Heun's method $(RK2.H)$ to solve the following initial value problems:

   (a) $y'(t) = te^{3t} - 2y^2$, $0 \le t \le 1$, $y(0) = 0$, $h = 0.2$

   (b) $y'(t) = t + (t - y)^2$, $0 \le t \le 2$, $y(0) = 1$, $h = 0.5$

6. Consider the following initial value problem:

$$(IVP) \begin{cases} \frac{dy}{dt} = t^2 + y^2 \, ; \ t \in [0, 1.5] \\ y(0) = 1 \end{cases}$$

   (a) Write first the discrete scheme of Euler's method, $(RK1)$, then use 2 steps of this scheme to approximate $y(0.25)$ and $y(0.50)$.

- Discrete Scheme

$$(RK1)\begin{cases} \dotfill \\ y_{i+1} = \dotfill \end{cases}$$

- Express all the computed results with a precision $p = 3$.

i	t_i	y_i	k_1	y_{i+1}
0	.	.	.	.
1	.	.	.	.

(b) Write first the discrete scheme of Heun's method, $(RK2.H)$, then use two steps of this scheme to approximate $y(0.75)$ and $y(1)$.

- Discrete Scheme

$$(RK2.H)\begin{cases} \dotfill \\ \dotfill \\ y_{i+1} = \dotfill \end{cases}$$

- Express all the computed results with a precision $p = 3$.

i	t_i	y_i	k_1	k_2	y_{i+1}
0	.	.	.	.	.
1	.	.	.	.	.

(c) Write first the discrete scheme of the midpoint rule, $(RK2.M)$, then use two steps of this scheme to approximate $y(1.25)$ and $y(1.50)$.

- Discrete Scheme

$$(RK2.M)\begin{cases} \dotfill \\ \dotfill \\ y_{i+1} = \dotfill \end{cases}$$

- Express all the computed results with a precision $p = 3$.

i	t_i	y_i	k_1	k_2	y_{i+1}
0	.	.	.	.	.
1	.	.	.	.	.

7. Repeat Exercise 5 using the midpoint method $(RK2.M)$

7.9   Computer Exercises

1. Test Algorithm 7.3: `function [t,Y]=myodeRK4Adaptive(T,h0,y0,tol)` on the following initial value problems:

   - $y'(t) = \sin(t)y^{1/2} + \cos(t)y,\ y(0) = 1.$
   - The Van der Pol equation:

   $$y'' - (1 - y^2)y' + y = 0,\ 0 < t \le 10,\ y(0) = 1,\ y'(0) = 0,$$

   after transforming it into a system of first-order equations.

2. Transform Algorithm 7.3 so as to have the control of the time step done using the following pairs of embedded Runge-Kutta methods:

   - $(RK_2(w))$ obtained by taking $w = \frac{3}{2}$ embedded in the third-order Heun scheme $(RK3.H)$.
   - The pair of Runge-Kutta schemes of order 4 and 5, whose coefficients are shown respectively in Table 7.11 and Table 7.12.
   - Test the resulting algorithms on the following initial value problems:

   (a) $y'(t) = ay(t) - be^{at}\sin(bt),\ t > 0;\ y(0) = 1.$

   (b) The Van der Pol equation:

   $$y'' - (1 - y^2)y' + y = 0,\ 0 < t \le 10,\ y(0) = 1,\ y'(0) = 0,$$

Answers to Odd-Numbered Exercises

Chapter 1

- **Exercise 1:**
  1.a $e \simeq (2.718)_{10} \simeq (10.10110.......)_2$.
  1.b $(0.875)_{10} = (0.111)_2$.
  1.c $(792)_{10} = (1100011000)_2$

- **Exercise 3:**
  3.a $(671.535)_8 = (441.681)_{10}$.
  3.b $(1145.32)_8 = (613.40625)_{10}$.

- **Exercise 5:**
  5.- $x = (0.6)_{10} = (0.\overline{46314})_8 = (0.\overline{1001})_2$.
  5.- $x = (0.6)_{10} = (0.\overline{1001})_2$.

- **Exercise 7:**
  7.a Incorrect
  7.b Correct
  7.c Correct
  7.d Incorrect
  7.e Correct

- **Exercise 9:**

t	$c(8)$	$f(23)$
0	10 000 101	000 000 000 001 $\underbrace{000 \ldots 000}_{11- \text{ zeros}}$

- **Exercise 11:**
  11.a $x = +0$
  11.b $x = -0$
  11.c $x = NaN$
  11.d $x = NaN$
  11.e $x = +1 \times 2^{-126}$
  11.f $x = +1.1111 \times 2^2$
  11.g $x = +1.0 \times 2^0$
  11.h $x = +1.10011001100110011001101 \times 2^{123}$

- **Exercise 13:**
  13.a 6.573972×10^{-1}
  13.b 2.979802×10^{83}
  13.c 3.301920×10^{81}
  13.d $8.128175418 \times 10^{5}$
  13.e 9.462402×10^{6}
  13.f 2.5281767×10^{3}
  13.g 3.506323×10^{3}
  13.h 3.3027656×10^{80}
  13.i 2.508630×10^{79}

- **Exercise 15:**
  15.a $x = -x_{\min}$.
  15.b $succ(x) = [80711111]_{16}$; $pre(x) = [80800001]_{16}$

- **Exercise 17:**
  17.a $b = [00480000]_{16}$
  17.b $succ(b) = [00480001]_{16}$
  17.c $b = [1802000000000000]_{16}$
  17.d $pre(b) = [1801FFFFFFFFFFFF]_{16}$

- **Exercise 19:**
  19.a

  First method: $f(x) = \begin{cases} \frac{\cos^2 x}{1+\sin x} & \text{if } x \simeq \frac{\pi}{2} + 2k\pi, \ k \in \mathbb{Z} \\ 1 - \sin x & \text{otherwise.} \end{cases}$

  Second method: $f(x) = \begin{cases} 1 - x + \frac{x^3}{3!} + \dots & \text{if } x \simeq \frac{\pi}{2} + 2k\pi, \ k \in \mathbb{Z} \\ 1 - \sin x & \text{otherwise.} \end{cases}$

  19.b

  First method: $f(x) = \begin{cases} \frac{\sin^2 x}{1+\cos x} & \text{if } x \simeq \frac{\pi}{2} + 2k\pi, \ k \in \mathbb{Z} \\ 1 - \cos x & \text{otherwise.} \end{cases}$

  Second method: $f(x) = \begin{cases} \frac{x^2}{2!} - \frac{x^4}{4!} + \dots & \text{if } x \simeq 0 \\ 1 - \cos x & \text{otherwise.} \end{cases}$

  19.c

  First method: $f(x) = \begin{cases} \cos 2x & \text{if } x \simeq \pm\frac{\pi}{4} + k\frac{\pi}{2}, \ k \in \mathbb{Z} \\ 2\cos^2 x - 1 & \text{otherwise.} \end{cases}$

  Second method: $f(x) = \begin{cases} \dots & \text{if } x \simeq \frac{\pi}{4} \\ 2\cos^2 x - 1 & \text{otherwise.} \end{cases}$

  19.d $f(x) = \begin{cases} 1 - x + \frac{x^2}{3!} + \dots & \text{if } x \simeq 0 \\ \dfrac{(\cos x - e^{-x})}{\sin x} & \text{otherwise.} \end{cases}$

  19.e $f(x) = \begin{cases} 2(\frac{x^2}{2!} + \frac{x^3}{3!} + \frac{x^6}{6!} + \dots) & \text{if } x \simeq 0 \\ e^x - \sin x - \cos x & \text{otherwise.} \end{cases}$

- **Exercise 21:**

  21.a First method: $f(x) = \begin{cases} f(x) = \ln \frac{x}{e} & \text{if } x \simeq 0 \\ \ln x - 1, & \text{otherwise.} \end{cases}$

  Second method: $f(x) = \begin{cases} f(x) = e^{-1}(x-e) - \dfrac{e^{-2}}{2}(x-e)^2 + \ldots & \text{if } x \simeq e \\ \ln x - 1, & \text{otherwise.} \end{cases}$

  21.b $f(x) = \begin{cases} 2\ln x & \text{if } x \simeq 1 \\ \ln x - \ln\left(\dfrac{1}{x}\right), & \text{otherwise.} \end{cases}$

  21.c $f(x) = \begin{cases} -\dfrac{1}{2} - \dfrac{x}{3} & \text{if } x \simeq 0 \\ x^{-2}(\sin x - e^x + 1), & \text{otherwise.} \end{cases}$

  21.d $f(x) = \begin{cases} \dfrac{e}{2!}(x-1)^2 + \dfrac{e}{3!}(x-1)^3 + \ldots & \text{if } x \simeq 1 \\ e^x - e, & \text{otherwise.} \end{cases}$

- **Exercise 23:**

  23.a $f(x) = \begin{cases} \dfrac{-1}{\sqrt{x^2-1}+|x|} & \text{if } x < 0,\ |x| >> \\ \\ x + \sqrt{x^2 - 1} & \text{otherwise} \end{cases}$

  23.b Directly with 3 significant digits, $f(-10^2) = 0$. Using remedy with 3 significant digits, $f(-10^2) = \frac{-1}{2\times 10^2} = -0.005000$.

- **Exercise 25:**

  25.a $f(x) = \begin{cases} 2 + \dfrac{2}{3!}x^2 + \dfrac{2}{5!}x^4 + \ldots & \text{if } x \simeq 0 \\ \\ \dfrac{e^x + e^{-x}}{x} & \text{otherwise} \\ . \end{cases}$

  25.b $f(0.1) = 2$
  25.c $f(0.1) = 2.003$
  25.d 4.83×10^{-2}; $1.672211587177143 \times 10^{-4}$

Chapter 2

- **Exercise 1:**
  1.a $f(x) = x - 2\sin x$
  The first bisector $y = x$ and the function $y = 2\sin x$ intersect at 3 points with respective abscissas:

$$root1 = 0, \ root2 > 0, \ root3 < 0$$

Therefore $root1=0$ is an exact root of $f(x) = x - 2\sin x$, while $root2$ and $root3$ can be approximated by the bisection method.
- $\frac{\pi}{2} < root2 < \frac{3\pi}{4}$, as $f(\frac{\pi}{2}) \times f(\frac{3\pi}{4}) < 0$

n	a_n	b_n	r_{n+1}	$f(r_{n+1})$
0	pi/2=1.5708	3pi/4=2.3562	1.9635	+
1	1.5708	1.9635	-1.7671	-
2	1.7671	1.9635	1.8653	-
3	1.8653	1.9635	1.9144	+
4	1.8653	1.9144	1.8899	-
5	1.8899	1.9144	1.9021	+
6	1.8899	1.9021	1.8960	+
7	1.8899	1.8960	1.8929	.

The bisection method took 7 iterations to compute $root2 \approx 1.8960$ up to 3 decimals.(The 8th confirms that the precision is reached).
- $\frac{-3\pi}{4} < root3 < \frac{-\pi}{2}$, as $f(\frac{-3\pi}{4}) \times f(\frac{-\pi}{2}) < 0$

n	a_n	b_n	r_{n+1}	$f(r_{n+1})$
0	-3pi/4= -2.3562	-pi/2=-1.5708	-1.9635	-
1	-1.9635	-1.5708	-1.7671	+
2	-1.9635	-1.7671	-1.8653	+
3	-1.9635	-1.8653	-1.9144	-
4	-1.9144	-1.8653	-1.8899	+
5	-1.9144	-1.8899	-1.9021	-
6	- 1.9021	-1.8899	-1.8960	-
7	-1.8960	-1.8899	-1.8929	.

The bisection method took 7 iterations to compute $root3 \approx -1.8960$ up to 3 decimals.(The 8th confirms that the precision is reached).
1.b $f(x) = x^3 - 2\sin x$
The cubic function $y = x^3$ and the function $y = 2\sin x$ intersect at 3 points with respective abscissas:

$$root1 = 0, \ root2 > 0, \ root3 = -root2 < 0.$$

Therefore $root1 = 0$ is an exact root of $f(x) = x^3 - 2\sin x$, while $root2$ and $root3$ can be approximated by the bisection method.
$1 < root2 < 1.5$, as $f(1) \times f(1.5) < 0$.
The same table as in (a) can be constructed to obtain the sequence of iterates:

$$\{1.2500, 1.1250, 1.1875, 1.2188, 1.2344, 1.2422, 1.2383, 1.2363\}.$$

Thus, the bisection method took 7 iterations to compute $root2 \approx 1.2363$ up to 3 decimals.(The 8th confirms that the precision is reached).
1.c $f(x) = e^x - x^2 + 4x + 3$
The exponential function $y = e^x$ and the parabola $y = x^2 - 4x - 3$ intersect at 1 point with negative abcissa : $root < 0$. Therefore the function $f(x) = e^x - x^2 + 4x + 3$ has a unique negative root, with:
● $-1 < root < 0$, as $f(-1) \times f(0) < 0$

n	a_n	b_n	r_{n+1}	$f(r_{n+1})$
0	-1	0	-0.5	+
1	-1	-0.5	-0.75	-
2	-0.75	-0.5	-0.6250	+
3	-0.75	-0.6250	-0.6875	+
4	-0.75	- 0.6875	-0.7188	+
5	-0.75	- 0.7188	- 0.7344	+
6	- 0.75	- 0.7344	-0.7422	-
7	- 0.7422	-0.7188	-0.7383	+
8	- 0.7422	-0.7305	-0.7363	-
9	- 0.7364	-0.7305	- 0.7354	+
10	- 0.7364	-0.7335	- 0.7349	-
11	- 0.7349	-0.7335	- 0.7351	-
12	- 0.7349	-0.7335	- 0.7350	

The bisection method took 12 iterations to compute $root1 \approx -1.8960$ up to 3 decimals.(The 13th confirms that the precision is reached).
1.d $f(x) = x^3 - 5x - x^2$
The cubic function $y = x^3 - 5x$ and the function $y = x^2$ intersect at 3 points with respective abscissas:

$$root1 = 0, \ root2 > 0, \ root3 < 0.$$

Therefore $root1 = 0$ is an exact root of $f(x) = x^3 - 5x - x^2$, while $root2$ and $root3$ can be approximated by the bisection method.
$2 < root2 < 3$, as $f(2) \times f(3) < 0$.
The same table as in (a) can be constructed to obtain the sequence of iterates approximating $root2$ up to 3 decimals:

$$\{2.5000, 2.7500, 2.8750, 2.8125, 2.7812, 2.7969, 2.7891, 2.7930\}.$$

Thus, the bisection method took 7 iterations to compute $root2 \approx 2.7930$ up to 3 decimals.(The 8th confirms that the precision is reached).
$-2 < root3 < -1$, as $f(-2) \times f(-1) < 0$.
One obtains the sequence of iterates approximating $root3$ up to 3 decimals:

$$\{-1.5000, -1.7500, -1.8750, -1.8125, -1.7812, -1.7969, -1.7891, -1.7930\}.$$

Thus, the bisection method took 7 iterations to compute $root3 \approx -1.7930$ up to 3 decimals.(The 8th confirms that the precision is reached).

- **Exercise 3:**
  Based on the bisection method, the theoretical number of iterations to approximate a root up to 4 decimal figures is $k = 11$.

  3.a $f(x) = x^3 - e^x$
  The computed sequence of iterations is:
  $r_1 = 1.500, r_2 = 1.7500, r_3 = 1.8750, r_4 = 1.8125, r_5 = 1.8438,$
  $r_6 = 1.8594,$
  $r_7 = 1.8516, r_8 = 1.8555, r_9 = 1.8574, r_{10} = 1.8564, r_{11} = 1.8569.$
  3.b $f(x) = x^2 - 4x + 4 - \ln x$
  The computed sequence of iterations is:
  $r_1 = 1.500, r_2 = 1.2500, r_3 = 1.3750, r_4 = 1.4375, r_5 = 1.4062,$
  $r_6 = 1.4219,$
  $r_7 = 1.4141, r_8 = 1.4102, r_9 = 1.4121, r_{10} = 1.4131, r_{11} = 1.4126.$
  3.c $f(x) = x^3 + 4x^2 - 10$
  The computed sequence of iterations is:
  $r_1 = 1.500, r_2 = 1.2500, r_3 = 1.3750, r_4 = 1.3125, r_5 = 1.3438,$
  $r_6 = 1.3594,$
  $r_7 = 1.3672, r_8 = 1.3633, r_9 = 1.3652, r_{10} = 1.3643, r_{11} = 1.3647.$
  3.d $f(x) = x^4 - x^3 - x - 1$
  The computed sequence of iterations is:
  $r_1 = 1.500, r_2 = 1.7500, r_3 = 1.6250, r_4 = 1.5625,$
  $r_5 = 1.5938,$
  $r_6 = 1.6094, r_7 = 1.6172, r_8 = 1.6211, r_9 = 1.6191, r_{10} = 1.6182, r_{11} = 1.6177.$
  3.e $f(x) = x^5 - x^3 + 3$
  The computed sequence of iterations is:
  $r_1 = 1.500, r_2 = 1.7500, r_3 = 1.8750, r_4 = 1.9375, r_5 = 1.9688,$
  $r_6 = 1.9844,$
  $r_7 = 1.9922, r_8 = 1.9961, r_9 = 1.9980, r_{10} = 1.9990, r_{11} = 1.9995.$
  3.f $f(x) = e^{-x} - \cos x$ The computed sequence of iterations is:
  $r_1 = 1.500, r_2 = 1.2500, r_3 = 1.3750, r_4 = 1.3125, r_5 = 1.2812,$
  $r_6 = 1.2969,$
  $r_7 = 1.2891, r_8 = 1.2930, r_9 = 1.2910, r_{10} = 1.2920, r_{11} = 1.2925.$

3.g $f(x) = \ln(1+x) - \frac{1}{x+1}$
The computed sequence of iterations is:
$r_1 = 1.500, r_2 = 1.7500, r_3 = 1.8750, r_4 = 1.9375, r_5 = 1.96888,$
$r_6 = 1.9844,$
$r_7 = 1.9922, r_8 = 1.9961, r_9 = 1.9980, r_{10} = 1.9990, r_{11} = 1.9995.$

- **Exercise 5:**
  $f(x) = \ln(1-x) - e^x$
  $f(x) = \ln(1-x)$ is monotone increasing on $(-\infty, 1)$; $y = e^x$ is monotone increasing on $(-\infty, +\infty)$, $\Rightarrow$ the 2 curves intersect at a unique point which is the root of f.
  $-1 < root < 0$, as $f(-1) \times f(0) < 0$

n	a_n	b_n	r_{n+1}	$f(r_{n+1})$
0	-1	0	-0.5	-
1	-1	-0.5	-0.75	+
2	-0.75	-0.5	-0.6250	-
3	-0.75	-0.6250	-0.6875	.

- **Exercise 7:**
  7.a Incorrect. For example for $n = 0$, $r > \frac{a_0 + b_0}{2}$.
  7.b Always correct since:

  $$\forall n, \; r \in (a_n, b_n), \text{therefore}, \; b_n - r \leq b_n - a_n = 2^{-n}(b_0 - a_0).$$

  7.c Incorrect on the basis that $r_{n+1} = \frac{a_n + b_n}{2}$ with r_n being either a_n or b_n but definitely not always b_n.
  7.d Incorrect on the basis that $r_{n+1} = \frac{a_n + b_n}{2}$ with r_n being either a_n or b_n but definitely not always a_n.

- **Exercise 9:**
  $f(x) = x^5 - x^3 - 3$, $1 < root < 2$ as $f(1) \times f(2) < 0$.
  $r_{n+1} = r_n - \frac{r_n^5 - r_n^3 - 3}{5r_n^4 - 3r_n^2}$; $r_0 = \frac{1+2}{2} = 1.5$. The first 3 iterates by Newton's method are: $r_1 = 1.4343$; $r_2 = 1.4263$; $r_3 = 1.4262$.

- **Exercise 11:**
  $x = \ln(3) \Rightarrow e^x - 3 = 0 \Rightarrow f(x) = e^x - 3$; $root = \ln(3)$, with $1 < root < 2$ as $f(1) \times f(2) < 0$
  $r_{n+1} = r_n - \frac{e^x - 3}{e^x}$, with $r_0 = \frac{1+2}{2} = 1.5$. The iterates of Newton's method approximating $root$ up to 5 decimals are:
  $r_1 = 1.16939$; $r_2 = 1.10105$; $r_3 = 1.10106$; $r_4 = 1.09862$; $r_5 = 1.1.09861$

- **Exercise 13:**
  $f(x) = x - \frac{e}{x}$; roots of f: $x = \pm\sqrt{e}$.

$-2 < Negative\ root < -1$, as $f(-2) \times f(-1) < 0$.
$r_{n+1} = r_n - \frac{r_n - \frac{e}{r_n}}{e/r_n^2}$; $r_0 = \frac{-1-2}{2} = -1.5$. The first 4 iterates by Newton's method are: $r_1 = -1.7584068$; $r_2 = -1.5166588$; $r_3 = -1.7498964$; $r_4 = -1.5285389$.

- **Exercise 15:**
  15.a $f(x) = \frac{1}{x} - 3$; $root = \frac{1}{3}$.
  $r_{n+1} = r_n - \frac{\frac{1}{r_n} - 3}{\frac{-1}{r_n^2}} = r_n(2 - 3r_n)$. Restriction: $0 < r_n < 2/3$.
  15.b
  (i) $r_0 = 0.5 < 2/3 \Rightarrow r_1 = 0.2500$; $r_2 = 0.3125$; $r_3 = 0.3320$; $r_4 = 0.3333$ this $\Rightarrow$ convergence to $root = 0.3333333...$
  (ii) $r_0 = 1 > 2/3 \Rightarrow r_1 = -1$; $r_2 = -5$; $r_3 = -85$; $r_4 = -21845 \Rightarrow$ divergence.

- **Exercise 17:**
  17.a $f(x) = \frac{1}{x^2} - 7$; negative $root = \frac{-1}{\sqrt{7}}$.
  $r_{n+1} = \frac{r_n}{2}(3 - 7r_n^2)$; restriction: $-\sqrt{(3/7)} < r_n < 0$, with formula not dividing by the iterate.

  17.b Let $r_0 = 0.45 \Rightarrow r_1 = 0.356063$; $r_2 = -0.376098$; $r_3 = -0.377951$;
  $r_4 = -0.377964$.

- **Exercise 19:** To compute $\sqrt{R}$, with $R > 0$, using Newton's method:
  19.a $r_{n+1} = \frac{1}{2}(r_n + \frac{R}{r_n})$. No restriction on initial condition.

  19.b $r_{n+1} = \frac{1}{2}r_n(3 - \frac{r_n^2}{R})$. Restriction on initial condition:
  $0 < r_n < \sqrt{3R}$.
  19.c $r_{n+1} = \frac{2Rr_n}{R+r_n^2}$. No restriction on initial condition, as for large values of x, $c(x) \approx x$.
  19.d $r_{n+1} = \frac{r_n}{2}(3 - \frac{r_n^2}{R})$. Restriction on initial condition:
  $0 < r_n < \sqrt{3R}$.
  19.e $r_{n+1} = 2R\frac{r_n}{R+r_n^2}$. No restriction on initial condition, as for large values of x, $c(x) \approx x$.
  19.f $r_{n+1} = \frac{r_n}{2}(3 - \frac{r_n^2}{R})$. Restriction on initial condition:
  $0 < r_n < \sqrt{3R}$.

- **Exercise 21:**
  21.a This function has 3 roots: $0 < root1 < 1$, as $f(0) \times f(1) < 0$, $1 < root2 < 2$, as $f(1) \times f(2) < 0$ and $-3 < root3 < -2$, as $f(-3) \times f(-2) < 0$.
  21.b Using the bisection method:

n	a_n	b_n	r_{n+1}	$f(r_{n+1})$
0	1	2	1.5	-
1	1.5	2	1.75	-
2	1.75	2	1.875	+
3	1.75	1.875	1.8125	-
4	1.8125	1.875	1.8438	+
5	1.8125	1.8438	1.8281	-
6	1.8281	1.8438	1.8359	+
7	1.8281	1.8359	1.8320	.

Using Newton's method, the first iterates computing the root up to 3 decimals are:
$r_1 = 1.5000; r_2 = 2.1429; r_3 = 1.9007; r_4 = 1.8385; r_5 = 1.8343.$

- **Exercise 23:** $f(x) = p(x) = c_2 x^2 + c_1 x + c_0$
  23.a Since in Newton's method:

$$|r_{n+1} - r| = \frac{1}{2}\left|\frac{f''(c_n)}{f'(r_n}\right|(r_n - r)^2 = \frac{1}{2}\frac{2|c_2|}{|p'(r_n)|}(r_n - r)^2| \le \frac{|c_2|}{d}(r_n - r)^2$$

i.e., $|r_{n+1} - r| = C(r_n - r)^2$ with $C = \frac{|c_2|}{d}$.
23.b Multiplying the last inequality by C and letting $e_n = C|r - r_n|$
yields $e_{n+1} \le e_n^2$.
For $n = 0$, $e_0 = C|r - r_0| < 1$ if and only if $|r - r_0| < \frac{1}{C} = \frac{d}{|c_2|}$.
Hence for such choice of r_0 $e_0 < 1$ implies $e_1 < e_0^2 < 1$ and by recurrence $e_n < 1$, i.e. the sequence $\{r_n\}$ belongs to the interval:

$$\left(r - \frac{1}{C}, r + \frac{1}{C}\right) \subseteq (a, b).$$

23.c If $e_0 = \frac{1}{2} < 1$ then $e_1 \le e_0^2$, $e_2 \le e_1^2 \le e_0^4 = e_0^{2^2}$. By recurrence, assuming $e_n \le e_0^{2^n}$, then $e_{n+1} \le e_n^2 \le (e_0^{2^n})^2 = e_0^{2^{n+1}}$.
Thus, $\frac{|r_n - r|}{|r_0 - r|} = \frac{e_n}{e_0} \le e_0^{2^n - 1}$. Therefore, the smallest n_p for which $\frac{|r_{n_p} - r|}{|r_0 - r|} \le 2^{-p}$ can be estimated using the inequalities:

$$e_0^{2^{n_p} - 1} \le 2^{-p} < e_0^{2^{n_p - 1} - 1}.$$

For $e_0 = \frac{1}{2}$, this is equivalent to:

$$n_p - 1 < \frac{\ln(p+1)}{\ln 2} \le n_p,$$

implying that $n_p = \lceil \frac{\ln(p+1)}{\ln 2} \rceil$.

- **Exercise 25:**
  The function $f(x) = x^3 - 2x + 2$ has a unique negative root :$-2 <$

root < -1, as $f(-1) \times f(-2) < 0$. The initial conditions are obtained by the bisection method applied twice on the interval $(-2, -1)$. This gives: $r_0 = -1.5000$, $r_1 = -1.7500$. The first 3 computed iterates using the secant method are:
$r_2 = -1.7737$, $r_3 = -1.7692$, $r_4 = -1.7693$

- **Exercise 27:**
  27.a The function $f(x) = e^x - 3x$ has a unique root : $0 < root < 1$, as $f(0) \times f(1) < 0$.
  The initial conditions are obtained by the bisection method applied twice on the interval $(0, 1) \Rightarrow: r_0 = 0.5$, $r_1 = 0.75$.
  The first computed iterates by the Secant method are:
  $r_2 = 0.631975$; $r_3 = 0.617418$; $r_4 = 0.619078$; $r_5 = 0.619061$.
  Therefore: 3 iterations are needed to compute *root* up to 5 decimals; the 4th one confirms reaching the required precision.
  27.b The function $f(x) = x - 2^{-x}$ has a unique root : $0 < root < 1$, as $f(0) \times f(1) < 0$.
  The initial conditions are obtained by the bisection method applied twice on the interval $(0, 1) \Rightarrow: r_0 = 0.5$, $r_1 = 0.75$.
  The first computed iterates by the Secant method are:
  $r_2 = 0.642830$; $r_3 = 0.641166$; $r_4 = 0.641185$; $r_5 = 0.641185$.
  Therefore: 3 iterations are needed to compute *root* up to 5 decimals; the 4th one confirms reaching the required precision.
  27.c The function $f(x) = -3x + 2\cos(x) - e^x$ has a unique root : $0 < root < 1$, as $f(0) \times f(1) < 0$.
  The initial conditions are obtained by the bisection method applied twice on the interval $(0, 1) \Rightarrow: r_0 = 0.5$, $r_1 = 0.25$.
  The first computed iterates by the Secant method are:
  $r_2 = 0.231462$; $r_3 = 0.229743$; $r_4 = 0.229731$; $r_5 = 0.229731$.
  Therefore: 3 iterations are needed to compute *root* up to 5 decimals; the 4th one confirms reaching the required precision.

Chapter 3

- **Exercise 1:**
  1.a

$$\begin{pmatrix} 3 & 4 & 3 & 5 \\ \boxed{1/3} & 11/3 & -2 & -5/3 \\ \boxed{2} & \boxed{-15/11} & -19/11 & -102/11 \end{pmatrix}.$$

By Back substitution: $x_3 = 5.3684 \Rightarrow x_2 = 2.4737 \Rightarrow x_1 = -7$.

  1.b

$$\begin{pmatrix} 3 & 2 & -5 & 0 \\ \boxed{4/3} & -26/3 & 26/3 & 0 \\ \boxed{1/3} & \boxed{-5/13} & 4 & 4 \end{pmatrix}.$$

By Back substitution: $x_3 = 1 \Rightarrow x_2 = 1 \Rightarrow x_1 = 1$

  1.c

$$\begin{pmatrix} 9 & 1 & 7 & 1 \\ \boxed{4/9} & 32/9 & 53/9 & -4/9 \\ \boxed{8/9} & \boxed{73/32} & -437/32 & 9/8 \end{pmatrix}.$$

By Back substitution: $x_3 = -0.082380 \Rightarrow x_2 = -0.025733 \Rightarrow x_1 = 0.17804$

- **Exercise 3:**
  3.a

$$IV = [1,2,3]; \ V = [2,1,3]; \ IV = [2,1,3]$$

$$\begin{pmatrix} \boxed{8/9} & 11/3 & 58/9 & -55/3 \\ 9 & 6 & -5 & 132 \\ \boxed{1/9} & \boxed{-2/11} & 118/11 & 72 \end{pmatrix}$$

By Back substitution: $x_3 = 6.7119 \Rightarrow x_2 = -16.797 \Rightarrow x_1 = 29.593$

  3.b

$$IV = [1,2,3]; \ IV = [2,1,3]; \ IV = [2,3,1]$$

,

$$\begin{pmatrix} \boxed{8/9} & \boxed{1/13} & 74/13 & 222/13 \\ 9 & 6 & -5 & 132 \\ \boxed{1/9} & 13/9 & 62/9 & 101/3 \end{pmatrix}$$

By Back substitution: $x_3 = 222/74 = 3.0411 \Rightarrow x_2 = 8.8040 \Rightarrow x_1 = 10.487$

  3.c

$$IV = [1,2,3]; \ IV = [2,1,3]; \ IV = [2,3,1]$$

$$\begin{pmatrix} \boxed{3/5} & \boxed{1/8} & -15/8 & 37/8 \\ 5 & 3 & 2 & 4 \\ \boxed{-1/5} & 8/5 & -13/5 & -1/5 \end{pmatrix}$$

By Back substitution: $x_3 = -37/15 = -2.4667 \Rightarrow x_2 = -233/60 = -3.8833 \Rightarrow x_1 = 4.11667$

- **Exercise 5:**
  5.a

$$IV = [1, 2, 3, 4]; \; IV = [2, 1, 3, 4]; \; IV = [2, 4, 3, 1]; \; IV = [2, 4, 1, 3]$$

Modified augmented matrix					Scales
$\boxed{1/7}$	$\boxed{1/12}$	65/12	7/12	2	6
7	6	7	9	0	9
$\boxed{3/7}$	$\boxed{-1/3}$	$\boxed{-8/65}$	−147/65	−49/65	4
$\boxed{5/7}$	12/7	−5	11/7	0	8

By Back substitution: $x_4 = \frac{1}{3} = 0.33333 \Rightarrow x_3 = \frac{1}{3} = 0.33333 \Rightarrow x_2 = \frac{2}{3} = 0.66667 \Rightarrow x_1 = \frac{-4}{3} = -1.3333$

5.b

$$IV = [1, 2, 3]; \; IV = [2, 1, 3]; \; IV = [2, 3, 1]$$

Augmented matrix				Scales
$\boxed{3/5}$	$\boxed{-3/7}$	−261/35	15	9
5	5	1	−20	5
$\boxed{0}$	7	5	0	7

By Back substitution: $x_3 = -\frac{175}{87} = -2.0115 \Rightarrow x_2 = \frac{-125}{87} = -1.4360 \Rightarrow x_1 = \frac{-188}{87} = --2.1609$

5.c

$$IV = [1, 2, 3, 4]; \; IV = [2, 1, 3, 4]; \; IV = [2, 1, 3, 4]; \; IV = [2, 1, 3, 4]$$

Modified augmented matrix					Scales
$\boxed{1/9}$	$\boxed{64/9}$	10/9	7/9	41/9	8
9	8	8	2	4	9
$\boxed{0}$	0	4	1	0	4
$\boxed{7/9}$	$\boxed{-29/64}$	$\boxed{105/288}$	−131/128	−131/64	9

By Back substitution: $x_4 = 2 \Rightarrow x_3 = -1/2 \Rightarrow x_2 = 1/2 \Rightarrow x_1 = 0$

- **Exercise 7:**
  7.a The augmented matrix of the system is: $A|b = \begin{pmatrix} 10^{-5} & 1 & 7 \\ 1 & 1 & 1 \end{pmatrix}$

The exact solution computed in high precision using Naive Gauss reduction or even Cramer's rule leads to:

$$x \approx 6, \ y \approx 7$$

7.b In $\mathbb{F}(10, 4, -25, 26)$, $(x = 0, y = 7)$.

$$\begin{pmatrix} 10^{-5} & 1 & 7 \\ \boxed{10^5} & -10^5 & -7*10^5 \end{pmatrix}$$

leading by back substitution to a wrong solution: $(x = 0, y = 7)$.

7.c $\begin{pmatrix} 1 & 1 & 1 \\ \boxed{10^{-5}} & 1 & 7 \end{pmatrix}$

leading by back substitution to the solution: $(x = -6, y = 7)$ that is very close to the exact one.

- **Exercise 9:**

  9.1.a $L = \begin{pmatrix} 1 & 0 & 0 & 0 \\ 1/4 & 1 & 0 & 0 \\ 1/4 & 3/5 & 1 & 0 \\ 1/2 & 0 & 5/9 & 1 \end{pmatrix}$; $U = \begin{pmatrix} 4 & 2 & 1 & 2 \\ 0 & 5/2 & 7/4 & 1/2 \\ 0 & 0 & 27/10 & 1/5 \\ 0 & 0 & 0 & 17/9 \end{pmatrix}$

  9.1.b - Determinant of A = Determinant of A = $4.\frac{5}{2}.\frac{27}{10}.\frac{17}{9} = 51$

  9.1.c

$$A^{-1} = \begin{pmatrix} 0.3921 & -0.2941 & 0.1569 & -0.2157 \\ -0.0392 & 0.5294 & -0.2157 & -0.0784, \\ -0.0196 & -0.2353 & 0.3922, & -0.0392 \\ -0.2353 & 0.1765 & -0.2941 & 0.5294 \end{pmatrix}$$

- **Exercise 11:**

  11.1.a

$$L = \begin{pmatrix} 1 & 0 & 0 \\ 1/6 & 1 & 0 \\ 1/3 & 1/81 & 1 \end{pmatrix} ; U = \begin{pmatrix} 6 & 8 & 9 \\ 0 & 8/3 & 7/2 \\ 0 & 0 & 25/16 \end{pmatrix}$$

$$P = \begin{pmatrix} 0 & 0 & 1 \\ 1 & 0 & 0 \\ 0 & 1 & 0 \end{pmatrix}$$

  11.1.b Determinant of A = $(-1)^2$.Determinant of U = $(6).(\frac{8}{3}).(\frac{25}{16}) = 25$

  11.1.c

  (i) The Lower triangular system $Ly = e_3$, gives $y = [0\,0\,1]^T$ by Forward substitution

  (ii) The Upper triangular system $Uc_2 = y$, gives $c_2 = [4/25 - 21/25 \ 16/25]^T$ by Backward substitution

  11.2.a

$$L = \begin{pmatrix} 1 & 0 & 0 & 0 \\ 1/6 & 1 & 0 & 0 \\ 2/3 & 1/2 & 1 & 0 \\ 2/3 & -1/31 & -1/39 & 1 \end{pmatrix} ; U = \begin{pmatrix} 6 & 6 & 4 & 2 \\ 0 & 6 & 16/3 & 26/3 \\ 0 & 0 & -13/3 & -8/3 \\ 0 & 0 & 0 & 253/39 \end{pmatrix}$$

$$P = \begin{pmatrix} 0 & 0 & 0 & 1 \\ 1 & 0 & 0 & 0 \\ 0 & 1 & 0 & 0 \\ 0 & 0 & 1 & 0 \end{pmatrix}$$

11.2.b Determinant of A $= (-1)^2$.Determinant of U $= (6).(\frac{8}{3}).(\frac{25}{16}) = 25$
11.2.c Solving successively:
(i) The Lower triangular system $Ly = e_3$, gives $y = [0\,0\,1]^T$ by Forward substitution
(ii) The Upper triangular system $U c_2 = y$, gives $c_2 = [4/25 - 21/25\,16/25]^T$ by Backward substitution.

- **Exercise 13:** B is not diagonally dominant as in the first row: $|8| < |-1| + |4| + |9|$.
  The 3 matrices A, B and C satisfy the Principal Minor Property as all their Principal submatrices have a non zero determinant.

- **Exercise 15:**
  15.1

$$T_n = \begin{bmatrix} a_1 & b_1 & 0 & 0 & \dots & 0 \\ c_1 & a_2 & b_2 & 0 & \dots & 0 \\ 0 & c_2 & a_3 & b_3 & \dots & 0 \\ \dots & \dots & \dots & \dots & \dots & \dots \\ \dots & \dots & \dots & \dots & \dots & \dots \\ \dots & \dots & \dots & c_{n-2} & a_{n-1} & b_{n-1} \\ 0 & \dots & 0 & 0 & c_{n-1} & a_n \end{bmatrix}$$

15.1.a At each reduction $k = 1 \to (n-1)$: 1 multiplier is computed: $c_k = c_k/a_k$ and 1 element is modified: $a_{k+1} = a_{k+1} - c_k.b_k$.
15.1.b

$$U = \begin{bmatrix} a_1 & b_1 & 0 & 0 & \dots & 0 \\ 0 & a_2 & b_2 & 0 & \dots & 0 \\ 0 & 0 & a_3 & b_3 & \dots & 0 \\ \dots & \dots & \dots & \dots & \dots & \dots \\ \dots & \dots & \dots & \dots & \dots & \dots \\ \dots & \dots & \dots & 0 & a_{n-1} & b_{n-1} \\ 0 & \dots & 0 & 0 & 0 & a_n \end{bmatrix} ;$$

$$L = \begin{bmatrix} 1 & 0 & 0 & 0 & \cdots & 0 \\ \boxed{c_1} & 1 & 0 & 0 & \cdots & 0 \\ 0 & \boxed{c_2} & 1 & 0 & \cdots & 0 \\ \cdots & \cdots & \cdots & \cdots & \cdots & \cdots \\ \cdots & \cdots & \cdots & \cdots & \cdots & \cdots \\ \cdots & \cdots & \cdots & \boxed{c_{k-2}} & 1 & 0 \\ 0 & \cdots & 0 & 0 & \boxed{c_{k-1}} & 1 \end{bmatrix}$$

15.1.c To compute the $(n-1)$ multipliers: $(n-1)$ flops are used and to modify the $(n-1)$ elements: $2(n-1)$ flops are used. $\Rightarrow$ Total number of flops: $3(n-1)$.

15.2

$$UQ_n = \begin{bmatrix} a_1 & b_1 & d_1 & 0 & \cdots & 0 \\ c_1 & a_2 & b_2 & d_2 & \cdots & 0 \\ 0 & c_2 & a_3 & b_3 & \cdots & 0 \\ \cdots & \cdots & \cdots & \cdots & \cdots & \cdots \\ \cdots & \cdots & c_{n-3} & a_{n-2} & b_{n-2} & d_{n-2} \\ \cdots & \cdots & \cdots & c_{n-2} & a_{n-1} & b_{n-1} \\ 0 & \cdots & 0 & 0 & c_{n-1} & a_n \end{bmatrix}$$

15.2.a At each reduction except the last, $k = 1 \rightarrow (n-2)$:
1 multiplier is computed: $c_k = c_k/a_k$ and 2 elements are modified:
$a_{k+1} = a_{k+1} - c_k.b_k$ and $b_{k+1} = b_{k+1} - c_k.d_k$.
At last reduction: 1 multiplier is computed: $c_{n-1} = c_{n-1}/a_{n-1}$ and 1 element is modified: $a_n = a_n - c_{n-1}.b_{n-1}$

15.2.b

$$U = \begin{bmatrix} a_1 & b_1 & d_1 & 0 & \cdots & 0 \\ 0 & a_2 & b_2 & d_2 & \cdots & 0 \\ 0 & 0 & a_3 & b_3 & \cdots & 0 \\ \cdots & \cdots & \cdots & \cdots & \cdots & \cdots \\ \cdots & \cdots & 0 & a_{n-2} & b_{n-2} & d_{n-2} \\ \cdots & \cdots & \cdots & 0 & a_{n-1} & b_{n-1} \\ 0 & \cdots & 0 & 0 & 0 & a_n \end{bmatrix} ;$$

$$L = \begin{bmatrix} 1 & 0 & 0 & 0 & \cdots & 0 \\ \boxed{c_1} & 1 & 0 & 0 & \cdots & 0 \\ 0 & \boxed{c_2} & 1 & 0 & \cdots & 0 \\ \cdots & \cdots & \cdots & \cdots & \cdots & \cdots \\ \cdots & \cdots & \cdots & \cdots & \cdots & \cdots \\ \cdots & \cdots & \cdots & \boxed{c_{k-2}} & 1 & 0 \\ 0 & \cdots & 0 & 0 & \boxed{c_{k-1}} & 1 \end{bmatrix}$$

15.2.c To compute the $(n-1)$ multipliers: $(n-1)$ flops are used and to modify the $[2(n-2)+1]$ elements: $[4(n-2)+2]$ flops are used. $\Rightarrow$ Total number of flops: $5n-7$.

15.3

$$LQ_n = \begin{bmatrix} a1 & b1 & 0 & 0 & \ldots & 0 \\ c1 & a2 & b2 & 0 & \ldots & 0 \\ d1 & c2 & a3 & b3 & \ldots & 0 \\ \ldots & \ldots & \ldots & \ldots & \ldots & \ldots \\ \ldots & \ldots & \ldots & \ldots & \ldots & \ldots \\ \ldots & \ldots & d_{n-3} & c_{n-2} & a_{n-1} & b_{n-1} \\ 0 & \ldots & 0 & d_{n-2} & c_{n-1} & a_n \end{bmatrix}$$

15.3.a At each reduction except the last $k = 1 \to (n-2)$: 2 multipliers are computed: $c_k = c_k/a_k$ and $d_k = d_k/a_k$, and 2 element are modified: $a_{k+1} = a_{k+1} - c_k.b_k$ and $c_{k+1} = c_{k+1} - d_k.b_k$
At last reduction: 1 multiplier is computed: $c_{n-1} = c_{n-1}/a_{n-1}$ and 1 element is modified: $a_n = a_n - c_{n-1}.b_{n-1}$.

15.3.b

$$U = \begin{bmatrix} a_1 & b_1 & 0 & 0 & \ldots & 0 \\ 0 & a_2 & b_2 & 0 & \ldots & 0 \\ 0 & 0 & a_3 & b_3 & \ldots & 0 \\ \ldots & \ldots & \ldots & \ldots & \ldots & \ldots \\ \ldots & \ldots & \ldots & \ldots & \ldots & \ldots \\ \ldots & \ldots & \ldots & 0 & a_{n-1} & b_{n-1} \\ 0 & \ldots & 0 & 0 & 0 & a_n \end{bmatrix}$$

$$L = \begin{bmatrix} 1 & 0 & 0 & 0 & \ldots & 0 \\ c_1 & 1 & 0 & 0 & \ldots & 0 \\ d_1 & c_2 & 1 & 0 & \ldots & 0 \\ \ldots & \ldots & \ldots & \ldots & \ldots & \ldots \\ \ldots & \ldots & \ldots & \ldots & \ldots & \ldots \\ \ldots & \ldots & \ldots & c_{n-2} & 1 & 0 \\ 0 & \ldots & 0 & d_{n-2} & c_{n-1} & 1 \end{bmatrix}$$

15.3.c To compute the $[2(n-2)+1]$ multipliers: $[2(n-2)+1]$ flops are used and to modify the $[2(n-2)+1]$ elements: $[4(n-2)+2]$ flops are used. $\Rightarrow$ Total number of flops: $6n - 9$.

- **Exercise 17:**

  - Column-Backward substitution:
    Total number of flops: $\sum_{j=2}^{n} 1 + (\sum_{j=2}^{n} \sum_{i=1}^{j-1} 2) + 1 = n^2$,
    as: $\sum_{i=1}^{j-1} 2 = 2(j-1)$ and $\sum_{j=2}^{n} \sum_{i=1}^{j-1} 2 = \sum_{j=2}^{n} 2(j-1) = n(n-1)$.

  - Row-Forward substitution:
    Total number of flops: $1 + (\sum_{i=2}^{n} \sum_{j=1}^{i-1} 2) + \sum_{i=2}^{n} 1 = n^2$,
    as: $\sum_{j=1}^{i-1} 2 = 2(j-1)$ and $\sum_{i=2}^{n} \sum_{j=1}^{i-1} 2 = \sum_{i=2}^{n} 2(i-1) = n(n-1)$.

Chapter 4

- **Exercise 1:**

  $D_3 = \{(0,7), (2,10), (3,25), (4,50)\}$.

  $l_0(x) = \frac{-(x-2)(x-3)(x-4)}{24}$; $l_1(x) = \frac{x)(x-3)(x-4)}{4}$; $l_2(x) = -\frac{x(x-2)(x-4)}{3}$; $l_3(x) = \frac{x(x-2)(x-3)}{8}$;

  $p_{0123}(x) = 7l_0(x) + 10l_1(x) + 25l_2(x) + 50l_3(x)$

- **Exercise 3:**

  $l_0(x) = \frac{(x-x_1)(x-x_2)(x-x_3)}{(x_0-x_1)(x_0-x_2)(x_0-x_3)}$; $l_1(x) = \frac{(x-x_0)(x-x_2)(x-x_3)}{(x_1-x_0)(x_1-x_2)(x_1-x_3)}$;

  $l_2(x) = \frac{(x-x_0)(x-x_1)(x-x_3)}{(x_2-x_0)(x_2-x_1)(x_2-x_3)}$; $l_3(x) = \frac{(x-x_0)(x-x_1)(x-x_2)}{(x_3-x_0)(x_3-x_1)(x_3-x_2)}$

  $p_{0123}(x) = y_0l_0(x) + y_1l_1(x) + y_2l_2(x) + y_3l_3(x)$

- **Exercise 5:**

  $D_4 = \{(1,-1), (2,-1/3), (2.5, 3), (3, 4), (4, 5)\}$

  5.a

i	x_i	y_i	$[.,.]$	$[.,.,.]$	$[.,.,.,.]$	$[.,.,.,.,.]$
0	1	-1				
			$2/3$			
1	2	$-1/3$		4		
			$20/3$		$-13/3$	
2	2.5	3		$-14/3$		$19/9$
			2		2	
3	3	4		$-2/3$		
			1			
4	4	5				

  5.b

  - Quadratic interpolating polynomial: $p_{123}(x) = -1 + \frac{20}{3}(x-2) - \frac{14}{3}(x-2)(x-2.5) \Rightarrow f(2.7) \approx p_{123}(x) = 4.3200$

  - Best Cubic interpolating polynomial; $p_{1234}(x) = p_{123}(x) + 2(x-2)(x-2.5)(x-3) \Rightarrow f(2.7) \approx p_{1234}(x) = 4.2360$

- **Exercise 7:**

  $p(x) = p_{01234}(x) + A(x-x_0)(x-x_1)(x-x_2)(x-x_3)(x-x_4)$

  $q(x) = p_{01234}(x) + B(x-x_0)(x-x_1)(x-x_2)(x-x_3)(x-x_4)$

  $\Rightarrow q(x) - p(x) = C(x-x_0)(x-x_1)(x-x_2)(x-x_3)(x-x_4)$. Substituting x by 4 in the identity above $\Rightarrow C = -\frac{443}{120} \Rightarrow q(x) = p(x) - \frac{443}{120}(x+1)(x)(x-1)(x-2)(x-3)$.

- **Exercise 9:**

  - Neville's polynomial:

  $p_{01}(x) = 2x + 1$; $p_{12}(x) = x + 2 \Rightarrow p_{012}(x) = \frac{(x-x_0)p_{12}(x) - (x-x_2)p_{01}(x)}{x_2 - x_0} = \frac{-x^2 + 5x + 2}{2}$.

- Newton's polynomial:
$$p_{012}(x) = [x_0] + [x_0 + x_1](x - x_0) + [x_0, x_1, x_2](x - x_0)(x - x_1) = \frac{-x^2 + 5x + 2}{2}$$

- **Exercise 11:**

  11. a $D_5 = \{(-2, 1), (-1, 4), (0, 11), (1, 16), (2, 13), (3, -4)\}$.

  $p_{01234}(x) = [x_0] + [x_0 + x_1](x - x_0) + [x_0, x_1, x_2](x - x_0)(x - x_1) + [x_0, x_1, x_2, x_3](x - x_0)(x - x_1)(x - x_2) + [x_0, x_1, x_2, x_3, x_4](x - x_0)(x - x_1)(x - x_2)(x - x_3) \Rightarrow$

  $p_{01234}(x) = 1 + 3(x + 2) + 2(x + 2)(x + 1) - (x + 2)(x + 1)x$, since $[x_0, x_1, x_2, x_3] = [x_0, x_1, x_2, x_3, x_4] = 0$

  11. b $q(x) = q_{012345}(x) = q_{012354}(x) = p_{01235}(x) + A(x - x_0)(x - x_1)(x - x_2)(x - x_3)(x - x_5)$

  $p(x) = p_{012345}(x) = p_{012354}(x) = p_{01235}(x) + B(x - x_0)(x - x_1)(x - x_2)(x - x_3)(x - x_5) \Rightarrow$

  $q(x) - p(x) = C(x - x_0)(x - x_1)(x - x_2)(x - x_3)(x - x_5)$.

  Substituting x by 2 in the identity above $\Rightarrow C = \frac{1}{8} \Rightarrow$

  $q(x) = p(x) + \frac{1}{8}(x + 2)(x + 1)(x)(x - 1)(x - 3)$.

- **Exercise 13:**

  13.a

i	x_i	y_i	$[\cdot, \cdot]$	$[\cdot, \cdot, \cdot]$	$[\cdot, \cdot, \cdot, \cdot]$
0	-2	-1			
			2		
1	-1	1		$1/2$	
			3		$-37/21$
2	0	4		$-17/3$	
			$-8/3$		
3	1.5	0			

  $p_{123}(x) = 1 + 3(x + 1) - \frac{17}{3}(x + 1)x$

  13.b $p(x) = p_{0123}(x) = p_{1230}(x) = p_{123}(x) + [x_0, x_1, x_2, x_3](x + 1)x(x_1.5) =$

  $p_{123}(x) - \frac{37}{21}(x + 1)x(x - 1.5)$.

  13.c Let $(x_A, y_A) = (-0.5, 2)$, then:

  $q(x) = q_{01A23}(x) = q_{0123A}(x) = q_{0123}(x) + C(x + 2)(x + 1)x(x - 1.5) \Rightarrow$

  $q(x) = p(x) + C(x + 2)(x + 1)x(x - 1.5)$ Substituting x by -0.5 in the identity above $\Rightarrow C = [2 - p(-0.5)]/ \Rightarrow$

  $q(x) = p(x) + (x + 2)(x + 1)x(x - 1.5)$

- **Exercise 15:**

i	x_i	y_i	$[x_i, x_{i+1}]$
0	0	1	$2/3$
1	1.5	2	$8/3$
2	2	6	-6
3	2.5	3	.

$$S(x) = \begin{cases} S_0(x) = 1 + \frac{2}{3}x, \ 0 \le x \le 1.5 \\ S_1(x) = 2 + \frac{8}{3}(x - 1.5), \ 1.5 \le x \le 2 \\ S_2(x) = 6 - 6(x - 2), \ 2 \le x \le 2.5 \end{cases}$$

- **Exercise 17:**

i	x_i	y_i	$[x_i, x_{i+1}]$	z_i
0	-1	3	-3	0
1	0	0	1	-6
2	1	1	1	8
3	2	2	.	-6

$$S(x) = \begin{cases} S_0(x) = -3(x + 1)^2 + 3, \ -1 \le x \le 0 \\ S_1(x) = 7x^2 - 6x, \ 0 \le x \le 1 \\ S_2(x) = -7(x - 1)^2 + 8(x - 1) + 1, \ 1 \le x \le 2 \end{cases}$$

- **Exercise 19:**

i	x_i	y_i	$[x_i, x_{i+1}]$	z_i
0	-1	0	1	1
1	0	1	-2	1
2	1/2	0	2	-5
3	1	1	-1	9
4	2	0	.	-11

$$S(x) = \begin{cases} S_0(x) = x - 1, \ 1 \le x \le 2 \\ S_1(x) = 1 + (x - 2) - 6(x - 2)^2, \ 2 \le x \le 2.5 \\ S_2(x) = -5(x - 2.5) + 14(x - 2.5)^2, \ 2.5 \le x \le 3 \\ S_3(x) = 3 + 9(x - 3) - 10(x - 3)^2, \ 3 \le x \le 4 \end{cases}$$

- **Exercise 21:**
  $a = 2; \ b = 1; \ c = 0; \ d = 1; e = -1.$

- **Exercise 23:**

i	x_i	y_i	z_i	w_i
0	1	0	8/3	0
1	2	1	$-10/3$	-8
2	3	0	$-304/63$	128/63
3	4	1	$-167/63$	20/3
4	5	0	$-148/63$	0

$$S(x) = \begin{cases} S_0(x) = \frac{8}{3}(x - 1) - \frac{8}{9}(x - 1)^3, \ 1 \le x \le 2.5 \\ S_1(x) = 1 - \frac{10}{3}(x - 2.5) - 4(x - 2)^2 + \frac{632}{189}(x - 2.5)^3, \ 2.5 \le x \le 3 \\ S_2(x) = -\frac{304}{63}(x - 3) + \frac{64}{63}(x - 3)^2 - \frac{292}{567}(x - 3)^3, \ 3 \le x \le 4.5 \\ S_3(x) = 1 - \frac{167}{3}(x - 4.5) + \frac{10}{3}(x - 4.5)^2 - \frac{20}{9}(x - 4.5)^3, \ 4.5 \le x \le 5 \end{cases}$$

- **Exercise 25:**

i	x_i	y_i	z_i	w_i
0	−0.2	0.7121	1.759	0
1	−0.1	0.8790	2.574	16.29
2	0.1	1.0810	3.1186	−10.844
3	0.2	1.1279	2.5764	0

$S(x) =$

$$\begin{cases} 0.7121 + 1.759(x + 0.2) + 13.575(x + 0.2)^3, & -0.2 \le x \le -0.1 \\ 0.8790 + 2.574(x + 0.1) + 8.145(x + 0.1)^2 + 4.5383(x + 0.1)^3, & -0.1 \le x \le 0.1 \\ 1.0810 + 3.1186(x - 0.1) - 5.422(x - 0.1)^2 - 9.0367(x - 0.1)^3, & 0.1 \le x \le 0.2 \end{cases}$$

- **Exercise 27:**

$$S(x) = \begin{cases} S_0(x) = 2x^2; & 0 \le x \le 1 \\ S_1(x) = 3x^2 - 2x + 1; & 1 \le x \le 2 \\ S_2(x) = 0.5x^3 + 4x - 3; & 2 \le x \le 3 \end{cases}$$

- **Exercise 29:**

1. As shown below, the given set of data D_4 verifies the following set of values:

i	x_i	y_i	z_i	w_i
0	0	1	−0.69088	0
1	0.25	0.7788	−1.2726	−4.654
2	0.75	0.4724	−2.0055	1.7226
3	1	0.3679	−1.7903	0

$h_1 = 0.25;\ h_2 = 0.5;\ h_3 = 0.25.$

- Using the Naive Gauss reduction, solve first the augmented system:

$$A|r = \begin{pmatrix} 0.75/3 & 0.5/6 & 0.2720 \\ 0.5/6 & 0.75/3 & 0.47347 \end{pmatrix} \Rightarrow w_2 = 1.7226;\ w_1 = $$

-4.654

- $z_0 = [x_0, x_1] - \frac{h_1}{6}(w_1 + 2w_0) = -0.69088.$
- Solve $z_{i+1} = z_i + \frac{h_{i+1}}{2}(w_i + w_{i+1})$, for i=0,1,2 $\Rightarrow z_1 = $ $-1.2726;\ z_2 = -2.0055;\ z_3 = -1.7903.$
- The equations of the Cubic spline $S(x)$ are as follows:

$$\begin{cases} 1 - 0.69088x - 3.1027x^3; & \text{if } 0 \le x \le 0.25 \\ 0.7788 - 1.2726(x - 0.25) - 2.327(x - 0.25)^2 \\ +2.1255(x - 0.25)^3; & \text{if } 0.25 \le x \le 0.75 \\ 0.4724 - 2.0055(x - 0.75) + 0.8613(x - 0.75)^2 \\ -1.1484(x - 0.75)^3; & \text{if } 0.75 \le x \le 1 \end{cases}$$

2. $\int_0^{0.25} e^{-x}\, dx = 0,2212 \approx \int_0^{0.25} S_0(x)\, dx = 0.225385.$

3. $f'(0.5) \approx S_1'(0.5).$

- **Exercise 31:**
  - From the first criterion of the Definition of Spline function of degree 4, , each of the $s_i(x)$ is determined by 5 parameters. Hence, full obtention of $s(x)$ requires $5n$ unknowns.
  - The second, third and fourth criteria impose now respectively $4(n-1)$ continuity conditions for s, s', s'', and s''' at the interior nodes, in addition to the $n+1$ interpolation conditions
  Hence for a total of $5n$ unknowns, one has a total of $4(n-1) + n + 1 = 5n - 3$ constraints. Obviously, to allow unique determination of the interpolating spline of degree 4, there appears to be a deficit of three constraints!

- **Exercise 33:** Upper Bounds on error terms for interpolating $f(x) = \frac{1}{1+x^2}$ over $[-5,5]$:

  1. Lagrange interpolation:

  $$\forall x \in [-5,5],\ |f(x) - p_{0\ldots 10}(x)| \leq \frac{10^{10}}{11!} \max_{x\in[-5,5]} |f^{(11)}(x)|$$

  2. Linear spline:

  $$\forall x \in [-5,5],\ |f(x) - S(x)| \leq Ch^2 \max_{x\in[-5,5]} |f^{(2)}(x)|,\ h = 1,$$

  3. Quadratic spline:

  $$\forall x \in [-5,5],\ |f(x) - S(x)| \leq Ch^3 \max_{x\in[-5,5]} |f^{(3)}(x)|,\ h = 1,$$

  4. Lagrange interpolation:

  $$\forall x \in [-5,5],\ |f(x) - S(x)| \leq h^4 \max_{x\in[-5,5]} |f^{(4)}(x)|,\ h = 1$$

  C a generic constant independent from h.

- **Exercise 35:**
  35.a

i	x_i	y_i	z_i	w_i
0	2	1	$-26/15$	0
1	3	0	$7/15$	$22/5$
2	4	1	$-2/15$	$-28/5$
3	5	-1	$-44/15$	0

$$S(x) = \begin{cases} S_0(x) = 1 - \frac{26}{15}(x-2) - \frac{11}{15}(x-2)^3, \ 2 \leq x \leq 3 \\ S_1(x) = \frac{7}{15}(x-3) + \frac{1}{5}(x-3)^2 - \frac{11}{7}(x-3)^3, \ 3 \leq x \leq 4 \\ S_2(x) = 1 - \frac{2}{15}(x-4) - \frac{14}{5}(x-4)^2 + \frac{14}{15}(x-4)^3, \ 4 \leq x \leq 5 \end{cases}$$

$f(3.4) \approx S_1(3.4) = 0.1181$

35.b

i	x_i	y_i	z_i	w_i
0	2	1	$-16/9$	2
1	3	0	$-4/9$	$2/3$
2	4	1	$-22/9$	$-14/3$
3	5	-1	$-43/9$	0

$$S(x) = \begin{cases} S_0(x) = 1 - \frac{16}{9}(x-2) + (x-2)^2 - \frac{2}{9}(x-2)^3, \ 2 \leq x \leq 3 \\ S_1(x) = -\frac{4}{9}(x-3) + \frac{1}{3}(x-3)^2 - \frac{8}{9}(x-3)^3, \ 3 \leq x \leq 4 \\ S_2(x) = 1 - \frac{22}{9}(x-4) - \frac{7}{3}(x-4)^2 - \frac{7}{9}(x-4)^3, \ 4 \leq x \leq 5 \end{cases}$$

$f(3.4) \approx S_1(3.4) = -0.1813$

Chapter 5

- **Exercise 1:**
  F.D. : $f'(0) \approx \frac{4.960-5}{0.1} = -0.400$
  C.D. : $f'(0.1) \approx \frac{4.842-5}{0.2} = -0.790$
  C.D. : $f'(0.2) \approx \frac{4.651-4.960}{0.2} = -1.545$
  C.D. : $f'(0.3) \approx \frac{4.393-4.842}{0.2} = -2.245$
  B.D. : $f'(0.4) \approx \frac{4.393-4.651}{0.1} = -2.580$

- **Exercise 3:**
  $h = 0.125 : \frac{\Delta_h f(0)}{h} = 2.1260$; $h = 0.25 : \frac{\Delta_h f(0)}{h} = 2.2580$; $h = 0.375 : \frac{\Delta_h f(0)}{h} = 2.4026$
  $h = 0.5 : \frac{\Delta_h f(0)}{h} = 2.5681$; $h = 0.625 : \frac{\Delta_h f(0)}{h} = 2.7646$

- **Exercise 5:**
  (i) $D = \phi(h) + c_1 h^{1/2} + c_2 h^{2/2} + c_3 h^{3/2} + \ldots$, i.e. $D = \phi(h) + O(h^{1/2})$
  (ii) $D = \phi(\frac{h}{2}) + c_1(\frac{h}{2})^{1/2} + d_2(\frac{h}{2})^{2/2} + d_3(\frac{h}{2})^{3/2} + \ldots$
  $\frac{\sqrt{2}(ii)-(i)}{\sqrt{2}-1} : D = [\frac{\sqrt{2}\phi(\frac{h}{2})-\phi(h)}{\sqrt{2}-1}] + d_2' h^2 + d_3' h^3 + \ldots$; i.e. $D = [\frac{\sqrt{2}\phi(\frac{h}{2})-\phi(h)}{\sqrt{2}-1}] + O(h^2)$

- **Exercise 7:**
  C.D. : $\psi_{\pi/3}(f(\pi/4)) = \frac{\cos(\pi/4+\pi/3)-\cos(\pi/4-\pi/3)}{2\pi/3} = -0.5847$;
  $\psi_{\pi/6}(f(\pi/4)) = \frac{\cos(\pi/4+\pi/6)-\cos(\pi/4-\pi/6)}{2\pi/6} = -0.6752$

$!f'(\pi/4) \approx \psi^1_{\pi/6}(f(\pi/4)) = \frac{4\psi_{\pi/6}(f(\pi/4))-\psi_{\pi/3}(f(\pi/4))}{3} = -0.7054;$
$f'(\pi/4) = -sin(\pi/4) = -0.7071;$ Relative Error=0.0024.

- **Exercise 9:**
  9.a $\Phi_{0.25}(f(0.25)) = \frac{3.233-2.122}{0.25} = 4.444$; $\chi_{0.25}(f(0.25)) = \frac{2.122-1}{0.25} = 4.488;$
  $\Psi_{0.25}(f(0.25)) = \frac{3.233-1}{0.5} = 4.466$
  9.b $\Psi_{0.25}(f(1)) = \frac{-1-4.455}{0.5} = -10.91;$ $\Psi_{0.5}(f(1)) = \frac{-1.255-3.233}{1} = -4.488;$
  $\Psi_1(f(1)) = \frac{-2-1}{2} = -1.5$
  $\Psi^1_{0.25}(f(1)) = \frac{4\Psi_{0.25}(.)-\Psi_{0.5}(.)}{3} = -13.051; \Psi^2_{0.25}(f(1)) = \frac{16\Psi^1_{0.25}(.)-\Psi^1_{0.5}(.)}{15} = -13.555$
  as: $\Psi^1_{0.5}(f(1)) = \frac{4\Psi_{0.5}(.)-\Psi_1(.)}{3} = -5.484$
  9.c F.D. : $f'(0) \approx \frac{2.122-1}{0.25} = 4.488$; B.D. : $f'(2) \approx \frac{-2-(-1.8)}{0.25} = -0.8$
  9.d F.D. : $f''(1) \approx \Phi_{0.25}(f(1)) = \frac{\Delta^2_{0.25}(.)}{(0.25)^2} = \frac{-1.255-2(-1)+5.566}{(0.25)^2} = 100.9760$
  F.D. : $f'''(1) \approx \Phi_{0.25}(f(1)) = \frac{\Delta^3_{0.25}(.)}{(0.25)^2} = \frac{5.566-2(4.455)+3.233}{(0.25)^2} = -1.7760$

- **Exercise 11:**
  11.a $\psi_{0.25}(f(0.5)) = \frac{y_8-2y_6+y_4}{(0.25)^2} = -0.61861280 ; \psi_{0.125}(f(0.5)) = \frac{y_6-2y_5+y_4}{(0.125)^2} = -0.70367360$
  $f''(0.5) \approx \psi^1_{0.125}(f(0.5)) = \frac{4\psi_{0.125}(.)-\psi_{0.25}(.)}{3} = -0.73202720.$
  11.b $f'''(1) \approx \chi_{0.125}(f(1)) = 6[x_{i-3}, x_{i-2}, x_{i-1}, x_i] = \frac{y_8-3y_7+3y_6-y_5}{(0.125)^3} = 0.57845760$

- **Exercise 13:**
  $\chi_{0.5}(f(1)) = \frac{y_8-y_4}{0.5} = 0.20223700; \chi_{0.25}(f(1)) = \frac{y_8-y_6}{0.25} = 0.12491040;$
  $\chi_{0.125}(f(1)) = \frac{y_8-y_7}{0.125} = 0.09103000$
  $\chi^1_{0.25}(f(1)) = \frac{2\chi_{0.25}(.)-\chi_{0.5}(.)}{1} = 0.04758380; \chi^1_{0.125}(f(1)) = \frac{2\chi_{0.125}(.)-\chi_{0.25}(.)}{1} = 0.0571496$
  $\chi^2_{0.125}(f(1)) = \frac{4\chi^1_{0.125}(.)-\chi^1_{0.25}(.)}{3} = 0.0603382$

- **Exercise 15:**
  $D^1_{0.125}(0) = \frac{4D_{0.125}(0)-D_{0.25}(0)}{3} = \frac{4(1.9624)-1.8880}{3}$, since:
  $D_{0.125}(0) = \frac{4(1.1108)-3(1)-1.1979}{0.125} = 1.9624$, and $D_{0.25}(0) = \frac{4(1.1979)-3(1)-1.3196}{0.25} = 1.8880$

- **Exercise 17:**
  $I \approx M(0.125) = (y_1 + y_3 + y_5 + y_7)0.25 = 1.2866$

- **Exercise 19:**
  19. a $h = \frac{1}{6} \Rightarrow$

$$I = \int_0^1 \frac{1}{1+x^2} dx \approx M(\frac{1}{6}) = [f(\frac{1}{6}) + f(\frac{3}{6}) + f(\frac{5}{6})]\frac{2}{6} = 0.78771$$

19. b $I = tan^{-1}1 = \pi/4 \Rightarrow |Error| = |\frac{\pi}{4} - 0.78771| = 0.0023118$

$f(x) = \frac{1}{1+x^2}; f''(x) = \frac{6x^2-2}{(1+x^2)^3}$, and $|f''(x)| \leq \frac{\max_{0 \leq x \leq 1} 6x^2-2}{\min_{0 \leq x \leq 1}(1+x^2)^3} = 4$

$\Rightarrow |Error| \leq \frac{4}{6^3} = 0.018518.$

- **Exercise 21:**

  $f(x) = e^{-x^2}$ and $f''(x) = e^{-x^2}(4x^2 - 2) \leq 2; h = \frac{1}{n}$

  $\Rightarrow Error = \frac{1}{6}\frac{1}{n^2}f''(c); |Error| \leq \frac{1}{3n^2} \leq \frac{10^{-4}}{2} \Rightarrow n \geq 10^2\sqrt{\frac{2}{3}} = 81.64 \Rightarrow$

  $n = 82.$

- **Exercise 23:**

  The definite integral $I = \int_a^b f(x)dx$ is approximated by:

  – $\sum_{k=1}^n (x_k - x_{k-1})f(x_{k-1})$ for "left composite rectangular" rule and

  – $\sum_{k=1}^n (x_k - x_{k-1})f(x_k)$ for "right composite rectangular" rule.

- **Exercise 25:**

  25. a $h = 0.5 \Rightarrow I \approx T(0.5)$

  $= \frac{[f(0)+2(f(0.5)+f(1)+f(1.5)+f(2)+f(2.5)+f(3)+f(3.5))+f(4)]}{2}.(0.5) = 21.8566$

  25. b Exact Value: $I = \frac{2^4-1}{\ln 2} = 21.6404. |Error| = 0.2162.$

  $$|Error| \leq \frac{1}{12}.(\frac{1}{2})^2.(\ln 2)^2.16 = 0.1601.$$

- **Exercise 27:**

  $I = \int_0^6 \sin(x^2)\, dx$

  $n + 1 = 55 \Rightarrow n = 54 \Rightarrow h = 1/9.$

  $f(x) = \sin x^2 \Rightarrow f^{(2)}(x) = 2\cos 2x; f^{(4)}(x) = -8\cos 2x;$

  $\bullet|Error_T| \leq \frac{6}{12}.(\frac{1}{9})^2.2 = 1.2345 \times 10^{-2}$

  $\bullet|Error_M| \leq \frac{1}{6}.(\frac{1}{9})^2.2 = 4.1152 \times 10^{-3}$

  $\bullet|Error_S| \leq \frac{1}{30}.(\frac{1}{9})^4.8 = 4.0064 \times 10^{-5}$

- **Exercise 29:**

  1st column: $T(1) = \frac{y_0+y_8}{2} = 1.2104; T(0.5) = (\frac{y_0+2y_4+y_8}{2})(0.5) = 1.2650;$

  $T(0.25) = (\frac{y_0+2(y_2+y_4+y_6)+y_8}{2})(0.25) = 1.2793;$

  $T(0.125) = (\frac{y_0+2(y_1+y_2+y_3+y_4+y_5+y_6+y_7)+y_8}{2})(0.125) = 1.2830;$

  2nd column: $R^1(0.5) = \frac{4T(0.5)-T(1)}{3} = 1.2832; R^1(0.25) = \frac{4T(0.25)-T(0.5)}{3} = 1.2841; R^1(0.125) = \frac{4T(0.125)-T(0.25)}{3} = 1.2842;$

  3rd column: $R^2(0.25) = \frac{16R^1(0.25)-R^1(0.5)}{15} = 1.284160000; R^2(0.125) = \frac{16R^1(0.125)-R^1(0.25)}{15} = 1.284206666;$

  4th column: $R^3(0.125) = \frac{64R^2(0.125)-R^2(0.25)}{63} = 1.284207407;$

- **Exercise 31:**

  Let $h = \{h_1, h_2, ..., h_n\}$ and $|h| = \max_k h_k$. then, the composite

trapezoidal rule for a non-uniform partition is obtained from: $T(h) = \sum_{k=1}^{n} T_k = \sum_{k=1}^{n} \frac{h_k}{2}(f(x_{k-1}) + f(x_k))$, where $h_k = x_k - x_{k-1}$.
As for the error term:

$$\int_a^b f(x)dx - T(h) = \sum_{k=1}^{n} \int_{x_{k-1}}^{x_k} f(x)dx - T_k.$$

Using:

$$\int_{x_k}^{x_{k+1}} f(x)dx = T_k + \frac{1}{2}\int_{x_k}^{x_{k+1}} (x - x_k)(x - x_{k+1})f''(c(x))dx, \; c(x) \in (x_k, x_{k+1}),$$

and the second mean value theorem:

$$\int_a^b f(x)dx - T(h) = -\frac{1}{12}\sum_{k=1}^{n} h_k^3 f''(c_k), \; h_k = x_k - x_{k-1}.$$

Hence using the intermediate value theorem:

$$|I - T(h)| \leq \frac{1}{12}|h|^2(b - a)\max_{x \in (a,b)} |f''(x)|.$$

- $I = \int_0^2 x^2 \ln(x^2 + 1)\, dx$; $h = 0.5$;
  1st method: $S(0.5) = \frac{[f(0)+4(f(0.5)+f(1.5))+2f(1)+f(2)]}{3}.(0.5) = 3.1092$
  2nd method: $S(0.5) = \frac{2}{3}M(0.5) + \frac{1}{3}T(1) = \frac{2}{3}(2.7078) + \frac{1}{3}(3.9120) = 3.1092$

- $I = \int_1^2 x^{-1}\, dx$; $h = 0.25$;
  $S(0.25) = \frac{[f(1)+4(f(1.25)+f(1.75))+2f(1.5)+f(2)]}{3}.(0.25) = 0.6932$;
  $f(x) = \frac{1}{x} \Rightarrow f^{(4)}(x) = \frac{24}{x^5}$; $\max_{1 \leq x \leq 2} |f^{(4)}(x)| = 24$;
  $|Error| \leq \frac{1}{180}.24.(0.25)^4 = 5.2083 \times 10^{-4}$.

- **Exercise 33:**
  $I = \int_0^2 x^2 \ln(x^2 + 1)\, dx$; $h = 0.5$;
  1st method: $S(0.5) = \frac{[f(0)+4(f(0.5)+f(1.5))+2f(1)+f(2)]}{3}.(0.5) = 3.1092$
  2nd method: $S(0.5) = \frac{2}{3}M(0.5) + \frac{1}{3}T(1) = \frac{2}{3}(2.7078) + \frac{1}{3}(3.9120) = 3.1092$

- $I = \int_1^2 x^{-1}\, dx$; $h = 0.25$;
  $S(0.25) = \frac{[f(1)+4(f(1.25)+f(1.75))+2f(1.5)+f(2)]}{3}.(0.25) = 0.6932$;
  $f(x) = \frac{1}{x} \Rightarrow f^{(4)}(x) = \frac{24}{x^5}$; $\max_{1 \leq x \leq 2} |f^{(4)}(x)| = 24$;
  $|Error| \leq \frac{1}{180}.24.(0.25)^4 = 5.2083 \times 10^{-4}$.

- **Exercise 35:**
  35.a $I = erf(x) = \frac{2}{\sqrt{\pi}}\int_0^x e^{-t^2}\, dt$; $\frac{|I-T(h)|}{|I|} \leq (0.5).10^{1-3}$; $f(t) = \frac{2}{\sqrt{\pi}}e^{-t^2}$; $f''(t) = \frac{2}{\sqrt{\pi}}\frac{4t^2-2}{e^{t^2}}$

where $|f''(t)| \leq \frac{4}{\sqrt{\pi}} \Rightarrow |Error| \leq \frac{1}{3\sqrt{\pi}} \cdot \frac{1}{n^2} \leq (0.42).10^{-2} \Rightarrow n^2 \geq 44.7 \Rightarrow$
$n \geq 6.6 \approx 7$.

Therefore the number of required partition points is 8.

35.b If the Romberg process has to be applied following the composite Trapezoidal rule, then: $n = 2^i = 8$, meaning that one should start with 9 partition points.

Chapter 6

- **Exercise 1:**
  $I - M(h) = \sum_{i=1}^{m} \frac{h_i^3}{3} f''(c_i) = f''(c) \sum_{i=1}^{m} \frac{h_i^3}{3}$
  $|I - M(h)| \leq (\frac{h^2}{3}|f''(c)|) = \frac{h^2}{6}|f''(c)|(b-a)$

- **Exercise 3:**
  $I = S(h) + ah^4 + O(h^6)$ leads to $I = \frac{16S(h/2)-S(h)}{15} + O(h^6) = S(h/2) + \frac{S(h/2)-S(h)}{15} + O(h^6)$

- **Exercise 5:**
  The estimate follows from the identities:
  $F(x) = \int_c^d f(x,y)dy$, $I = \frac{h}{2}\sum_{i=1}^{m}(F(x_{i-1}) + F(x_i)) - \frac{h^2}{12}f_{xx}(\xi,\eta)$ and
  $F(x_i) = \frac{k}{2}\sum_{j=1}^{n} f(x_i, y_{j-1}) + f(x_i, y_j) - \frac{k^2}{12}f_{yy}(x_i, \zeta_j)$ and through repeated applications of the intermediate value theorem.

- **Exercise 7:**
  7.a $\int_2^4 \int_1^2 \ln(2xy)\, dy\, dx = \frac{1}{2}\int_2^4 [\ln(2.5x) + \ln(3.5x)]dx = \frac{1}{2}[\ln(6.25) + \ln(12.25)] = 2.1691$
  7.b $\int_2^3 \int_2^4 (x^2 + y^3)\, dy\, dx = \int_2^3 (2x^2 + 58.5)\, dx = 71.1250$

- **Exercise 9:**
  9.a $\int_0^1 \int_0^1 e^{y-x}\, dy\, dx = \int_0^1 \int_0^1 e^y e^{-x}\, dy\, dx = \int_0^1 7.0157 e^{-x}\, dx = 18.1071$
  9.b $\int_0^\pi \int_0^\pi y \cos x\, dy\, dx = \int_0^\pi 2\pi \cos x\, dx = 0$

- **Exercise 11:**
  11.a
  (Midpoint)$\int_{-1}^1 \int_1^2 \int_0^1 y\, dz\, dy\, dx = \int_{-1}^1 \int_1^2 y\, dy\, dx = \int_{-1}^1 1.5\, dx = 1.5$
  (Trapezoid)$\int_{-1}^1 \int_1^2 \int_0^1 y\, dz\, dy\, dx = \int_{-1}^1 \int_1^2 y\, dy\, dx = \int_{-1}^1 \frac{3}{2}\, dx = \frac{3}{2}$
  11.b
  (Midpoint)$\int_{-1}^1 \int_0^1 \int_1^2 xyz\, dx\, dy\, dz = \int_{-1}^1 \int_0^1 1.5yz\, dy\, dz = \int_{-1}^1 0.75z\, dz = 0$
  (Trapezoid)$\int_{-1}^1 \int_0^1 \int_1^2 xyz\, dx\, dy\, dz = \int_{-1}^1 \int_0^1 3yz\, dy\, dz = \int_{-1}^1 \frac{3}{2}z\, dz = 0$

Chapter 7

- **Exercise 1:**
  1.a $|f(t, y_1) - f(t, y_2)| = |\sin(t)|.|y_1 - y_2| \leq |y_1 - y_2|$.
  Hence, $\forall t, y$, the function $f(t, y)$ is Lipshitz and the IVP has a unique solution for all values of initial conditions.
  1.b $|f(t, y_1) - f(t, y_2)| = |e^{t/2}|.|e^{-y_1/2} - e^{-y_2/2}| = |e^{t/2}|.|e^{c/2}|.|y_1 - y_2|$, $c \in (-y_1, -y_2)$ or $c \in (-y_2, -y_1)$. Hence:
  For $a \leq y_0 \leq b$, a, b arbitrary and $0 \leq t \leq T$, T arbitrary:

  $$|f(t, y_1) - f(t, y_2)| \leq e^{(T-a)/2}|y_1 - y_2|.$$

  The function f satisfying a Lipshitz condition, the IVP has a unique solution for $y_0 \in (-\infty, \infty)$, $0 \leq t \leq T < \infty$. **1.c** $|f(t, y_1) - f(t, y_2)| = |\frac{2t^2}{1+t^4}|.|y_1 - y_2| \leq L|y_1 - y_2|$, where $L = \max_{\forall t} |\frac{2t^2}{1+t^4}|$.
  Hence, $\forall t, y$, the function $f(t, y)$ is Lipshitz and the IVP has a unique solution for all values of initial conditions.

- **Exercise 3:**
  Picard's iteration for $y' = -4y + t$, $0 \leq t \leq 1$, $y(0) = 1$.

  $$y^1(t) = 1 - 4t + \frac{t^2}{2}; \; y^2(t) = 1 - 4t\frac{15}{2}t^2 - \frac{2t^3}{3}$$

- **Exercise 5:**
  Results of Heun's method to solve:
  5.a $y'(t) = te^{3t} - 2y^2$, $0 \leq t \leq 1$, $y(0) = 0$, $h = 0.2$

ti	yi	k1	k2
0.00	0.00000E+00	0.00000E+00	3.64424E-01
0.20	3.64424E-02	3.61768E-01	1.30437E+00
0.40	2.03057E-01	1.24558E+00	3.22087E+00
0.60	6.49702E-01	2.78556E+00	5.90574E+00
0.80	1.51883E+00	4.20484E+00	8.94822E+00
1.00	2.83414E+00	4.02086E+00	1.74433E+01

5.b $y'(t) = t + (t - y)^2$, $0 \leq t \leq 2$, $y(0) = 1$, $h = 0.5$

ti	yi	k1	k2
0.00	1.0000E+00	1.0000E+00	1.5000E+00
0.50	1.6250E+00	1.7656E+00	3.2735E+00
1.00	2.8848E+00	4.5524E+00	1.4903E+01
1.50	7.7486E+00	4.0545E+01	6.7909E+02
2.00	1.8766E+02		

- **Exercise 7:**
  Results of Modified Euler's method to solve:
  7.a $y'(t) = te^{3t} - 2y^2, \ 0 \le t \le 1, \ y(0) = 0, \ h = 0.2$

ti	yi	k1	k2
0.00	0.00000E+00	0.00000E+00	1.34986E-01
0.20	2.69972E-02	3.62966E-01	7.29869E-01
0.40	1.72971E-01	1.26821E+00	2.06109E+00
0.60	5.85190E-01	2.94489E+00	4.16865E+00
0.80	1.41892E+00	4.79188E+00	6.18614E+00
1.00	2.65615E+00	5.97530E+00	8.65107E+00

7.b $y'(t) = t + (t - y)^2, \ 0 \le t \le 2, \ y(0) = 1, \ h = 0.5$

ti	yi	k1	k2
0.00	1.0000E+00	1.0000E+00	1.2500E+00
0.50	1.6250E+00	1.7656E+00	2.4829E+00
1.00	2.8665E+00	4.4837E+00	8.7433E+00
1.50	7.2381E+00	3.4426E+01	2.0041E+02
2.00	1.0744E+02		

Bibliography

[1] K.E. Atkinson. *An Introduction to Numerical Analysis.* John Wiley & Sons, Inc. New York, NY, USA, 1989.

[2] G. Birkhoff and G. Rota. *Ordinary Differential Equations.* John Wiley & Sons, Inc. New York, NY, USA, 1989.

[3] P. Bogacki and L.F. Shampine. A 3(2) pair of Runge-Kutta formulas. *Appl. Math. Letters*, pages 1–9, 1989.

[4] R. Burden and J.D. Faires. *Numerical Analysis, 9th edition.* Brooks, Cole, USA, 2010.

[5] J.C. Butcher. *Numerical Methods for Ordinary Differential Equations.* John Wiley & Sons, Inc. New York, NY, USA, 2008.

[6] Chaitin, G.H. Randomness and mathematical proof, *Scientific American* 232, No. 5, May 1995,pp. 47-52.

[7] S. Chapra and R. Canale. *Numerical Methods for Engineers, 6th edition.* McGraw Hill, 2010.

[8] Chatelin, F. CERFACS Publications, 1993.

[9] P. Cheney and D. Kincaid. *Numerical Analysis and Mathematics of Scientific Computing, 7th edition.* Brooks/Cole, Boston, MA, USA, 2013.

[10] P. Ciarlet. *Analyse Numérique Matricielle et Optimisation.* Masson, Paris, 1985.

[11] M. Crouzeix and Mignot A. *Analyse Numérique des équations Différentielles.* Masson, Paris, 1984.

[12] J.R. Dormand and P.J. Prince. A family of embedded Runge-Kutta formulae. *J. Comp. Appl. Math.*, pages 19–26, 1980.

[13] H. Edelsbrunner. *Geometry and Topology for Mesh Generation.* Cambridge University Press, Cambridge, UK, 2001.

[14] Erhel, J. Erreur de Calcul des ordinateurs, IRISA Publications, 1990.

[15] A. Gilat and V. Subramaniam. *Numerical Methods for Engineers and Scientists.* John Wiley & Sons, Inc. New York, NY, USA, 2008.

[16] G.H. Golub and C.F. Van Loan. *Matrix Computations, 2nd edition*. John's Hopkins University, 2001.

[17] H. Gould, J. Tobochnik, and W. Christian. *An Introduction to Computer Simulation Methods. 3rd edition*. Pearson-Addison-Wesley, 2006.

[18] E. Hairer, S. Nørsett, and G. Wanner. *Solving Ordinary Differential Equations I: Nonstiff Problems*. Springer Verlag, 1993.

[19] N. Higham. *Accuracy and Stability of Numerical Algorithms, 2nd edition*. SIAM, 2002.

[20] Ø. Hjelle and M. Dæhlen. *Triangulations and Applications*. Mathematics and Visualization Series, Springer 2006, 2006.

[21] E. Isaacson and H.B. Keller. *Analysis of Numerical Methods, 4th edition*. John Wiley & Sons, Inc. New York, NY, USA, 1966.

[22] A. Iserles. *A First Course in the Numerical Analysis of Differential Equations, 2nd edition*. Cambridge University Press, Cambridge, UK, 2009.

[23] J.M. Muller. *Arithmétique des Ordinateurs*. Masson, Paris, 1982.

[24] Niederreiter, H. Quasi-Monte Carlo methods and pseudo-random numbers, *Bull. Amer. Math. Soc.* Volume 84, Number 6 (1978), 957-1041.

[25] J. Ortega and W. Poole. *An Introduction to Numerical Methods for Ordinary Differential Equations*. Pitman, Mansfield, MA, USA, 1981.

[26] A. Quateroni, F. Saleri, and P. Gervasio. *Scientific Computing with Matlab and Octave, 3rd edition*. Springer, Berlin Heidelberg, 2010.

[27] M.H. Schultz. *Spline Analysis*. Prentice Hall, Englewood Cliffs, 1973.

[28] L.F. Shampine and M.K. Gordon. *Computer Solutions of Ordinary Differential Equations*. W.H. Freeman and Co., San Francisco, 1975.

[29] W.A. Smith. *Elementary Numerical Analysis*. Harper and Row, New York, 1979.

[30] I. Sobol. *A Primer for the Monte Carlo Method*. CRC Press, Boca Raton, USA, 1994.

[31] G. Strang. *Linear Algebra and its Applications, 3rd edition*. Philadelphia, PA: Saunders, 1988.

[32] G. Strang. *Introduction to Linear Algebra*. Wellesley-Cambridge Press, 1993.

Index